Metamorphism

A Study of the Transformations
of Rock-masses

by the late
ALFRED HARKER
M.A., LL.D., F.R.S.
Sometime Fellow of St John's College and Emeritus Reader
in Petrology in the University of Cambridge

CHAPMAN AND HALL
LONDON

First published in 1932
by Methuen & Co. Ltd
Second edition 1939
Third edition 1950

Reprint 1974 published by
Chapman and Hall Ltd
11 New Fetter Lane, London EC4P 4EE

Original copyright 1932 with A. Harker

© 1974 Chapman and Hall Ltd

Printed in Great Britain
at the University Printing House, Cambridge

SBN 412 12940 x

Distributed in the U.S.A.
by Halsted Press, a Division
of John Wiley & Sons Inc., New York

Metamorphism

PREFACE TO THE FIRST EDITION

THE following pages reproduce in substance a course of lectures delivered at Cambridge. The lack of any published work planned on similar lines has encouraged me to offer them to a wider circle of students.

Metamorphism is here conceived, not as a status, but as a process; viz. a progressive change in response to changing conditions of temperature and stress. In the first part of the book high temperature alone is the ruling condition. This is the case of 'contact metamorphism', which has usually been treated as an isolated phenomenon. Here, on the contrary, a study of purely thermal metamorphism is regarded as the natural line of approach to the more complex problem in which the dynamic factor enters in conjunction with the thermal. This general case (regional metamorphism) is the main subject of Part II. In discussing it I have given special prominence to the controlling influence of shearing stress, as distinguished from uniform pressure, which also has its part. Another factor to which I attach a certain importance is the mechanical generation of heat by the crushing of rocks. In the final chapter the various retrograde changes which may partly undo the work of metamorphism are only briefly noticed.

The subject being metamorphism, not metamorphosed rocks, detailed petrographical description has been reduced to a minimum, but its place is partly supplied by a large number of figures drawn from the microscope. Considerations of space forbid the insertion of chemical analyses: a useful collection of these is contained in a recent publication of the Geological Survey.

In choosing examples mainly from British sources, I am far from undervaluing the work of the many distinguished Continental petrologists who have contributed to our knowledge of metamorphism. Rather has it been my design to show that this country enjoys peculiar advantages as a field for research, and that British workers have not wholly neglected the opportunities so liberally offered.

A. H.

St. John's College, Cambridge
October, 1932

v

NOTE ON THE SECOND EDITION

THE late Dr. Harker completed the revision of the text of his book only a few weeks prior to his death. His intention to provide a second preface was not to be fulfilled, and it is at his wish that I have undertaken the revision of the proofs for the press. In the new edition the original plan, and so far as possible the original text, have been preserved, but opportunity has been taken to make certain minor changes and corrections, and to incorporate reference to some later researches.

C. E. TILLEY.

CAMBRIDGE,
August, 1939.

CONTENTS

PART I

THERMAL METAMORPHISM

vii

CONTENTS

METAMORPHISM

PART I

THERMAL METAMORPHISM

CHAPTER I

THERMAL METAMORPHISM: GENERAL CONSIDERATIONS

Scope of Subject—Plan of Treatment—Agents of Metamorphism—Conditions Controlling Thermal Metamorphism—Attainment of Chemical Equilibrium.

SCOPE OF SUBJECT

THE term ' metamorphism ', i.e. change of form, is understood in geology as having reference to molecular and atomic configuration, as well as to visible shapes and relations ; and it comprises therefore both mineralogical and structural rearrangements in rock-masses. Every branch of geology is, in fact, largely concerned with the phenomena of change in the material world, and it is evident that a study of rock-metamorphism, in the fullest sense of the word, would embrace a large part of the whole subject-matter of petrology. As conceived in the following pages, and in the usage of most petrologists, its scope is much less comprehensive, and it will be proper therefore to define at the outset the limitations to be observed.

The fundamental principle to be assumed as axiomatic is that the internal changes which take place in a rock are a response to changes in external conditions, and are to be interpreted as an effort to re-establish equilibrium under the changed conditions. The *conditions* which are relevant in this connexion are two : viz. temperature and stress. The province of metamorphism, as here understood, can then be defined as follows. We take as point of departure the various types of rocks known to us, such as can be handled and subjected to direct examination ; i.e. rocks at ordinary atmospheric temperature and sensibly free from stress. We shall endeavour to follow the changes induced in like rocks where they have been exposed in nature to more or less elevated temperature and more or less intense stress. Since the two conditions are, at least theoretically, independent, we may distinguish *thermal* metamorphism, consequent upon rise of tempera-

1

ture without important intervention of the stress factor, and *dynamic* metamorphism, related to mechanical stress operating without significant rise of temperature. We have further to consider the general case, involving the joint action and interaction of the two factors, thermal and dynamic. For reasons which will appear, this most general kind of metamorphism is typically developed throughout some extensive tract of the earth's crust, and it is conveniently styled *regional* metamorphism. It will be in accordance with scientific method to study in turn the two simpler special cases before essaying the more complex problems presented by regional metamorphism.

It has been tacitly assumed that the conditions of equilibrium are sufficiently determined by temperature and stress. This is true with regard to internal reactions within a defined and isolated system. A rock undergoing metamorphism comes under this designation only on the understanding that its bulk-composition remains unchanged during the process. If there be addition or subtraction of material, a new factor, and from our immediate point of view a disturbing factor, is introduced, and must be duly taken into account. Metamorphism, or change of form, is then complicated by metasomatism, or change of substance.

The domain of metamorphism, as here understood, excludes all those processes, mainly superficial, which fall under the head of ' weathering ', comprising hydration, oxidation, carbonation, and other changes of the nature of degradation. Depending on reactions between the rocks on the one hand and the atmospheric water and gases on the other, they involve a large element of metasomatism. Their results are very noticeable by an observer, whose own habitat is confined to that narrow zone of our planet where lithosphere and atmosphere meet. This class of changes has indeed its importance in relation to our subject, but chiefly as making part of a great cycle of operations, of which metamorphism is the complementary part. The subject of rock-weathering is large enough and important enough in its applications to be treated as a distinct branch of geology. Another department which lies outside our scheme is that which deals with mineral-veins and ore-deposits, where again the element of metasomatism enters largely. This is commonly treated, for good reasons, as a separate subject of study. It has to do largely with the sulphides and oxides of the heavy metals, which play only a minor part in general petrology ; and, further, the manner of treatment is partly guided by economic and other extraneous considerations.

In thus limiting a field which still remains a wide one, we are following the general practice of British geologists. American writers on metamorphism, in particular Van Hise and his school, have occupied a much wider province. Adopting the more restricted definition, we shall find that metasomatic processes play a much less important part. In most cases of metamorphism that will be discussed we shall be able to assume without sensible error that the rock, and even any small part of the rock, suffers no change in total composition during the process. The exceptions to this generalization will be considered as they arise.

PLAN OF TREATMENT

Metamorphism and its effects in rock-masses being the subject, there are two alternative lines of approach. They correspond with two different points of view, and the choice between them is a choice between two very different plans of treatment. These two methods may be distinguished for convenience as the *descriptive* and the *genetic*.

On the one hand, we may study metamorphosed rocks simply as specimens which come under our notice, describing and classifying them with reference to their actual characters—chemical, mineralogical, textural, and structural. This does not, it is true, preclude further inquiry into the manner in which those characters have been acquired ; but such considerations do not, in this descriptive treatment, enter into the study in a way which affects its development. As applied to the crystalline schists and allied products of regional metamorphism, this manner of procedure is seen as the last surviving relic of Wernerian geology. It has been followed especially by the German school, to whom we owe much of our knowledge of these rocks. The standard work here is that of Grubenmann,[1] which should be in the hands of every student. The first part of the book does indeed treat of the principles of metamorphism ; but no attempt is made to apply these principles in detail, and the systematic part which follows is on purely descriptive lines. A division into twelve groups is made on the basis of chemical composition, and three grades of metamorphism are distinguished, and are correlated with different zones of depth in the earth's crust. The conception of metamorphism as a progressive process, however, is quite obscured, and indeed in each group the most highly metamorphosed rocks are placed first. The products of purely dynamic metamorphism are recognized in appendices to certain of the

[1] *Die Kristallinen Schiefer* (1906) ; 2nd ed. (1910). The author had planned a third edition, with greatly enlarged scope, the completion of which has devolved upon his successor Prof. Niggli ; but only the first instalment of this work has yet appeared ; *Die Gesteinsmetamorphose*, by U. Grubenmann and P. Niggli, I. *Allgemeiner Teil* (1924).

groups, but ' contact ' (i.e. purely thermal) metamorphism is definitely excluded. The crystalline schists are regarded as constituting a distinct great class of rocks, co-ordinate with the two other classes, igneous and sedimentary, their relation to these being consequently ignored.

While the descriptive method is rooted in the conception of *meta-morphic rocks* as a distinct *class of rocks* defined by certain characters, that which I have called the genetic method starts from the idea of *metamorphism* as a certain *class of changes*, which may affect rocks of any kind and alter their characters. In its origin it attaches itself to the Huttonian doctrines as developed by Lyell, to whom the word metamorphism is due, and with the tenets of that school the notion of gradual progressive change was entirely in harmony. That from this logical starting-point no serious advance was made, is to be ascribed in part to that unfortunate neglect of the chemical and petrographical side which was long the reproach of British geology ; but in truth little could have been accomplished on these lines with the knowledge then at command. The rapid development of physical chemistry in later years, and the successful application of the experimental method to the problems of petrology, have greatly altered the situation. On the foundation now being laid there may not improbably be built up in the future a complete theory of metamorphism on a rational and genetic basis.[1]

The first serious attempt to discuss the process of metamorphism in relation to first principles was made by Van Hise,[2] ; but the scope of his work is far wider than that here adopted, and a considerable part of his massive volume is devoted to such subjects as weathering, cementation, and ore-deposits. The later work of Leith and Mead [3] is conceived on a plan no less comprehensive.

An important contribution to the theory of thermal metamorphism has been made by Goldschmidt [4] in a memoir dealing primarily with the Oslo district. He makes direct application of the Phase Rule to determine the possible associations of minerals in a metamorphosed rock of given total composition. A rock completely metamorphosed in presence of a pervading solvent is regarded simply as a ' condensed ' system of n components, and it is deduced that the number of distinct minerals which can exist together in equilibrium is then n, or at an

---

[1] Harker, Anniversary Address to the Geological Society, *Quart. Journ. Geol. Soc.*, vol. xxiv (1919), pp. lxiii–lxv.

[2] *A Treatise on Metamorphism, Monog. xlvii U.S. Geol. Sur.* (1904).

[3] *Metamorphic Geology : a Text-book* (1915).

[4] *Die Kontaktmetamorphose im Kristianiagebiet, Vidensk. Skr.* (1911). See also *Zeits. Anorg. Chem.*, vol. lxxi (1911), pp. 313–22.

invariant-point $n + 1$. The assumption of true chemical equilibrium cannot be universally admitted, and other assumptions underlying the author's reasoning have been criticized. To demur to Goldschmidt's argument, however, is not necessarily to combat his specific conclusions, which are generally supported by the observed facts ; and the classification of different types of ' hornfels ' to which he is led is of service as an ideal scheme. Nevertheless, since its application is only to rocks which have suffered total reconstruction, it throws no light upon metamorphism regarded as a progressive process.

Although we possess at present no complete theory of metamorphism based directly upon accepted principles, it is possible at least to prepare the way for such systematic treatment by marshalling observations and inferences with this ideal constantly in view. Such is the design of the present work, which to that extent may profess to aim at a rational or genetic treatment of the subject. Laboratory experiment, which has so greatly enlarged our understanding of the genesis of igneous rocks, is already being applied to some questions important in relation to metamorphism ; while the known laws of physics and chemistry, based ultimately upon experiment, are always at our service.[1] Help from these quarters comes to supplement and reinforce the results of geological and petrographical inquiry ; and it is by combining all these resources that we may best hope to gain an insight into the processes of rock-metamorphism. A philosophical treatment of the subject as a whole is not yet among things possible.

THE AGENTS OF METAMORPHISM

We read in the older text-books of geology that rocks are metamorphosed by the agency of heat and pressure, to which is commonly added the presence of water. The part played by water will be discussed later : regarded strictly, it is not to be reckoned among the controlling conditions of metamorphism, but makes part of the material in which metamorphism operates. For the rest, since we are concerned at present, not with the question of energy, but with the conditions of equilibrium, we may interpret ' heat ' to mean rise of temperature. The remaining factor calls for more particular consideration.

The term ' pressure ', as loosely used by the older writers, ignores a distinction which it is of the first importance to observe. On the one hand, the pressure at every point within a body may be the same in all directions. Since this is the only type of pressure-distribution that can be maintained in a fluid, it is conveniently distinguished as

[1] For a useful summary of these see Johnston and Niggli, *Journ. Geol.*, vol. xxi (1913), pp. 481–516, 588–624.

hydrostatic pressure, and in discussing the behaviour of liquids and gases the word pressure can usually be employed without ambiguity. This is true also of the customary operations of the laboratory, and the chemist can generally regard temperature and (hydrostatic) pressure as completely determining the conditions to which a given system is subjected. In metamorphism, however, we have to do with changes which proceed in the heart of a solid rock, and a solid is capable of sustaining pressure which, at a given point, is different in different directions. This is tantamount to saying that a solid, unlike a fluid, can sustain *shearing stress.* A simple analysis shows that any non-uniform pressure at a point within a solid body is equivalent to a simple (hydrostatic, pressure together with certain shearing stresses.[1] This mode of presentation is adopted in mechanics in discussing the correlation of stress and strain, because the ' modulus of compression ', which connects uniform pressure with voluminal compression, and the ' modulus of rigidity ', which connects shearing stress with deformation, are two independent constants. It is not less necessary in discussing molecular and atomic rearrangements within a body effectively solid, for here too the influence of simple pressure and of shearing stress must be carefully discriminated. The influence of hydrostatic pressure upon various transformations and chemical reactions can be expressed in terms of simple laws, but concerning the influence of shearing stress in this field much yet remains to be learnt. This is the more to be regretted, since shearing stress is undoubtedly a factor of great moment in controlling mineralogical changes in metamorphism, as it manifestly is also in respect of structural rearrangement.

In the following pages *pressure,* without qualification, will be understood to mean pressure of the simple hydrostatic type. The term *stress* properly comprises both pressure and shearing stress, but there will be little risk of misunderstanding if it is often employed for brevity in place of shearing stress.

The mathematician's analysis of stresses and strains is strictly applicable only to a homogeneous and isotropic body, and it is easy to see that uniform compression and pressure are not theoretically possible in a crystalline rock. If a cube of granite be subjected to uniform pressure from without upon its six faces, its mechanical status is different from that of a cube of glass under like external forces. Quartz and felspar are not equally compressible, and the compressibility of each crystal is different in different directions. The granite therefore cannot yield without some internal deformation and the

---

[1] Thomson and Tait, *Treatise on Natural Philosophy,* art. 682, and see below, Chapter X.

setting up of shearing stress. The same result follows from expansion or contraction with change of temperature. The point to be observed concerning these internal stresses is that, besides being of no great magnitude, they have no common direction, since the crystals lie in all ways indifferently. Consequently stress is in great measure annulled by mutual compensation. The small balance can be relieved by slight slipping of one crystal against another, by the opening of cleavage-cracks, by 'gliding' in the crystals of some minerals and bending in others, without appreciable change in the structure of the rock as a whole. Experiment shows indeed that the mechanical behaviour of a crystalline rock is, in the gross, much like that of an isotropic body, and that it conforms rather closely with Hook's law.[1]

Of a very different order are the shearing stresses set up in rockmasses in response to the powerful external forces which arise in connexion with orogenic movements. The magnitude of these stresses is limited only by the crushing strength of the rocks, and in fact this limit is very often reached. Moreover, since the stresses here have a common direction, imposed by the external force-distribution, there is no mutual compensation. Relief can come only from very radical changes in the rock, which almost always involve mineralogical as well as structural rearrangement. The mineralogical changes induced under stress are dependent also on temperature, but we shall see that they differ in general from those changes which would take place at like temperatures in the absence of the stress factor.

Since the absence or presence of any important shearing stress is of prime significance in determining the mineralogical changes which follow when rocks are subjected to rise of temperature, and since it must evidently be of capital importance also in relation to the setting up of new structures, the distinction so implied will properly determine the plan of treatment of the whole subject. The first part of this volume will deal accordingly with thermal metamorphism, not complicated by the stress factor ; the latter part will be devoted to metamorphism in which shearing stress enters as a ruling condition, with or without significant rise of temperature.

CONDITIONS CONTROLLING THERMAL METAMORPHISM

Shearing stress being absent or negligible, the factors which control thermal metamorphism are temperature and pressure (in the hydrostatic sense). Logically the two are co-ordinate, but in most cases

[1] Adams and Coker, *An Investigation into the Elastic Constants of Rocks, More Especially with Reference to Cubic Compressibility*, Carnegie Inst., Washington (1906).

a very moderate change of temperature is as effective as a very great change of pressure. For most purposes, therefore, we may regard the former as the dominant factor and pressure as merely modifying the influence of temperature. The grounds for this assertion will appear, if we examine very briefly and in general terms how various physical and chemical changes are dependent upon these two ruling conditions.

In the first place, a given mineral, supposed for simplicity to possess a stoichiometric composition, has a definite temperature-range of stability. When one of the limiting temperatures is passed, the mineral must undergo change, if equilibrium is to be maintained : alternatively, it may persist as a metastable form. The limiting temperatures include melting-points, inversion-points of dimorphous compounds, and dissociation-points. Dry fusion has no part in metamorphism, but an inversion-point in reversible dimorphism is strictly analogous to a melting point, and its dependence on pressure is expressed by the simple relation :—

$$\frac{dT}{dp} = (v_\alpha - v_\beta)\frac{T}{L},$$

where T is the inversion-temperature reckoned from absolute zero, p is the pressure, L is the latent heat of inversion, and the expression in parentheses is the difference of specific volume between the higher and lower forms. The change from lower to higher is therefore promoted or retarded by increased pressure, according as it is accompanied by contraction or expansion. If the volume-change is small, pressure will have little effect, except in the possible case of the heat-change being also very small. Inversion may be greatly promoted, in the sense of its *rate* being accelerated, by the presence of some other body ; but the inversion-temperature is not altered, unless this other body enters in solid solution. In irreversible or monotropic dimorphism there is no inversion-point : the lower form is merely metastable, and exists only by reason of an infinitesimally slow rate of inversion. Rise of temperature, by accelerating the rate, may bring about the change, but not at any determinate temperature. Here, too, a catalyser may greatly promote the change to the stable form.

Solution again is in all respects analogous to melting. The solubility of a given mineral in a given liquid is a function of temperature and pressure. It is increased or diminished by rise of temperature, according as heat is absorbed or liberated in the act of solution. The former is the more usual case, but we possess few data for rock-forming minerals. Increased pressure augments or diminishes solubility, according as the volume-change and the heat-change in solution are of opposite signs or like sign ; but the little that is known from experiment suggests that, for solution of a solid in a liquid, pressure is not usually of great moment.

In considering chemical reactions as controlled by temperature and pressure, we must distinguish between a balanced reaction and one which proceeds to completion. Suppose, first, that the total composition of the system remains unchanged, and consider a balanced reaction of the type :

$$A + B \rightleftharpoons C + D,$$

where the letters represent single molecules of the four phases involved. The equilibrium arrangement is determined by the mass-action equation :

$$[A][B] = K[C][D],$$

where $[A]$ stands for the concentration of A, etc., and K, the reaction-constant, is a function of temperature and pressure. Its dependence upon these is such that a rise of temperature drives the reaction in that direction which involves absorption of heat, and an increase of pressure drives it in the direction that involves diminution of volume. If only solid and liquid phases be present, the volume-change is seldom very considerable, and only a very great pressure will have any sensible influence. It is otherwise when a gaseous phase is involved. Pressure will then have a very pronounced effect, in the sense of resisting the reaction by which the gas is liberated.

Suppose now, on the other hand, that one of the four phases can pass out of the system. There can then be no balance, but the reaction will proceed continually in one direction until that body is eliminated. Thus, if D be a gas, and if the circumstances be such that it can escape, the reaction will be driven in the direction from left to right until A and B are exhausted and C alone remains. The result will be the same if the reaction is one between bodies in solution and D is insoluble, so that it passes out of solution as soon as it is formed. Again, if all four bodies be soluble in different degrees, a like situation is reached as soon as the least soluble body arrives at the point of saturation.

It may be laid down summarily, that the changes which are promoted by rise of temperature are those which involve absorption of heat, and the changes which are promoted by increase of pressure are those which involve diminution of volume. If the latter of these two general laws is more frequently cited than the former, it is perhaps because the volume-change is more easily calculated than the heat-change. The ' Volume Law ' in its application to metamorphism may be illustrated by an example borrowed from Becke. In the equation :

$$Mg_2SiO_4 + CaAl_2(SiO_4)_2 = CaMg_2Al_2(SiO_4)_3$$

$$\underset{\text{Forsterite}}{} \quad \underset{\text{Anorthite}}{} \quad \underset{\text{Garnet}}{}$$

the molecular volumes on the left side are $43\cdot9 + 101\cdot1 = 145\cdot0$, and on the right $125\cdot8$, giving a diminution of volume to the amount of 13 *per cent*. Here the molecular volumes are computed from the specific gravities of the minerals in the laboratory, and would be greater at higher temperatures, but would doubtless still show a difference of the same order. It follows that, if the reaction indicated is a possible one, it will be very sensibly aided by high pressure. It must be remembered, however, that the Volume Law merely formulates the effect of pressure as considered apart from other factors. We shall see later that its importance has sometimes been exaggerated by attributing to uniform pressure effects which are really connected with unequal pressure and shearing stress. It is doubtless very generally true that rocks are denser after metamorphism than before, but this cannot be credited wholly to the cause in question. Expulsion of volatile substances will tend to the same result.

ATTAINMENT OF CHEMICAL EQUILIBRIUM

There are other considerations to be weighed before we can with confidence apply the data of chemistry to concrete problems in rock-metamorphism. The investigations of the physical chemist are usually directed to determining the *equilibrium* configuration of a given system under varying conditions of temperature and pressure. But, while the changes induced in a rock in metamorphism are always in the direction of restoring equilibrium, we are not entitled to assume that equilibrium is necessarily established. It is certain that it is not always realized even when a rock has been totally reconstituted, and to the lower grades of metamorphism the limitation applies with greater force. This does not remove metamorphism from the province of chemical science, but it does counsel caution in the application of simple laws to complex cases. In another field, that of the crystallization of molten rock-magmas, we know that the imperfect attainment of equilibrium has far-reaching consequences, but it has not been found impossible to include these consequences in the general scheme of petrogenesis. Various features of metamorphosed rocks, notably the comparative rarity of zoned crystals, seem to indicate that equilibrium is here more promptly attained, or more closely approached, than in a crystallizing magma. The magma has the advantage of freer diffusion, but this is more than counterbalanced by the fact that crystallization there proceeds with falling temperature, while the reverse is the case in metamorphism. It is a capital principle, to be abundantly illustrated in our study of metamorphism, that equilibrium is reached or approached far more promptly with rising temperature or increasing stress than with falling temperature or declining stress. Doubtless many of the chemical reactions characteristic of rock-metamorphism are theoretically reversible, but for the most part they are not in fact reversed when the conditions which induced them have passed away. Minerals belonging to the higher grades,[1] instead of reverting to other products, more usually remain to indicate, as it were, the high-water mark of metamorphism. Were it otherwise, any study of the subject on the petrographical side would be impossible, since it would have little material to work on.

The matter can be discussed in terms somewhat more definite. In addition to the data of temperature, pressure, and concentration, which theoretically determine a certain chemical reaction, we must take account also of the *rate of reaction*, and must recognize that the rate is in some cases excessively slow. If it be so slow as to be negligible the sensible result is that the reaction does not take effect. The comparison here is between the rate of reaction and the time during which the conditions favourable to it are maintained : an unlimited lapse of time after those conditions have ceased will be of no avail. The rate itself is dependent upon the conditions, and especially upon the temperature. According to Johnston and Niggli, it may be doubled by a rise of 10°, while a rise of 100° may perhaps increase it a thousand-fold and 200° a million-fold. For this reason it is chiefly in the lower grades of metamorphism that complication arises from the non-adjustment of equilibrium. Concerning the influence of stress upon the rates of reactions we know little, but it is probable that simple pressure is without sensible effect. More important is it to observe that a rate of reaction may be greatly accelerated by the presence of some body which apparently does not itself take part in the reaction, or at least does not enter into the resulting products. We are probably to infer that it plays an essential part in some intermediate reaction and is finally set free. Whatever be the true nature of this ' catalytic ' action, it certainly has its importance in metamorphism. If by its means a rate, otherwise insensible, becomes sensible, the practical effect is that the catalyser induces a reaction which would not take place in its absence.

The change in the imposed conditions, temperature and pressure,

[1] The word ' grade ' will be used always with reference to temperature.

which determines metamorphism, is a gradual and continuous change. If chemical equilibrium were constantly maintained, mineralogical reconstruction would likewise proceed steadily and continuously, keeping pace with the changing conditions. Something like this we may suppose realized in the higher grades of metamorphism, where rates of reaction are accelerated by high temperature. In the lower grades this cannot in general be true, for the process does not commonly start from equilibrium. The given rock, which we take as our point of departure, has in fact a past history. If, for instance, it be an igneous rock, some of the high-temperature minerals which compose it are not truly stable under the actual conditions, but survive only in virtue of that chemical inertia which has been indicated. If it be a sedimentary rock, the several minerals may be individually stable, but not in true equilibrium with one another. They remain unchanged merely because, under the actual conditions, the rate of any possible reaction is so small as to be sensibly *nil*. This is the case known as 'false equilibrium'.

The conception of metamorphism to be kept constantly in mind is that of something *progressive*. In response to rising temperature the substance of a given rock passes through a certain sequence of transformations, the stage actually reached depending upon the highest temperature attained. In the original rock, however, there are, in the most usual case, some constituents which are more susceptible of change than others, in the sense of being affected at an earlier stage of the rise of temperature. They may dissociate, or react with one another, or merely recrystallize. In an early grade therefore such a rock is only partially metamorphosed. It consists partly of new and recrystallized minerals, partly of *residual minerals* still intact. With continued rise of temperature these are in turn drawn into the sphere of the processes of metamorphism ; so that in any advanced grade the rock may be regarded as totally reconstituted. The only noteworthy exception is that of the highly refractory mineral zircon, which can even survive complete dissolution of a rock which contained it.

The changes which take place in the earlier stages of metamorphism depend, then, upon the *initial mineralogical constitution* of the rock ; but in an advanced grade all is determined by the *total chemical composition*, in conjunction with the given conditions of temperature and pressure, the past history of the rock being no longer relevant. When a new mineral has once come into being, or an old one has recrystallized, it is still, potentially at least, a party to all that happens thereafter with further rise of temperature. It may be called in a sense alive, in that it responds freely to suitable stimulus from without. Through the medium of a common solvent—a matter which we have yet to consider —the several minerals present are maintained in chemical equilibrium with one another by a large number of balanced reactions among them. These reactions are balanced at any given temperature : rise of temperature displaces the balance, and may bring into play new reactions.

By rejuvenation of the several minerals, together with reactions between them, a rock, once reconstituted, is still being continually made over again with advancing metamorphism.   The student should guard against importing into the discussion of metamorphism the conception of an ' order of crystallization ', which is a prominent feature of igneous rocks.   Even in the crystallization of a molten magma, conducted with falling temperature, it would be an error to suppose that a mineral, once crystallized, passes out of the province of chemical reactions : we know that it is sometimes resorbed or dissociated at a later stage.   In metamorphism, which proceeds with rising temperature, such readjustment is not an exceptional but a universal incident, and the several constituents of a metamorphosed rock, apart from residual minerals, if any, are in effect of simultaneous crystallization. We have next to consider the mechanism by which this continual building anew of the rock is effected.

CHAPTER II

THERMAL METAMORPHISM : GENERAL CONSIDERATIONS

(continued)

The Rôle of Solution in Metamorphism—Limit set to Diffusion—Aureoles of Thermal Metamorphism—Some Illustrative Areas—Vitrification of Shales and Sandstones.

THE RÔLE OF SOLUTION IN METAMORPHISM

WHEN in discussing the processes of metamorphism, we speak of a chemical reaction between two minerals, such as calcite and quartz, we are using an elliptical expression. No sensible reaction can in general be verified at the contact of two crystalline bodies.[1] We are to suppose that the bodies in question enter into solution, and there suffer dissociation and reassociation, the new products finally passing out of solution. Even recrystallization of a single mineral, where no chemical reaction is implied, must usually be brought about by solution. The presence of some *solvent medium* pervading the rocks is therefore to be presumed as an essential part of the mechanism of metamorphism of any kind.

It is no less important to observe, however, that the solvent must be present in general only in *very exiguous quantity*. The kind of solution to which we make appeal is a *local and temporary solution*. Bodily dissolved, a rock would lose its identity, yielding not a metamorphosed product but a totally new rock. In thermal metamorphism at least, the preservation of various residual structures, such as the banding in sediments or the ophitic and other characteristic peculiarities of igneous rocks, shows that the rocks have in fact maintained their identity throughout the process. We are then to conceive a rock which suffers metamorphism as being worked over *gradually and piecemeal* by the very small quantity of solvent present, which is continually set free to act upon new portions of the rock. If metamorphism is a

---

[1] An interesting paper by N. W. Taylor and F. J. Williams treats of 'Reactions between Solids in the System $CaO-MgO-SiO_2$'. Here, however, the lime and magnesia were introduced as carbonates, yielding abundant carbon dioxide to act as a solvent ; *Bull. Geol. Soc. Amer.*, vol. xlvi (1935), pp. 1121-36.

slow process, this is due, not only to the tardy rate of some of the reactions involved, but to the small total amount of disposable solvent, which must therefore be used over and over again. When we say that mineral substances enter into solution, take part there in chemical reactions, and pass out of solution in new forms, we are not to conceive that the metamorphism of the rock as a whole falls into these distinct stages ; but merely that such is the sequence of operations at any one spot in the rock, and is realized successively at different spots.

This is no imaginary picture. Its truth can be verified in that type of spotted slates (' Knotenschiefer ' or ' Fruchtschiefer ') which often figures as the lowest grade of thermal metamorphism in argillaceous sediments. When a rock has been completely transformed in the manner sketched, all trace of the earlier stages of the process is obliterated ; so that in general all that belongs to solution is a closed chapter. In the case cited, however, the process has been left incomplete, local solution having taken place but not the correlative recrystallization. We then have the opportunity of observing the course of metamorphism as arrested at an early stage. The type of spotted slate in question has been studied by Hutchings [1] and others. The essential constituent of the spots is an amorphous, isotropic substance of a pale yellow colour, which can be regarded only as a glass (see page 24, Fig. 1, B). It may enclose minute new crystals, e.g. of rutile, which recrystallizes very readily. The glass, as such, is structureless, but has sometimes given rise to indistinctly cryptocrystalline matter, or is beginning to develop a finely flaky structure, with feeble depolarization. If devitrification has gone farther, there results a minutely crystalline mosaic which can be partly resolved into mica and quartz.

The interpretation of these phenomena can scarcely be in doubt. The first step in the metamorphism was *local solution*, beginning at many *isolated points* within the rock. This should have been followed by recrystallization, setting free the solvent to attack new portions of the rock-mass ; but the crystallization of silicate-minerals demands time. In the actual circumstances the temperature attained has been high enough to initiate local solution, but the duration of the high-temperature conditions did not suffice for the complementary process of recrystallization. The dissolved spots passed therefore into a glassy or largely glassy state, just as an igneous magma will do with rapid cooling. In this glass, we must suppose, the small quantity of solvent is itself incorporated.

The principal solvent which officiates in the metamorphism of rocks

[1] *Geol. Mag.*, 1894, pp. 43–5, 64–8.

is doubtless the omnipresent water. Most inorganic substances are in some degree soluble in water, and in general the solubility is increased by rise of temperature. Not a few characteristic minerals of metamorphism—micas, chlorites, epidotes, amphiboles, idocrase, etc.—have hydroxyl or basic hydrogen as part of their constitution. This affords direct evidence of the presence of water during the metamorphism, though not a measure of the amount present. It teaches us, too, that the function of water is not always limited to the part of simple solvent, since it may also participate in chemical reactions. Even when no part of it enters into the final products, it may possibly have taken part in intermediate reactions. If we compare rocks in different grades of metamorphism, we see that water enters to a less extent into the constitution of the new minerals at the higher temperatures. This may be compared with the crystallization of an igneous magma, conducted likewise in presence of water but with falling temperature. There the earlier products of crystallization are all anhydrous ; such minerals as hornblendes and micas come later ; and minerals rich in water, such as analcime, appear only in the closing stages.

While the chemical action of water is essential to the production of particular minerals of metamorphism, its solvent action is universal, and may be regarded as its prime function. We find therefore no place for a special type of ' hydrothermal metamorphism ', as distinguished by some geologists. With water as the chief solvent are associated other substances, which have a less general distribution, and figure usually in much smaller quantity. Here are to be reckoned borates, fluorides, chlorides, carbon dioxide, and others of less importance. Being more potent solvents than water, they may perhaps play a part by no means negligible, even when present in very minute quantity. It may be supposed, too, in view of their greater chemical activity, that they take part in essential chemical reactions to a greater extent than water does. Direct evidence of their action is seen where some part of the boron, fluorine, etc., has become fixed in certain new minerals ; but the absence of such material trace does not preclude the possibility that these active bodies have had a share in the metamorphism, whether as catalysers or merely as solvents.

The critical temperature of water is 374° C., and this figure will not be much raised by a small admixture of other volatile substances. The critical pressure is for pure water about 200 atmospheres, equivalent to about 2,500 feet of rocks. It appears then that, while, under the ordinary conditions of thermal metamorphism, the solvent medium may possibly remain in the liquid state up to a temperature in the neighbourhood of 400°, we must suppose it to be gaseous at higher temperatures, and therefore in any really advanced grade of metamorphism. The solvent power of liquid water falls off rapidly as the critical point is approached, but there seems to be little information concerning the properties of gaseous water above that point. It is true that many experiments are on record in which various substances have been heated with water in sealed vessels,[1] and from these it appears that numerous minerals have been deposited, as if from aqueous solution, at temperatures well above 400°. It must be remembered, however, that water and the associated volatile bodies, besides acting as solvents and fluxes, may take part in chemical reactions of a cyclical kind, i.e. may officiate as catalysers. There is experimental evidence that gases above the critical temperature have the power of dissolving non-volatile bodies, and that the solubility increases with rising temperature.[2] Although such investigations have not been extended to the rock-forming minerals, it is shown that silica is soluble in gaseous water, at least at temperatures above 700° C.

Concerning the source of those special volatile substances which may co-operate with the water a few words will suffice. We cannot in general suppose them to be derived from the material of the rocks which suffer metamorphism. It is true that tourmaline, for instance, is a widespread constituent of ordinary sediments ; but the boric acid there contained is not only in very small relative quantity, but is already locked up in a highly stable combination. Detrital tourmaline recrystallizes readily in metamorphism, but with nothing to suggest that it enters into special relations with other minerals present. One class of rocks there is, which in this respect stands upon a peculiar footing. Partly calcareous sediments undergoing metamorphism are capable in certain circumstances of liberating abundant carbon dioxide, and may be said to provide a competent solvent from their own substance. This is one of the features which invest the metamorphism of such rocks with exceptional characteristics. Setting aside this special case, it appears from the mineralogical evidence that the distribution of volatile substances other than water is of a local kind. Moreover, it is localized in evident relation to igneous activity. Metamorphic minerals containing boron or fluorine or chlorine are found in general only near igneous intrusions belonging to the epoch of the metamorphism. At the immediate contact such distinctive minerals are sometimes very abundant, and are clearly related to a pneumato-

---

[1] For a useful summary of such experiments see Morey and Niggli, *Journ. Amer. Chem. Soc.*, vol. xxxv (1913), pp. 1086–1130.

[2] Greig, Merwin, and Shepherd, *Amer. Journ. Sci.* (5), vol. xxv (1933), pp. 61–73 ; Ingerson, *Econ. Geol.*, vol. xxix (1934), pp. 454–70.

lytic replacement, which has usually affected igneous and metamorphic rocks in common. We may infer with confidence that the more active solvents and mineralizers in metamorphism, carbon dioxide excepted, are of direct magmatic origin.

This is also true in the main of the water itself, for any extensive penetration of surface-water into the heated interior crust of the earth is not an admissible hypothesis. Free circulation is confined to very moderate depths, and capillarity necessarily ceases at the critical temperature.[1] In any but quite superficial rock-masses the contained water must be attributed to the same ultimate source as the other volatile bodies. Water, however, being present much more abundantly in igneous magmas than the rest, comes to have a much more general distribution in rocks at large. There is evidence that its degree of concentration varies locally, and that in a manner directly related to igneous intrusions ; but this variation is superposed upon a general distribution, which has been attained cumulatively during preceding ages.

That part of the earth's crust which is the theatre of metamorphism is to be conceived therefore as everywhere permeated by a medium consisting of water with other volatile substances. In all places where the temperature is above the critical point, and at lower temperatures where pressures are not high, this pervading medium is in the gaseous state. In general extremely tenuous, it may attain a more notable concentration in the neighbourhood of igneous intrusions ; and there also the more active volatile substances, elsewhere quite subordinate to water, may acquire enhanced importance.

LIMIT SET TO DIFFUSION

The physical properties of gases at high temperatures and under great pressures are very imperfectly known ; but we may most probably conceive the general solvent medium as possessing a high degree of viscosity, with a density such as we associate normally with liquids.[2] Sparingly distributed through the rocks, it is held in capillary and subcapillary passages and in solid solution in the minerals of the rocks themselves. In the general case, anything of the nature of free circulation is quite precluded. In the shallowest levels of the earth's crust, where such circulation is possible, water acts as a carrier promoting hydration, oxidation, dolomitization, and many other processes which lie outside the province of metamorphism as here understood. In dynamic metamorphism, too, where it has been effected under a

[1] Johnston and Adams, *Journ. Geol.*, vol. xxii (1914), pp. 1–15.
[2] Compare Arrhenius, *Geol. Fören. Stock. Förh.*, vol. xxii (1900), p. 395.

thin cover, there is evidence of the carriage of material in solution, but this we may regard as a special case. In general the conditions prohibit flowing movement of the solvent medium, and any redistribution of material must be effected, not by molar, but by molecular flux, that is by *diffusion*. This is true equally of the solvent medium itself and of rock-substance which passes into solution, but there are important circumstances which discriminate the two cases.

We possess no knowledge of the numerical constants of diffusion in viscous media at high temperatures, but the form assumed by the fundamental law of diffusion itself makes it evident that redistribution by this process must always be very slow. The propagation of any sensible concentration beyond a very short distance will demand a very prolonged lapse of time. The diffusion of the volatile solvent through the earth's crust can become effective, despite the extreme slowness of the process, because it is perennially in progress, and the actual distribution at any epoch in the world's history has been gradually brought about during ages preceding. The diffusion of dissolved rock-substance presents a very different case. In general the solubility of a mineral in any sensible degree is dependent upon an elevated temperature, besides being augmented, as we shall see, by stress. Solution therefore, and consequently diffusion of dissolved material, are subject to the primary conditions which determine metamorphism. Here diffusion is not only a slow process, but is operative only during a limited time, the ' diffusivity ' rising with temperature and stress to a maximum and falling off as these conditions decline. Migration or interchange of material within a rock undergoing metamorphism is consequently confined to very narrow limits.

The study of metamorphosed rocks amply confirms this conclusion, and makes it appear that the mineral formed at any given point depends upon the composition of the rock within a very small radius about that point.[1] The limit of effective diffusion thus indicated is commonly a small fraction of an inch. So, for example, in a banded sediment composed of thin layers of different nature—argillaceous, gritty, calcareous—metamorphism does not confuse all to one average type. The several narrow bands remain distinct, each represented by its own association of new minerals, even when a dozen alternations are included in the field of the microscope. It is needless to refer in this place to many other phenomena, which point to the same conclusion. The dynamic element in metamorphism, while introducing some modification, does not invalidate this generalization.

Diffusivity, as already remarked, is a function of temperature,

[1] Harker, *Journ. Geol.*, vol. i (1893), pp. 574–78.

increasing with rise of temperature. For this reason the latitude of migration of material, always small, becomes somewhat enlarged in the higher grades of metamorphism. Moreover, if we may legitimately picture the temperature-conditions as steadily waxing and then waning, it follows that, the higher the maximum temperature reached, the longer does the temperature remain above any assigned figure. One result is that the individual crystals of new minerals, or of old ones recrystallized, can attain larger dimensions.

Increase of grain size, if we suppose the total quantity of a mineral to remain unchanged, implies that this given quantity is by some means shared out among a smaller number of individual crystals. How this is effected is a question which demands closer consideration. It is to be remarked that an important function of the general solvent is to preserve a balance, not only between the different minerals present, but also between the several crystals of any one mineral. Crystals of the same kind but of different sizes, in presence of their saturated solution and within the range of effective diffusion, constitute a sensitive system, in that a slight cause may suffice to bring about corrosion of some crystals with correlative addition of material to others. Such a cause is found in *surface-tension*. Since the pressure due to surface-tension is proportional to the curvature of the surface, a small crystal is under greater stress than a larger one. Increased stress, as we shall have occasion to point out later, causes increased solubility. Material is therefore dissolved from the smaller crystals and deposited upon larger ones in their neighbourhood, until the smaller have disappeared. If a mineral is only sparingly present in the rock, the process will cease when the distance apart of the crystals exceeds the latitude of effective diffusion. If the mineral is so abundant that the crystals are necessarily close together, redistribution of material will continue, tending always to increasing coarseness of grain and also to uniformity of grain-size. The even-grained character is most noticeable in rocks of very simple constitution, such as marbles and quartzites. They illustrate also the coarser texture which goes with advancing metamorphism, and indeed in the highest grades a rock of any kind often shows a coarseness of grain comparable with that of a plutonic igneous rock.

The action described, depending upon surface-tension, is that to which Rinne [1] has given the name *Sammelkristallisation*, and he attributes to it an important part in various geological processes. In metamorphism it figures as one factor among others which go to determine the micro-structure of a reconstituted rock. It will suffice

---

[1] *Gesteinskunde*, 3rd ed. (1908), p. 167 ; 6–7th ed. (1921), p. 187.

here to note one way in which the ideal simplicity of the process may be disturbed. The approach to an equilibrium arrangement which we have pictured, depends upon the assumption that there is a fixed quantity of a given mineral in the rock. In the more general case there are reactions in progress, by which some minerals are being continually generated at the expense of others, and there can evidently be no finality so long as this continues.

AUREOLES OF THERMAL METAMORPHISM

The phenomena of thermal metamorphism are best studied in the rocks surrounding a large plutonic intrusion of a type not closely connected with orogenic movements. Here the heat, which is the proximate cause of the metamorphism, is of course drawn from the earth's internal store, but has been carried by an ascending molten magma. If, for a rough estimate, we suppose the temperature of intrusion to range from 1,000° to 600° C., according to the nature of the magma, and assume 0·3 as the mean specific heat of the solid igneous rock, we have from 300 to 180 calories as the measure of the heat given out in cooling ; to which must be added about 100 calories for the latent heat of fusion. From a large body of plutonic rock, therefore, an enormous amount of heat will be set free, and the whole of this passes by conduction into and through the surrounding rocks, raising their temperature in its passage.

This is not the only way in which rock-masses may become raised to an elevated temperature. As part of geological events of a large order, there can be a *direct* invasion of the earth's internal heat, conceivable as a general rise of the isothermal surfaces within the crust throughout a large tract. Igneous intrusion is likely to be an incident of this movement, but the intrusion cannot be regarded here as the cause of the high temperature. The rise of temperature affects the solid rocks, not only over a great areal extent, but to a great depth, and a very large increase of volume is therefore implied. The expansion cannot take effect uniformly, being free only in the upward direction, and so powerful shearing stresses are set up in the rocks. This case, therefore, falls under the head of what we distinguish as regional metamorphism. Even where the heating of rocks is brought about merely by an intrusion, if this is of very large dimensions, there must be a certain measure of shearing stress set up in the rocks of the aureole. In strictness, therefore, pure thermal metamorphism cannot be developed on a very large scale.

Under different and more local conditions the dynamic factor may figure, not as the consequence, but as the immediate cause of the

M.—2

heating of rocks, viz. by the mechanical generation of heat in crushing. This is a case to be discussed later. It is evident that here too the element of shearing stress removes the effects beyond the province of that simple type of metamorphism which will first engage our attention.

This simple type is sometimes styled ' local ' in contradistinction to ' regional ' metamorphism ; but pure dynamic metamorphism and effects resulting from the mechanical generation of heat have equally a local distribution. The term in common use among Continental geologists is ' contact-metamorphism ', although the phenomena may be exhibited at a distance of miles from any igneous contact. Inasmuch as the effects are due, not to contact, but to heat and high temperature, the term *thermal metamorphism* seems more appropriate.

The belt of metamorphosed rocks surrounding a plutonic intrusion, convèniently styled a *metamorphic aureole*, has a width which depends upon more than one factor, but mainly upon the size of the intrusive mass. Since not all kinds of rocks are equally susceptible of change, we may expect the visible effects to extend farther outward in some rocks than in others, but in many instances this selective action is in fact little apparent. The aureole of the granite boss of Shap, in Westmorland, comprises grits, flags, slates, calcareous shales and tuffs, pure and impure limestones, basalts, andesites, rhyolites, and various pyroclastic rocks ; but, if the first definite formation of new minerals be taken to mark the outer limit of the aureole, this can be drawn at about 1,200 or 1,300 yards from the granite-contact in very different rocks. The width so indicated is roughly equal to the semidiameter of the granite boss. The larger granite masses of Cornwall have aureoles up to two or three miles in breadth, and here the selective action is sometimes evident. At some places in Cornwall, indeed, effects of thermal metamorphism have been recorded even farther from any visible granite,[1] but the possible underground extension of the intrusions is to be taken into account. An intrusive mass like the Skiddaw granite in Cumberland, which shows only limited exposures but underlies the neighbouring rocks at a low inclination, has, for this reason, what appears a disproportionately large aureole of metamorphism. Its outer boundary, too, influenced by the varying surface-relief of the country, follows an irregular course on the map. In short, it is to be borne in mind that the aureole seen is merely the section by the actual ground-surface of a three-dimensional aureole.

The degree of metamorphism experienced at any place within an

[1] Effects have sometimes been ascribed to this cause, which belong truly to an earlier and quite independent regional metamorphism.

aureole must be supposed dependent, for a given type of rock, upon the highest temperature attained at that point ; but with the proviso that the high temperature was maintained long enough for the possibly slow process of metamorphism to be completed. Accordingly we can note, in a general sense at least, an advance in metamorphism from the outer limit of the aureole up to the actual contact. The highest grade of metamorphism, attained close to the contact, stands in relation with the temperature of intrusion of the plutonic magma, which is highest for the most basic types and lowest for those of acid or specially alkaline composition.

There is, however, another factor which is of importance. The intruded magma not only supplies heat to the surrounding rocks, but may also furnish more or less copiously the water and other volatile bodies which play so essential a part in all metamorphism. It is noticeable how generally muscovite-bearing granites and pegmatites are surrounded by an important aureole of metamorphism. This is not due to a high temperature of intrusion, but to the richness of these acid magmas in gaseous constituents. If we look to the actual mineralogical changes induced, we shall see that the highest grades of metamorphism are found especially near basic and ultrabasic rocks, intruded at high magmatic temperatures. The transference of water, in the gaseous state, from an igneous magma into and through the surrounding rocks, while promoting metamorphism, may set no obvious mark upon the resulting products. It is otherwise when volatile bodies of much greater chemical activity pass in quantity from an igneous magma into the rocks immediately contiguous. Metamorphism proper then becomes complicated by pneumatolysis, involving an important amount of metasomatism. The discussion of this pneumatolytic element in metamorphism will be conveniently deferred, until we have dealt with the effects of simple thermal metamorphism in different classes of rocks. The special effects, it should be observed, are confined to the neighbourhood of an igneous contact, and are essentially dependent upon that situation. They might approximately be designated ' contact-metamorphism ', had not that term already acquired a less suitable connotation.

If a metamorphic aureole embraces a varied succession of rocks, the progressive advance of metamorphism can be appreciated only in a general way, since there is no obvious term of comparison between the different rock-types. If the rocks are of one general type, the progress of metamorphism can be followed step by step ; and, if there be frequent alternations of different rocks, the same end can be attained by confining attention to one type. In favourable circumstances it is then possible to divide the aureole into successive *zones of meta-morphism*, possessing distinctive characters and representing successive grades of metamorphism. This was first attempted by Rosenbusch [1] in the metamorphosed Palæozoic slates known as the Steiger Schiefer round the granite mass of Barr-Andlau in the Vosges (Fig. 1). He distinguished three zones, from without inward : (l) *Knotenschiefer* or spotted slates, (ii) *Knotenglimmerschiefer* or spotted mica-schists, and (iii) *Hornfels*, rocks totally reconstructed. Other aureoles of

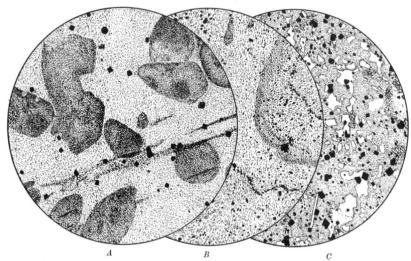

A   B   C

FIG. 1.—STAGES OF METAMORPHISM IN THE STEIGERSCHIEFER OF BARR-ANDLAU,
VOSGES ; × 25.

A. Typical ' Knotenschiefer '. The spots are marked by aggregations of the dis-seminated carbonaceous matter, now reduced to graphite. The only distinct new mineral is a little magnetite, formed at the expense of haematite flakes.
B. A somewhat more advanced stage. The fine micaceous material, making the bulk of the rock, is recrystallized, but only on a minute scale. The spots here have a different significance, being places where recrystallization is incomplete and much amorphous matter remains (p. 15).
C. Andalusite-cordierite-hornfels. The rock is now visibly recrystallized through-out, and is composed of magnetite, quartz, mica, andalusite, and (altered) cordierite.

metamorphism have been divided into zones on somewhat similar lines. When these have been defined with reference to general descriptive characters, such as the coming in and disappearance of spotted struc-tures, the divisions are necessarily arbitrary. It may be possible, however, to introduce an element of precision by taking as criterion of the outer limit of a zone the first appearance of some distinctive new mineral. The assumption required to make this valid is that such mineral demands a certain minimum temperature for its forma-

[1] *Die Steiger Schiefer und ihre Contactzone, Abh. Geol. Spezialkarte Elsass-Lothr.*, vol. i (1877), pp. 80–393.

tion, and is formed, in a rock of suitable composition, wherever that temperature has been reached. This would make the boundary as laid down an isothermal line, in the special sense that the maximum temperature reached in metamorphism was the same for points along that line. That for a given mineral there is a definite temperature of formation is doubtless true (with some allowance for varying pressure) in certain cases, e.g. where dimorphism of the reversible kind enters, and may be sufficiently near to the truth to be a serviceable assumption in some other cases.

SOME ILLUSTRATIVE AREAS

In the more special study of thermal metamorphism which follows, illustrations will be taken by preference from British examples ; and it will be convenient to enumerate here, for purposes of reference, the principal metamorphic aureoles in this country of which we possess some knowledge. Those surrounding the Cornish and Dartmoor granites will be found described in various Memoirs of the Geological Survey of England and Wales.[1] The rocks metamorphosed are the Devonian slates, including calcareous slates, and the Culm Measures, besides sills of dolerite and spilitic lavas.

In the North of England, the Skiddaw granite [2] metamorphoses the Skiddaw slates and grits, while the aureole of the Eskdale granite is in the Ordovician volcanic series, and that of the Shap granite [3] takes in a varied succession of Ordovician and Silurian rocks, sedimentary and volcanic. The Cheviot granite has metamorphosed the surrounding andesites of the Old Red Sandstone.[4] The Galloway granites,[5] intruded among Lower Palæozoic strata, are surrounded by well-marked belts of metamorphism, the rocks affected comprising grits, slates, and shales (some carbonaceous), cherts, and impure calcareous beds, besides the Ballantrae volcanic group.

[1] *Land's End* (1907), pp. 20–30 ; *Newquay* (1906), pp. 46–50 ; *Padstow and Camelford* (1910), pp. 63–77 ; *Bodmin and St. Austell* (1909), pp. 80–104 ; *Tavistock and Launceston* (1911), pp. 75–83 ; *Dartmoor* (1912), pp. 44–56 ; *Ivybridge* (1912), pp. 82–9. On Dartmoor see also Busz, *Neu. Jb. Min.*, B.Bd. xiii (1900), pp. 90–139, and *Geol. Mag.*, 1896, pp. 492–3.

[2] Harker, *Geol. Mag.*, 1894, pp. 169–70, and *Naturalist*, 1906, pp. 121–3 ; Rastall, *Quart. Journ. Geol. Soc.*, vol. lxvi (1910), pp. 116–40.

[3] Harker and Marr, *Quart. Journ. Geol. Soc.*, vol. xlvii (1891), pp. 292–327, and xlix (1893), pp. 359–71 ; Hutchings, *Geol. Mag.*, 1891, pp. 459–63.

[4] Kynaston, *Trans. Edin. Geol. Soc.*, vol. vii (1901), pp. 18–26.

[5] Teall in *The Silurian Rocks of Britain*, vol. i, *Scotland* (*Mem. Geol. Sur. U.K.*, 1899), pp. 632–51 ; Miss M. I. Gardiner, *Quart. Journ. Geol. Soc.*, vol. xlvi (1890), pp. 569–80 ; C. I. Gardiner and Reynolds, *ibid.*, vol. lxxxviii (1932), pp. 26–31.

The Caledonian plutonic intrusions of the Scottish Highlands break through members of the Dalradian and Moine series, which were already in the condition of crystalline schists. In the bordering tract, where the schists were in a low grade, the metamorphic aureoles are well marked ; viz. about the Ben Cruachan mass,[1] the Garabal-Glen Fyne complex,[2] and that of Càrn Chois, near Comrie.[3] The Deeside area is of interest for the striking effects produced by the Cairngorm and Lochnagar granites, especially in the Dalradian limestones.[4] The nepheline-syenites of the Assynt district of Sutherland have metamorphosed the Cambrian dolomites,[5] and like effects are to be observed near Tertiary intrusions of gabbro and granite in Skye.[6] In the latter area may be studied also the thermal metamorphism of igneous rocks, both lavas and dykes[7] ; and like phenomena are exhibited in great variety about the corresponding plutonic intrusions in the Isle of Mull.[8]

The Foxdale granite, in the Isle of Man, has a metamorphic aureole in the Manx slates.[9] Some striking effects of metamorphism are shown by the Ordovician sediments bordering the Leinster granites at numerous localities in Counties Dublin and Wicklow, but there are at present few published accounts dealing with this area. It may be remarked in conclusion that interesting illustrations of thermal metamorphism, though limited in extent, are sometimes to be found near large sills and dykes of dolerite, such as the Whin Sill of Teesdale,[10] the Plas Newydd dyke in Anglesey,[11] and the dolerites of Portrush[12] and Larne[13] in Antrim.

The literature dealing with numerous European districts is too

---

[1] *Geology of Oban and Dalmally* (*Mem. Geol. Sur. Scot.*, 1908), pp. 139–52.

[2] Cunningham-Craig, *Quart. Journ. Geol. Soc.*, vol. lx (1904), pp. 25–6 ; Clough, *Geology of Cowal* (*Mem. Geol. Sur. Scot.*, 1897), pp. 98–101.

[3] Tilley, *Quart. Journ. Geol. Soc.*, vol. lxxx (1924), pp. 22–70.

[4] *Geology of Braemar, Ballater, and Clova* (*Mem. Geol. Sur. Scot.*, 1912), pp. 104–9 ; Hutchison, *Trans. Roy. Soc. Edin.*, vol. lvii (1933), pp. 557–92.

[5] Teall, in *The Geological Structure of the North-West Highlands* (*Mem. Geol. Sur. Gr. Brit.*, 1907), pp. 453–62.

[6] Harker, *Tertiary Igneous Rocks of Skye* (*Mem. Geol. Sur. U.K.*, 1904), pp. 144–52.

[7] *Ibid.*, pp. 50–4, 318–19.

[8] *Tertiary and Post-Tertiary Geology of Mull* (*Mem. Geol. Sur. Scot.*, 1924) ; M'Lintock, *Trans. Roy. Soc. Edin.*, vol. li (1915), pp. 25–31.

[9] Watts, in *Geology of the Isle of Man* (*Mem. Geol. Sur. U.K.*, 1903), pp. 106–8.

[10] Hutchings, *Geol. Mag.*, 1895, pp. 122–31, 163–9, and 1898, pp. 69–82, 123–31.

[11] Harker, *ibid.*, 1887, pp. 413–14.

[12] Cole, *Proc. Roy. Ir. Acad.*, vol. xxvi (B) (1906), pp. 56–66.

[13] Tilley, *Min. Mag.*, vol. xxii (1929), pp. 77–86.

voluminous to be cited here. We may mention as of special interest the works of various writers on the Harz, Brittany, and the Predazzo district, as well as the researches of Rosenbusch in the Vosges, of Lacroix in the Pyrenees, of Beck and others in Saxony, and of Brögger and Goldschmidt in the Oslo district.

VITRIFICATION OF SHALES AND SANDSTONES

As already intimated, thermal metamorphism is not wholly confined to the aureoles of large plutonic intrusions. Any striking effects due to dykes or sills must, however, be regarded as exceptional. In the instances cited above the rocks invaded were of a kind peculiarly susceptible to metamorphism, such as impure calcareous sediments. In general the metamorphism bordering a minor intrusion is very limited, both in extent and in kind. It is seen in such changes as slight induration of argillaceous rocks, decoloration of red sandstones, and incipient marmorization of limestones.

There is, however, one special case worthy of notice, viz. the *vitrification* of argillaceous or arenaceous sediments for a few inches from their contact with a dyke or sill. We have already seen (p. 15) how a partially metamorphosed rock may be locally vitrified because cooling was too rapid to permit recrystallization. That the effect was there confined to isolated spots was due to the very small quantity of solvent present in the rock. At an igneous contact, however, it is possible that a sufficient supply of solvent, viz. water, may be supplied directly from a magmatic source, and the rock may become bodily vitrified. The effect extends only a few inches from the contact, and it is rare, because it requires a concurrence of favourable conditions. The high initial temperature demanded is realized only in basic or ultra-basic intrusions, and magmas of this kind are those least rich in water. Further, to ensure a relatively rapid cooling, the intrusion must be one of no great dimensions, and must be apart from any regular aureole.

Vitrification is found also in another case, viz. where fragments of some sedimentary rock have been enclosed in a basic lava or dyke or sill.[1] Here, however, complication is often introduced by some intermingling of the magma itself with the fused rock.[2]

Such superficial and local effects as the calcining of limestones and the charring of carbonaceous deposits at contact with a lava-flow do not call for particular notice. It is sufficient to observe how they

---

[1] The name *buchite* is applied to such vitrified rocks, or sometimes more specifically to vitrified sandstones.

[2] Thomas has described interesting examples from sills in the Isle of Mull: *Quart. Journ. Geol. Soc.*, lxxviii (1922), pp. 229–59.

differ from the phenomena of metamorphism effected within the earth's crust. Nor is it part of our plan to discuss solfataric effects, in which there enters commonly a large element of metasomatism. It is of interest to remark, however, that thermal metamorphism of the ordinary type has sometimes been produced locally by the passage of heated gases, without any intrusion of magma. Good illustrations of this are seen in the Isle of Rum.[1] The Torridon sandstone there exposed is traversed in many places by vertical crush-bands, ranging from mere fissures to fifty feet in width. Some of these have been injected in Tertiary times with basaltic magma, much modified by dissolving sandstone fragments ; others contain no igneous material. The latter, as well as the former, give proof of thermal metamorphism which, though so narrowly limited, is of a high grade ; and this can be attributed only to the passage of highly heated gases through the shattered rock.

[1] Harker, *Geology of the Small Isles* (*Mem. Geol. Sur. Scot.*, 1908), pp. 60–7.

CHAPTER III

STRUCTURES OF THERMALLY METAMORPHOSED ROCKS

Metamorphism Contrasted with Magmatic Crystallization—Residual Structures —Crystal Growth in the Solid—Force of Crystallization—Characteristic Structures of Thermal Metamorphism—Significance of Inclusions in Crystals.

METAMORPHISM CONTRASTED WITH MAGMATIC CRYSTALLIZATION

ROCKS which have become totally recrystallized in metamorphism present some obvious features of likeness to crystalline rocks of direct igneous origin. Many of the component minerals are common to the two classes. In textural and structural characters also, i.e. in respect of the size, shape, and disposition of the several constituents and their visible relations with one another, resemblances are easily perceived. Consciously or unconsciously, the student is liable to give to these distinctive characters in metamorphosed rocks the same significance that they bear in igneous rocks. It is of vital importance therefore to make it clear at the outset that such similarities of texture and structure do not import any real analogy. The determining factors are wholly different in the two cases ; and although metamorphism may give rise to peculiarities which mimic well-known structures of igneous rocks, such as the ophitic and the porphyritic, the interpretation of them is in no wise the same.

We have to observe in the first place that the crystallization or recrystallization of minerals in metamorphism proceeds, not in a fluid medium, but *in the heart of a solid rock*, an environment which cannot fail to modify greatly their manner of growth. Further, the rock may be, during the process of recrystallization, in a condition of shearing stress imposed by external forces, and in the typical crystalline schists this additional factor has had a very important influence. These two postulates, the one universal and the other conditional, suffice to differentiate metamorphism fundamentally from the crystallization of a molten magma, and that in a manner which must make itself evident in distinctive structural characters.

There is another fundamental distinction to be remarked, which is of even greater moment. The crystallization of an igneous rock-magma proceeds with falling temperature, and under such conditions a continuous readjustment of chemical equilibrium is seldom, if ever, realized in so complex a system. It must often happen that crystals which have separated at a higher temperature are no longer in equilibrium with a magma which has cooled and become changed in composition, and are, or ought to be, attacked by it. Bowen [1] in particular has discussed what he styles the Reaction Principle in Petrogenesis, and enforced its wide application. Although his conclusions may appear too sweeping, and the details of his formal scheme be open to criticism, the validity of the general argument can scarcely be questioned. None the less, a survey of the actual characters of igneous rocks in general, taking note of the complex mineralogical constitution of many common types and the frequency of zoned crystals, makes it evident that the reactions demanded by equilibrium have in most cases taken effect only very partially or not at all. In so far as this is true, the completed igneous rock as we see it represents a merely metastable arrangement. Be this as it may, the several constituent minerals belong to different stages of the process of consolidation. There is, for a given rock, a definite order of crystallization—including of course simultaneous crystallization of two or more minerals—and this order is plainly written in the visible structure of the rock.

In interpreting the structures of metamorphosed rocks the conception of an order of crystallization, with all that it implies, is to be totally discarded. The reconstruction of a rock in metamorphism proceeds *with rising temperature*, and all the facts go to show that, at least in any advanced stage of the process, adjustment of equilibrium in general keeps pace somewhat closely with the rise of temperature. Minerals formed at an earlier stage are, potentially at least, parties to the reactions which succeed at higher temperatures. If they do not change their composition or give place to other new minerals, they are still to be conceived as continually rejuvenated from stage to stage. The crystals are, as a rule, homogeneous, without zoning. Where crystals of two different new minerals come together, the boundary between them is not determined by priority of formation, for there is in this sense no priority, all the constituent new minerals of the rock being in effect of simultaneous crystallization.

This rough statement of the case will suffice to show that the textural and structural characters of metamorphosed rocks are determined by factors quite different from those which are operative in rocks of igneous origin. The subject is indeed one of some complexity, and it has been confused by the common practice of treating ' contact '

[1] *Journ. Geol.*, vol. xxx (1922), pp. 177-98.

(i.e. thermal) and regional metamorphism as two wholly unrelated classes of phenomena.

RESIDUAL STRUCTURES

Before discussing the characteristic new features developed in metamorphism, it is to be remarked that there is often, in the earlier stages at least, a survival of structures proper to the original rocks. These *residual structures*, aptly named by Sederholm ' palimpsest ' structures, may vary from a clear transcript of the original to a mere shadowy reminiscence. Where internal differential movement is involved, pre-existing structures are likely to be soon obliterated, but in purely thermal metamorphism they may be still traceable in an advanced grade, despite mineralogical reconstruction of the rock. Their preservation is due to the very narrow limit set to diffusion in the lower grades of metamorphism ; and their gradual fading out is a consequence of the more enlarged amplitude of diffusion which goes with higher temperature. It follows that there is a direct relation between the scale of magnitude of any original structure and its possible persistence with advancing metamorphism. The appearance of a clastic origin is very quickly lost in a fine-textured sediment, but remains evident longer in a coarse grit, while a pebbly structure is still to be detected in a high grade of metamorphism, after pebbles and matrix alike have been totally reconstituted (Figs. 26, *A* ; 134, *B*). The larger structures of igneous rocks, such as porphyritic and amygdaloidal, may still be indicated in outline when all the original minerals have been recrystallized or replaced by new minerals (Figs. 42, *A* ; 43). We can of course reason back from these facts of observation. Residual structures afford direct proof of the narrow limits of diffusion of material in metamorphism, and can be used to form estimates of those limits. To the geologist these relics have an obvious value as giving indications of the original nature of a rock, now represented, it may be, by an entirely new mineral-aggregate.

CRYSTAL GROWTH IN THE SOLID

We come now to those *new structures* which are set up in the process of metamorphism, and are exemplified in great variety in rocks which have suffered reconstruction under the conditions already glanced at. The critical study of this subject is a thing of recent years, and much is still to be learnt before a complete presentation of the matter can be attempted on systematic lines. So striking a feature as the foliation of crystalline schists naturally attracted attention at an early time, and its significance was discussed by Darwin in the middle

of the last century. Later investigation, with the aid of the microscope, made petrographers aware that, not only crystalline schists,
but rocks recrystallized in thermal metamorphism have peculiarities
of micro-structure which distinguish them from igneous rocks. Salo-
mon in particular, describing the rocks of the inner ring of metamorphism about the tonalite of the Adamello Mountains, recognized
a class of structures to which he gave the name ' contact-structures '.
Among these he distinguished the simple mosaic type and various
sieve-like, skeletal, and spongy arrangements arising from the interpenetration or inclusion of one mineral by another. He also used
the epithet ' contact ' as prefix to the name of a rock-type : thus
a ' contact-pyroxenite ' is a rock with the mineralogical composition
of a pyroxenite, but with a micro-structure of a kind proper to a
thermally metamorphosed rock. To connect such structural peculiarities definitely with the conditions of crystallization in a solid medium
with rising temperature represents, however, a further step.

What long delayed a clearer understanding of the micro-structure
of metamorphosed rocks was the tacit assumption that the rocks
themselves play a merely passive part in the process of metamorphism.
Despite the splitting of rocks by frost, the formation of crystals of
pyrites in slate, of selenite in clays, and various other familiar phenomena, the powerful mechanical force which can be exerted by growing
crystals received in this connexion only a tardy recognition. Its
importance was first clearly recognized in 1903 by F. Becke,[1] in a
paper dealing generally with the mineralogical composition and
structure of the crystalline schists. The principles there plainly but
very briefly set forth have been more fully developed by others, and
especially by Grubenmann in his well-known work, also treating ex-
plicitly of the crystalline schists. It is a consequence of the complete
divorce which Continental geologists have made between regional
and ' contact ' metamorphism, that the writers named do not distinguish those features which depend merely upon crystal-growth in
the solid, and are therefore common to metamorphism of both kinds,
from those which are related to a state of internal shearing stress
maintained by external forces. For the same reason others, who
have followed the lead given by Becke's exposition, have generally
been slow to perceive how much of what he lays down for crystalline
schists is no less true of thermally metamorphosed rocks also.[2] In
accordance with the plan of the present work those peculiarities

[1] *Comptes Rendus ix Cong. Geol. Intern. Vienne*, 1903 (1904), pp. 553-70.
[2] Compare Erdmannsdörffer, *Centr. f. Min.*, 1909, pp. 501-3 ; *Jahrb. k.
preuss. geol. Land.*, vol. xxx (1909), pp. 341-52.

which belong to the crystalline schists, being related to crystallization under special stress-conditions, are reserved for later consideration. We are concerned at present with the *structures of thermally meta-morphosed rocks* only, depending on crystallization in the solid, free from any externally imposed shearing stress. In discussing these we can still accept Becke as our guide.

A mineral separating from an igneous magma grows, like a crystal of salt suspended in its saturated solution, by the tranquil addition of layer upon layer under no more restraint than is implied in viscosity and surface-tension. In metamorphism there is no such freedom. The growing crystal must make a place for itself against a solid resist-ance, and is to be conceived as *forcibly thrusting its way outward from its starting-point.* In consequence of the constraint so imposed, metamorphosed rocks of all kinds come to have a peculiar structure, or class of structures, for which we adopt Becke's term *crystalloblastic*, connoting the idea of sprouting or shooting (βλαστάνω).

The very great mechanical force that can be exercised by a grow-ing crystal, which encounters resistance is illustrated by a simple experiment devised by Becker and Day.[1] They showed that a crystal of alum, growing in a saturated solution, can lift a heavy weight, and in so doing sustains a pressure of many pounds to the square inch. Indeed, geological observation and experiment alike go to prove that growing crystals are capable of exerting forces of the same order of magnitude as their own crushing strength.

In the experiment cited the material to build up the alum crystal was drawn from the surrounding medium at large ; so that the lifting of the weight was to provide space for a certain amount of new material. In metamorphism, however, a crystal grows at the expense of part of the rock-substance, which it replaces, and there is in general no increase of total volume but often a certain diminution. In so far, therefore, as the crystal is built up at the cost of immediately con-tiguous material, it will experience no constraint in respect of volume but a definite constraint in respect of shape. In other words, shearing stresses will be set up.[2] Actually, as we have seen, interchange of material can take place within certain narrow limits ; and such readjustment, proceeding concurrently with crystal-growth, will in some measure reduce or relieve the stress. Despite such accommoda-tion, it is evident that, as reconstruction proceeds in the rock, and the

[1] *Proc. Wash. Acad. Sci.*, vol. vii (1905), pp. 283–8 ; *Journ. Geol.*, vol. xxiv (1916), pp. 313–33.

[2] The alum crystal was likewise subjected to shearing stress, being loaded above but free at the sides.

newly-formed crystals come to impinge upon one another, shearing stresses must inevitably be set up, and potentially stresses of considerable intensity. Now there are several ways in which these potential stresses may be actually relieved to a greater or less extent, and we shall see that these various devices are reflected in the multiform peculiarities of structure met with in metamorphosed rocks. In fine, the key to the interpretation of these structures is to regard them as *contrivances for evading or minimizing the internal stresses set up by crystal-growth in a solid medium.*

FORCE OF CRYSTALLIZATION

Accepting Becke's 'force of crystallization' (Kristallisationskraft) as a definite property of a growing crystal, we must expect to find, first, that it is a *vector property*, differing in intensity in different crystallographic directions ; secondly, that it is a *specific property*, differing in different mineral species ; and, thirdly, that, like other properties of crystals, it is a *function of temperature*. Each of these principles has its importance in relation to the structures of metamorphosed rocks.

In proportion as the force differs sensibly in different crystallographic directions, and according as such force is called into play by resistance, the effect should be apparent in the shape assumed by the completed crystal. This is in all cases of a very simple kind. There is here little of the variety of habit shown by some pyrogenetic minerals ; still less the richness in facets seen on crystals grown from aqueous solution in geodes and fissures. Crystal-boundaries of any kind are much less general than in igneous rocks. Where they are found, the principal, and commonly the only, faces present are those which are parallel to well-marked cleavages ; and we may infer that the force of crystallization is greater along than across a cleavage. A large proportion of the minerals of metamorphism fall accordingly under one or other of two types ; some having the tabular or flaky habit which goes with a single perfect cleavage, as in the micas ; others the columnar or acicular habit which goes with two good cleavages, as in the amphiboles. Very commonly the mica-flakes show ragged edges, and the prisms of amphibole are devoid of terminal planes. Twinning is decidedly less frequent in metamorphic than in pyrogenetic minerals, and in particular the plagioclase felspars are often free from albitelamellae.

The tabular and columnar forms are familiar, for the appropriate minerals, in crystals grown freely from a molten magma, but the characteristic habit is very decidedly exaggerated in crystals developed in a solid rock. There is another point of difference. In igneous rocks the unequal development in different directions is strongly marked only in microlites, and tends to disappear in full-grown crystals : in metamorphosed rocks the peculiarity of habit, often little noticeable at first, becomes more pronounced as the crystals continue to grow in the face of resistance.

Such unequal growth in the heart of a solid rock must manifestly set up internal shearing stresses. If, for a mineral making up a considerable part of the rock, the crystals had a common orientation, these stresses would be additive. By a different arrangement, in

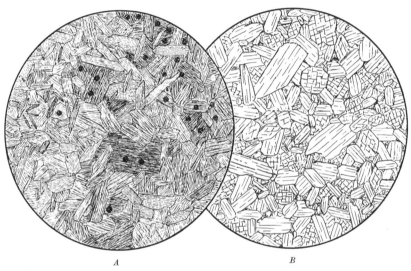

A   *B*

FIG. 2.—DECUSSATE STRUCTURE ; × 23.

A. Biotite-Hornfels, De Lank, near the Bodmin Moor granite, Cornwall. The dark spots are ' haloes ' surrounding radioactive inclusions. See also Figs. 10, *B*, etc.
    B. Wollastonite-rock, Moonbi, New South Wales. See also Figs. 30, *B*, etc.

which corresponding axes of contiguous crystals lie in diverse directions, the stresses can be made in great measure to cancel one another by mutual accommodation. This is the arrangement typically exhibited in thermally metamorphosed rocks, and the micro-structure described, a criss-cross or *decussate* structure, is for such rocks highly characteristic. The component crystals lie in all directions ; *not* at random, by the operation of a mathematical law of chance, but as part of a definite mechanical expedient for minimizing internal stress. This peculiar structure is most noticeable in a rock which is composed largely of minerals with a flaky or a columnar habit (Fig. 2). The interlacing of the little scales or prisms imparts to a ' hornfels ' of this type a remarkable toughness and a distinctive kind of fracture.

The decussate structure is, however, by no means universal, even in rocks containing abundant mica. Our argument has assumed that the rock had originally no structure of the kind that implies weakened cohesion in a particular direction. The conditions are different in the earlier stages of metamorphism of a shale or slate which possessed a marked fissility, due either to bedding-lamination or to superinduced cleavage. As their fissile property attests, such rocks offer much less resistance along the bedding or cleavage than across it. Until this difference is obliterated by total reconstruction of the rock, new

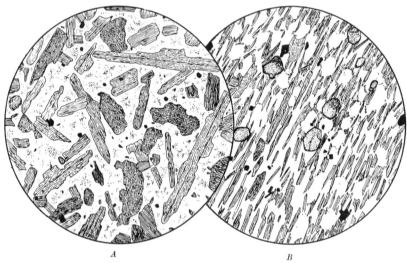

FIG. 3.—STRUCTURES OF HIGHLY METAMORPHOSED SEDIMENTS ; × 23.

A. Biotite-Hornfels, inclusion in nordmarkite, Grorud, Norway. The flakes of biotite have the typical decussate arrangement.
B. Garnet-Biotite-Schist, Ordovician slate near a granite-contact, Glendalough, Co. Wicklow. Here the flakes still retain a parallel arrangement which marks the original lamination.

minerals are generated and grow in a medium having peculiar mechanical properties. The flakes naturally push their way in the direction of least resistance, and so acquire a regular parallel orientation. The parallel structure in this case has in fact precisely the same significance as the decussate structure in the other case ; i.e. it is a contrivance to elude the setting up of internal shearing stress. The parallel orientation may persist as a residual structure into a higher grade of metamorphism, but is ultimately lost. Indeed, schistosity arising in this way is essentially a residual effect, perpetuating and sometimes accentuating an initial property of the rock. The schistosity which characterizes so many types of rocks in regional metamorphism is, on the contrary, a new property with a direction imposed by external forces. The immunity from externally provoked stress which we have postulated in thermal metamorphism is, however, not absolute ; and in a broad aureole bordering a large intrusive mass a tendency to parallel orientation from this cause is sometimes observable in highly metamorphosed rocks (Fig. 3, B).

Parallel arrangement of mica-flakes as a residual structure is related, not necessarily to the general lie of the bedding, but to the direction of lamination at the spot, which may have been affected

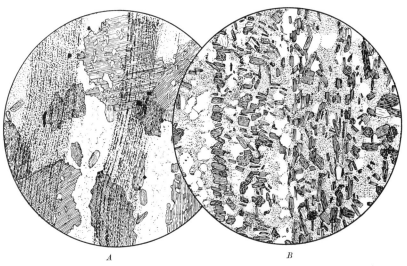

A   *B*

FIG. 4.—METAMORPHOSED SLATES PRESERVING BANDED STRUCTURES ; × 25.

A. Manx Slate near an aplite dyke, Crosby, Isle of Man. Shows large irregular flakes of biotite in a matrix of fine sericite and quartz. Trains of inclusions, marking the original lamination, pass undisturbed through the biotite.
B. Highly metamorphosed Devonian slate enclosed in the St. Austell granite, Cornwall : essentially of biotite, some muscovite, altered cordierite, and quartz. The mica-flakes have a definite orientation, in some bands parallel but in others transverse to the original bedding.

by contortion on a small scale. Such disturbance has sometimes been localized in certain narrow bands which were unusually yielding, giving rise there to an arrangement of new-formed mica-flakes transverse to the bedding (Fig. 4, B).

CHARACTERISTIC STRUCTURES OF THERMAL METAMORPHISM

Despite mutual accommodation, crystals starting from neighbouring points and thrusting out, each from its own centre, necessarily come into competition in a struggle for space, or more accurately for shape. In a completely reconstituted rock the shape of each crystal has been determined by its encounters with its immediate neighbours. If all are of the same mineral, as in a pure marble or quartzite, they compete on equal terms, and no one is able to assert its natural crystal outline against its neighbours. The compromise results in a mosaic which, as seen in section, is made up of polygonal or partly inter-locking elements of approximately equal size with sutural junctions (Fig. 20, below). A rock composed of two or more minerals shows a like structure, if the several minerals do not differ sensibly in their force of crystallization. Where any pronounced difference in this respect exists, the simplicity of relations is modified accordingly, the stronger mineral being able, in greater or less degree, to insist upon its natural development at the expense of its weaker neighbour.

This, it will be observed, is essentially different from the mutual relations between the minerals of an igneous rock, determined by the order in which they have crystallized. That is not a fixed order for all igneous rocks, but depends upon the relative proportions of the several minerals present as well as upon their specific properties ; so that in one rock augite is idiomorphic towards felspar and in another the reverse. In metamorphism, as has already been laid down, the conception of an order of crystallization finds no place. Rather, in a rock totally reconstituted, we may assume a practically simultaneous crystallization of all the minerals involved (p. 30). If we accept this reasoning in its entirety, and suppose that the rock retains no trace of its past history, it follows that the mutual relations of the different crystals must be fixed solely by specific properties of the minerals themselves. The determining factor is, in short, the force of crystal-lization, which is found to differ widely in different minerals. The strongest, such as rutile and spinel, constantly make good their claims against all competitors, and usually exhibit good crystal-boundaries ; the weakest minerals in this sense, such as the potash-felspars, are defeated in the struggle, and never develop their natural forms ; for any mineral in general the issue depends upon its own force of crystal-lization as compared with that of the contiguous minerals, its immedi-ate competitors. Since the terms ' idiomorphic ' and ' xenomorphic ' carry a connotation which is proper to igneous but alien to meta-morphosed rocks, we shall adopt in their stead the terms used by Becke, primarily in connexion with the crystalline schists, *idioblastic* and *xenoblastic*. A stronger mineral is idioblastic against a weaker one in contact with it, and the relative force of crystallization of the several minerals in a rock can therefore be determined from their observed mutual relations. Becke, and after him Grubenmann, have shown that the constituent minerals of the crystalline schists can in this way often be ranged in an order of relative strength, to which they give the name *crystalloblastic series*. The minerals of thermally metamorphosed rocks can likewise be at least approximately ranged according to the same property, i.e. in a list beginning with the strongest and ending with the weakest. Since some of the minerals are found only in certain classes of rocks, it is more convenient to make two lists.

In Argillaceous and Arenaceous Rocks

Rutile, pleonaste ;
Garnet, sillimanite, tourmaline ;
Magnetite and ilmenite, andalusite ;
Muscovite, biotite, chlorite ;
Plagioclase and quartz, cordierite ;
Orthoclase and microcline.

In Calcareous and Igneous Rocks

Sphene, spinel minerals, pyrites ;
Wollastonite, lime-garnet, apatite ;
Magnetite and pyrrhotite, zoisite and epidote ;
Forsterite and chondrodite, hypersthene and diopside, chalybite and dolomite ;
Scapolite, albite, muscovite, biotite and phlogopite ;
Tremolite, idocrase and calcite ;
Plagioclase, quartz, orthoclase and microcline.

The order here set down will be found to hold good very generally, but is not to be accepted as a rigidly fixed standard. It is probable that the relative crystallizing strength of different minerals may be modified by factors which are not easily taken into account, such as pressure and the concentration of a solvent ; but the principal cause of exceptions or apparent exceptions to the rule is to be found in the effects of retrograde metamorphism. If a mineral has been produced as a pseudomorph during the declining phase of metamorphism, its visible form and relations are determined, not by its own properties, but by those of the mineral which it has replaced.

In works dealing with the crystalline schists an accepted nomenclature is in use for describing the varied micro-structures there encountered. Some of the terms are equally applicable to simple thermal metamorphism, inasmuch as the structures which they denote arise merely from crystallization in the solid, and are not necessarily related to shearing stress impressed from without. The simplest type is the *granoblastic*, answering to the mere mosaic arrangement which has been noticed. It is necessarily modified when minerals of pronounced tabular or columnar habit make a noteworthy part of the rock, and, as we have seen, more than one variety of structure can then be recognized. In the tough close-textured type of rock properly styled ' hornfels ' [1] the decussate arrangement prevails, while the mica-schists are characterized by a parallel disposition of the mica-flakes. Some more special structures have received convenient names. In the *porphyroblastic* some one mineral makes crystals of conspicuously larger size than the other constituents of the rock (Figs. 4, *A* ; 5, *B*).

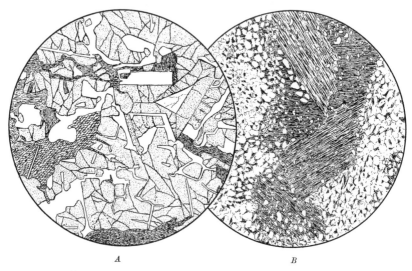

A   *B*

FIG. 5.—SPECIAL STRUCTURES IN METAMORPHOSED ROCKS ; × 23.

A. Diablastic structure : an intergrowth of andradite garnet, colourless axinite, and green hornblende in a metamorphosed dolerite near Falmouth.
B. Poeciloblastic structure in porphyroblasts in biotite-cordierite-hornfels, Rietpoort, N. of Parys, Transvaal.

The term *diablastic* comprehends various intergrowths of two or more minerals, whether parallel or radiate or on some less regular plan (Fig. 5, *A*). In the *poeciloblastic* or ' sieve '-structure, which often goes with the porphyroblastic, the crystals of one mineral enclose numerous smaller crystals of another (Fig. 5, *B*). The form of these words should remind the student that, for instance, the poeciloblastic structure in metamorphosed rocks has only a superficial resemblance to the poecilitic in igneous rocks, not a real analogy.

---

[1] This word, however, is often employed in so wide a sense as to lose any precise meaning, and some writers have affronted the English language by using ' hornfels ' as a verb.

We may go farther than this. These terms themselves, as applied to metamorphosed rocks, are to be understood as having merely a descriptive, not a genetic signification, for the structure denoted by a given name may arise in more than one way. Doubtless the principle of minimizing internal stress may often furnish the key. Competing crystals of different minerals may lessen the stress by a compromise which takes the form of interpenetration ; or a large growing crystal encountering small crystals of other minerals may find it easier to swallow them than to thrust them aside. Very often, however, the explanation is to be sought in some earlier stage of the progress of metamorphism. Where two minerals are seen intimately intergrown, or one mineral is crowded with inclusions of another, it is probable that the two have often been formed as joint-products of some reaction, and have not yet been able to disentangle themselves.

SIGNIFICANCE OF INCLUSIONS IN CRYSTALS

In the earlier stages of metamorphism, however, inclusions are of a residual nature, representing material which could not be incorporated in the new-formed minerals. The original lamination of a finely banded sediment is often indicated by trains of minute inclusions of this kind (Fig. 4, A). If they are of a mineral such as graphite, which does not take part in chemical reactions, they may persist into an advanced grade, though they are liable to be disturbed by growing crystals of a strong mineral.

Growing crystals endeavour to clear themselves by expelling foreign inclusions of any kind, but their power to do so depends upon their inherent force of crystallization. A very weak mineral, like cordierite, may remain almost to the last crowded with inclusions. In a mineral of moderate strength, like scapolite, the elimination of inclusions often goes hand-in-hand with the development of crystal shape (Fig. 6, A and B). It should be remarked, however, that the force of crystallization is effective only when it is called into play by resistance, as growth proceeds. So even a strong mineral, such as garnet, often shows a nuclear portion full of residual inclusions, while the marginal part of the crystal is clear. This contrasts with the peripheral arrangement of inclusions so common in the crystals of igneous rocks. The dependence of force of crystallization upon temperature also has its application here. In a very high grade of metamorphism all but the weakest minerals tend to be free from inclusions, always excepting such as are being currently generated by high-temperature reactions.

We have seen that the force of crystallization, upon which the expulsion of inclusions depends, is a vector property. An instructive case is that in which a growing crystal has been able to brush aside foreign material, but not completely to eject it ; the result being that trains of inclusions remain caught in the crystal along certain directions in which the force was least effective. The chiastolite variety of andalusite is a familiar example, and comparable phenomena are to be observed less frequently in cordierite,[1] staurolite,[2] and garnet.[3] Of these four minerals, garnet stands very high in the crystalloblastic series and cordierite very low, while the other two are of medium

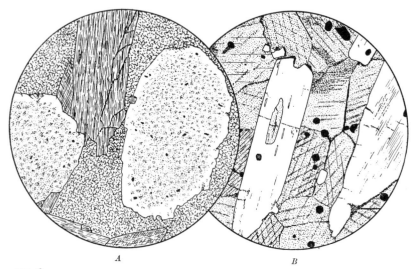

A   *B*

FIG. 6.—SCAPOLITE IN METAMORPHOSED LIMESTONES, from the Pyrenees ; × 25.

A. Pouzac, Hautes Pyrénées : showing scapolite (dipyre), with numerous minute inclusions, and green actinolite.
B. Arnave, Ariège : a coarsely crystalline limestone enclosing clear crystals of scapolite and grains of pyrrhotite. In this higher grade the scapolite is cleared of inclusions, and has developed crystal outlines.

strength. What they have in common is an exceptional inequality of the forces in different crystallographic directions.

The peculiar features of *chiastolite* have been often discussed.[4]

[1] Kikuchi, *Journ. Sci. Coll. Tokyo*, vol. iii (1890), pp. 313–34.

[2] Penfield and Pratt, *Amer. J. Sci.* (3), vol. xlvii (1894), pp. 87–9.

[3] Renard, *Bull. Mus. Roy. Belg.*, vol. i (1882), pp. 18–19, and plate I, fig. 1 ; Karpinsky, *Mél. Phys. Chim. St. Pétersbourg*, vol. xii (1887), pp. 639–45.

[4] Rohrbach, *Zeits. Deuts. Geol. Ges.*, vol. xxxix (1887), pp. 632–8 ; Becke, *Tsch. Min. Pet. Mitt.*, vol. xiii (1892), p. 256 ; Sederholm, *Geol. Fören. Stockh. Förh.*, vol. xviii (1896), p. 390 ; Flett, *Geol. Lower Strathspey (Mem. Geol. Sur. Scot.*, 1902), pp. 54–6 ; Mawson, *Mem. Roy. Soc. S. Austr.*, vol. ii (1911), pp. 189–210.

The dark area in the centre (Fig. 7) represents the nucleus of the crystal, unable at that stage to free itself from inclusions. Subsequent growth has been effected by a thrusting outward, which was most effective in the directions perpendicular to the prism-faces. Much of the foreign matter brushed aside in this growth accumulated on the edges of the prism, and was enveloped by the growing crystal. The arms of the dark cross represent thus the traces of the prism-edges as the crystal grew. Finally, re-entrant angles may be left, but more usually these are filled in by the latest growth of the crystal, remaining

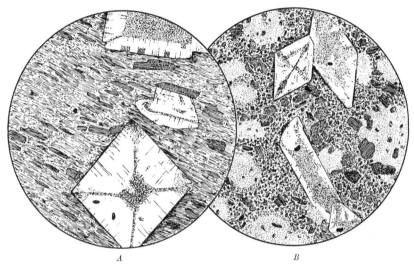

A   B

FIG. 7.—METAMORPHOSED SKIDDAW SLATES WITH CHIASTOLITE,
CUMBERLAND ; × 23.

A. This rock has not yet lost the parallel arrangement of the new-formed flakes of biotite.
B. Chiastolite-Cordierite-Biotite-Hornfels, representing a higher grade of metamorphism. The cordierite is still in ill-defined round grains, which contain inclusions of biotite.

full of inclusions. A longitudinal section of the crystal would show the arrangement of inclusions as modified by the thrusting forth in the direction of the long axis. The streaming out of trains of inclusions away from the centre presents almost a visible picture of the operation of the forces of crystallization. It is in marked contrast with the concentric zones of inclusions common in the crystals of igneous rocks, Chiastolite affords also an impressive indication of the magnitude of the forces developed in crystal growth in the solid, for these forces have often been sufficient to shatter the crystal itself. The cracks follow the prismatic cleavage, but only one of the two cleavages is opened in each sector (Fig. 7, *A*). This feature is most clearly exhibited when the crystals enclose an unusually large quantity of foreign matter.[1] A very prevalent feature of *cordierite* is the pseudo-hexagonal habit with six-fold twinning, and it is in connexion with this that a regular arrangement of inclusions is sometimes found. Kikuchi has given the name ' cerasite ' to this variety of cordierite. The inclusions are grouped along hexagonal pyramids with apex at the centre of the crystal and bases on the basal planes (Fig. 8, *A*). In *staurolite* [2] a

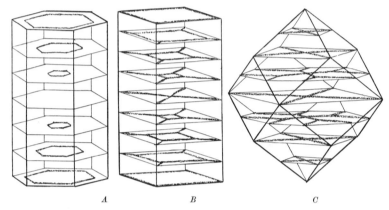

A   B   C

FIG. 8.—REGULAR ARRANGEMENT OF INCLUSIONS IN CRYSTALS, *Illustrated by a Series of Sections Parallel to the Basal Plane.*
A. Cordierite.   *B.* Staurolite.   *C.* Garnet.

regular pattern is not often seen. The inclusions are ranged along rhombic pyramids having apices at the centre and bases coincident with the basal planes of the rhombic crystal, and also, outside these pyramids, along median planes parallel to the two pinacoids (Fig. 8, *B*). In *garnet* too any special arrangement is not common. It is seen in manganese-garnets from the Ardenne and the Urals, and may sometimes be observed in the garnet, likewise manganiferous, of the New Galloway aureole. The inclusions are here distributed along twelve pyramids with apices at the centre and bases on the dodecahedral faces [3] : in other words, along the traces of the edges of the growing dodecahedral crystal (Fig. 8, *C*).

[1] Compare Sederholm, *Geol. För. Förh.*, vol. xviii (1896), pp. 390–3, with figures.

[2] This is a mineral especially characteristic of crystalline schists, but is included here for the sake of completeness.

[3] Karpinsky, *loc. cit.* Renard states erroneously that they lie along the principal planes of symmetry. See also Raisin, *Quart. Journ. Geol. Soc.*, vol. lvii (1901), p. 64, fig. 3.

CHAPTER IV

THERMAL METAMORPHISM OF NON-CALCAREOUS SEDIMENTS

Constitution of Argillaceous Sediments—Lower Grades of Metamorphism in Argillaceous Sediments—Medium Grades of Metamorphism in Argillaceous Sediments—Higher Grades of Metamorphism in Argillaceous Sediments.

CONSTITUTION OF ARGILLACEOUS SEDIMENTS

BEFORE attempting to follow the course of metamorphism in rocks of various kinds, it will be proper to make some inquiry into the constitution of the rocks prior to metamorphism. How that constitution has been acquired is a question which lies outside our province, and will be dealt with only summarily. Sedimentary deposits are derived, by one or more removes, from crystalline rocks. They are the result of destructive action, partly mechanical and partly chemical, including biochemical, together with a redistribution of the various products. This redistribution, too, is in part mechanical, in part chemical in its operation ; for it includes the sorting of finer from coarser detritus by running water, the separation of soluble from insoluble constituents, and secretion by organic agency. The solid particles derived from the chemical destruction of crystalline minerals are in general much smaller than the grains which result from mere mechanical disintegration ; and consequently a separation between finer and coarser material is roughly a separation between decomposed and undecomposed. The one goes to make clays and the other sandstones. The argillaceous sediments represent then, as a first approximation, the products of chemical degradation of the parent rocks, or rather the insoluble residue of those products. There is, however, a variable admixture of mechanical detritus, chiefly of the finest washings of quartz.

Note first that, while some small part of the lime from felspars and pyroxenes may be fixed in secondary epidote, etc., the bulk of the lime goes into solution as carbonate. Redistribution of dissolved material differs from mechanical transport in that there is no limit to its range, and most of the carbonate goes to make limestones in

45

distant areas. Apart therefore from an actual admixture of cal-careous material, argillaceous sediments are notably poor in lime. Of magnesia, on the other hand, derived from the destruction of olivines, pyroxenes, amphiboles, and biotites, only a small part is dissolved as carbonate, and this constituent passes generally into minerals of the serpentine and chlorite groups. The original silicate compounds contained iron in the ferrous state replacing magnesia and sometimes in the ferric state replacing alumina. There is less of this replacement in the resulting serpentines and chlorites ; so that much of the iron is set free as oxide, its final form being a hydroxide which for convenience we may name limonite. This too is the ultimate representative of most of the iron-ore minerals of the parent rocks. The titaniferous iron-ores yield in addition granular sphene and perhaps rutile, but the minute rutile-needles which are so characteristic a feature of slate-rocks come from the destruction of biotite (Fig. 69, *B*, below).

The alteration-products of the alkali-felspars contribute largely to the composition of argillaceous sediments, and it is unfortunate that their true nature is involved in much obscurity. If we set aside the china-clays, which have a special mode of origin connected with pneumatolysis, the alkali-content shown in analyses of ordinary clay-rocks proves that kaolin is a constituent of minor importance. It may be suggested that, when it is present, it comes rather from albite than from orthoclase, since the low ratio of soda to potash in most argillaceous rocks shows that soda is to a greater extent than potash removed in solution. None the less we shall see that finely granular albite is present in many sediments of this class, while orthoclase seems to be always destroyed. However this may be, it is certain that the excessively fine flaky substance which makes a large part of all ordinary clay-rocks is of a micaceous nature. Since even the well-crystallized white micas present considerable difficulty to the systematic mineralogist, the constitution of minute flakes which cannot be isolated for analysis remains still more problematical. If we are justified in distinguishing two ideal molecules :

$$\text{Muscovite, } H_2KAl_3(SiO_4)_3,$$
$$\text{Phengite, } H_4K_2Al_4Si_8O_{25}$$

then the ratio $K_2O : Al_2O_3$ given by the bulk-analyses of average argillaceous sediments points decidedly to a phengitic composition.[1] The study of clays still in a plastic condition is complicated by the fact that some part of their substance is in the colloid state. This,

[1] See also analyses of ' sericite,' Niggli, *Schw. Min. Pet. Mitt.*, 1933, p. 84.

however, is only a temporary circumstance, and the rocks which suffer metamorphism may be supposed made up of definite, if sometimes imperfectly known, minerals.

Under certain physical conditions, of which a tropical climate is one, not only the felspars but all other silicate compounds may be totally decomposed, the silica liberated passing into solution together with the whole of the alkalies, lime and magnesia. The alumina and iron remain as hydroxides, at first largely in the colloid state. With them are mingled quartz-grains, if the rock was a quartzose one, and rutile. Such deposits (laterites) represent the extreme result of the chemical degradation of silicate-rocks. According to the nature of those parent rocks, they may be almost purely aluminous (bauxite) or highly ferruginous (lithomarge). There are also accumulations of iron-oxide, normally in a hydrated form, due to direct deposition by the agency of bacteria, and these may have an admixture of siliceous and other impurities. These peculiar types of deposit, including also the kaolin-clays, will require separate treatment in the sequel. The more ordinary argillaceous sediments, notwithstanding a considerable variety of composition, can be dealt with collectively.

In the usual and less drastic course of decomposition only part of the lime, magnesia, alkalies, and silica is carried off in solution, to reappear elsewhere as limestones, salt-deposits, and cherts. Their loss, however, is enough to raise notably the proportion of the alumina which remains, and *richness in alumina* is accordingly the salient chemical characteristic of argillaceous sediments as a class. Other noteworthy features are the poverty in lime (in deposits not distinctly calcareous) and the strong preponderance of potash over soda.

In the destructive chemical changes which have been noted we see a partial breaking down (and even in laterization a complete breaking down) of higher silicate compounds to form silicates of lower type and non-silicates, the process often involving hydration or peroxidation or carbonation. Broadly we may recognize a readjustment to atmospheric conditions, i.e. to low temperature and unlimited access of water and the atmospheric gases. From the standpoint of energy-content the reactions involved are *exothermic*, and the quantity of heat evolved and dissipated in the complete chemical degradation of a crystalline rock is very great. The chemical changes in thermal metamorphism are essentially *endothermic*. Besides dehydration, reduction of peroxides, and elimination of carbon dioxide, there is in most of the reactions to be noted below a building up of higher types of silicates from lower, or of silicates from non-silicates, with the aid of the detrital quartz present. To this extent rising temperature brings about a reversal of the changes which we have enumerated as proceeding at low temperatures. It is evident, however, having regard to the bulk-composition of argillaceous deposits, that the new minerals produced can correspond only partially with those of igneous rocks. In particular, we must expect to meet some silicates notably rich in alumina.

One other point is to be emphasized. Almost all the constituents of an argillaceous sediment result from chemical reactions completed at low temperatures. They exist as very minute particles, intimately commingled with one another and in presence of an abundant solvent. It may be assumed that under these favourable conditions a near approach to true chemical equilibrium has been attained, an assumption seldom to be justified in other rocks of composite origin. This being so, metamorphism *starts from equilibrium* ; so that there is from the beginning a gradual readjustment in response to rising temperature, and the earlier stages exhibit here a clearer gradation than is to be recognized in any other class of rocks. Especially is this noteworthy in contrast with the case of those partially calcareous sediments which are excluded from present consideration, and will be treated in a later chapter.

LOWER GRADES OF METAMORPHISM IN ARGILLACEOUS SEDIMENTS

The earliest effects of rising temperature are felt by some of the minor and non-essential constituents of the sediment. Many black shales contain a noteworthy quantity of organic matter, and this is quickly affected by heating. Under a low pressure it may be wholly expelled, but more commonly it is reduced to graphite. The minute black particles have a strong tendency to gather into clots and patches, and one type of ' spotted ' structure, common in the lowest grade of metamorphism is due to this aggregation of the dark pigment of the rock (Fig. 1, *A*). In some other black shales the finely divided dark matter is limonite and this is reduced to granules or minute octahedra of magnetite. In cleaved slates the iron-oxide is usually in the form of haematite, and this too is reduced to magnetite. The minute rutile needles, so abundant in many argillaceous sediments, give place to rather larger and stouter crystals. It is safe to assert also that, if any colloid matter still existed in the original sediment, it becomes reconstituted at a very early stage.

At a somewhat higher temperature the general body of the rock begins to be affected. The process begins, as has been said, by temporary solution initiated at many isolated points throughout the mass (p. 15). If it go no farther, there will result a ' spotted slate ' of the type already described. Such rocks have been styled Knotenschiefer, Fleckschiefer, and Fruchtschiefer, names which, however, have been applied also to spotted structures of a different nature. If, on the other hand, as we now suppose, solution is duly followed by recrystallization, and the process spreads gradually through the mass of the rock, we reach a definite landmark of metamorphism in the production of one or more new minerals.

Very often the first important new product is *biotite*, and this mineral is seldom absent from metamorphosed argillaceous sediments in most succeeding grades. Although it is not possible to write down a precise equation representing the reaction, it is clear that the biotite is formed from the chlorite, ' sericite ', iron-ore, and rutile of the original sediment. If much of the iron-ore enters, some free silica must be taken up too, and we see that practically all the constituents of the rock are already involved at this early stage. Of the exact composition of the biotite produced in thermal metamorphism we possess scarcely any knowledge. In the only recorded analysis [1] of such a mica the molecular ratio $K_2O : Al_2O_3$ is 0·50, which suggests that the sericite was of a phengitic composition and the chlorite of a kind poor in alumina. In the same analysis the ratio $FeO : MgO$ is 1·54, and, judged by its deep colour, the biotite of metamorphosed rocks in general is notably ferriferous. At its first appearance this mineral is in numerous very small elements, shapeless or sometimes rounded, but it soon develops the characteristic flakes, which may have either the criss-cross or the parallel arrangement (Fig. 3, above). Relatively large flakes with a poeciloblastic structure are not uncommon (Fig. 4, *A*).

In sediments containing any noteworthy amount of kaolin (or possibly of bauxite or gibbsite) biotite is not the first new mineral to appear, but is preceded by one of the distinctively aluminous silicates *andalusite* and *cordierite*, or by both together. Andalusite comes simply from the decomposition of kaolin :

$$H_4Al_2Si_2O_9 = Al_2SiO_5 + SiO_2 + 2H_2O ;$$

but in the more chloritic sediments the ferro-magnesian mineral cordierite forms instead, and is in fact the commoner of the two. Unlike the nascent biotite, these aluminous silicates figure from the first as individuals of relatively large dimensions. The formation of such large crystals in a low grade, and in a matrix which is still practically unchanged, is connected with their composition and their capacity for enclosing an unusual amount of foreign matter. The

[1] Lang, *Nyt. Mag. Nat.*, vol. xxx (1886), p. 318 (Oslo district).

growth of isolated crystals in a solid rock is resisted only in so far as the rock-substance cannot be either incorporated in the crystals or enveloped by them. If the bulk composition were identical with that of the growing crystals, there would be no resistance, and the actual conditions are favourable in proportion as that ideal case is approached. The crowded minute inclusions are at this early stage merely the undigested part of the rock-substance, that residuum which could not be incorporated in the crystals themselves. At first both minerals have rounded or oval shapes. Andalusite, capable of exerting a

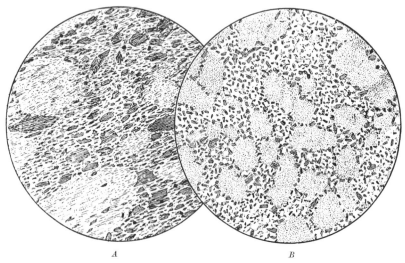

FIG. 9.—METAMORPHOSED ARGILLACEOUS ROCKS WITH ALUMINOUS SILICATES ;
× 25.

These minerals are seen as oval spots, containing abundant minute inclusions but relatively free from the biotite developed in the general matrix.
A. Skiddaw Slates, Glenderamackin Valley, Cumberland ; with large andalusites.
B. Coniston Flags, near the Shap granite, Westmorland ; with numerous round grains of cordierite.

considerable force of crystallization, speedily acquires crystal outline, always that of the simple prism, and pushes aside its enclosed foreign matter into the familiar chiastolite pattern. The feebler cordierite very seldom attains to any crystal shape, and is unable to clear itself from inclusions.

It is not to be supposed that these aluminous silicates are formed only when kaolin was present. They are also produced jointly with biotite by reactions involving ' sericite ', chlorite, and iron-ore. In this case they come in at a rather later stage, and may continue to form, together with biotite, until some part of the requisite material is exhausted. As before, their first appearance is in round grains. These are conspicuous by contrast, being relatively free from the biotite-flakes abundantly developed in the rest of the rock, and they thus characterize still another type of ' spotted slate ' (Fig. 9). The andalusite makes well-defined crystals, clear except for the chiastolite cross (Fig. 7), and at a later stage this also disappears (Fig. 10). The shapeless grains of cordierite still carry abundant inclusions, but these have now become converted to recognizable minerals, such as biotite and quartz (Fig. 11, B). The cordierite at this stage often shows its

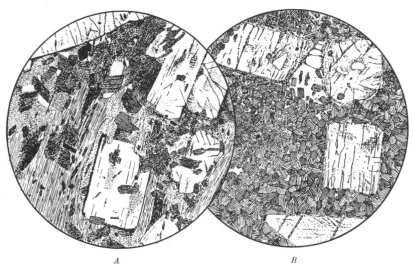

A B

FIG. 10.—ANDALUSITE-BEARING ROCKS ; × 25.

A. Andalusite-Mica-schist, Killiney, near the Dublin granite. Besides biotite, this rock contains abundant large flakes of muscovite with parallel disposition.
B. Andalusite-Biotite-Hornfels, Perran Sands, near the Cligga Head granite, Cornwall. The biotite illustrates the typical decussate structure.

characteristic polysynthetic twinning. It is to be observed that this mineral is not merely a magnesian but a ferro-magnesian silicate, having the general formula $(Mg,Fe)_2Al_4Si_5O_{18}$. The molecular ratio FeO : MgO is sometimes higher than unity, and has probably no limit. The highly ferriferous variety imparts to a rock of which it makes the chief bulk a deep violet colour, with an unmistakable resinous lustre. In a thin slice it is seen making a general mosaic in which the other constituents are embedded, and has often a distinctly bluish tint.

MEDIUM GRADES OF METAMORPHISM IN ARGILLACEOUS SEDIMENTS

At the stage now reached nothing of the original material remains unchanged. What has not been used to make new minerals has recrystallized. The remaining quartz, according to its greater or less amount, figures in a fine-grained mosaic or appears as new granules enclosed in cordierite. If white mica is left, it is in new flakes, much larger than the original minute scales of 'sericite', and it may be assumed to have now the composition of muscovite, allowing for some replacement of potash by soda. If we exclude rocks at granite-contacts, where pneumatolysis may be suspected, muscovite is a much less common mineral among the products of thermal metamorphism than in the crystalline schists. Of other minerals, some magnetite

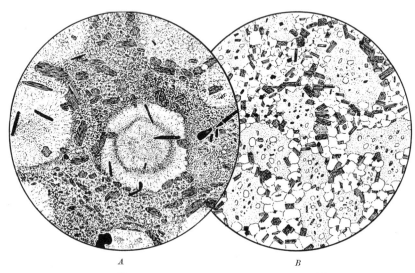

A   B

FIG. 11.—CORDIERITE-BIOTITE-HORNFELS ; × 25.

A. Metamorphosed Skiddaw Slates, Bowscale Tarn. The cordierite crystals here are exceptionally well developed, one basal section showing the hexagonal outline and also the peculiar arrangement of inclusions (p. 44). The opaque mineral is ilmenite.
B. Metamorphosed Culm, Leusdon Common, near the Dartmoor granite. This represents a higher grade of metamorphism. The cordierite contains recognizable inclusions of biotite and quartz.

usually remains, though in much-reduced quantity. Besides contributing to the making of biotite and cordierite, it has sometimes reacted with rutile to produce flakes of ilmenite (Fig. 11, *A*). Most of the original rutile has gone into biotite, but a few crystals may be left even in an advanced grade of metamorphism. The other two forms of titania are also known, but in this mode of occurrence are decidedly rarities and perhaps anomalies. Brookite [1] has been

[1] Beck, *Neu. Jb. Min.*, 1892, vol. i, pp. 159–60 ; *Tscherm. Min. Petr. Mitt.* vol. xiv (1893), p. 314 (Elbe Valley).

recorded in a comparatively low grade of metamorphism, and anatase [1] in a more advanced grade. They do not take the place of rutile at any determinate stage, but occur capriciously and sporadically, and it is probable that both minerals are here merely metastable forms.

That a rock is totally reconstituted does not imply finality, for, with further rise of temperature, new reactions will come into play, giving rise to new minerals. In following the further course of metamorphism, we have to observe that it takes one line or another in accordance with initial differences of composition. Much depends upon the ratio of alumina to magnesia, and therefore upon the relative proportions of the sericitic and chloritic constituents in the original sediment. It is mainly from these that the new minerals have been formed in the earlier stages of metamorphism already discussed, and which of the two is first exhausted is now the important criterion. An excess of chlorite on the one hand or of white mica on the other will lead to different results in the higher grades of metamorphism.

Suppose, first, that there is an excess of chlorite after all the white mica has disappeared. This chlorite does not remain unchanged, or merely recrystallized, through higher grades of metamorphism, but takes part in further reactions. These are (i) conversion of any andalusite present to cordierite ; (ii) if this has not disposed of all the chlorite, formation of some new ferro-magnesian silicate, either non-aluminous or at least with a ratio $RO : R_2O_3$ greater than in cordierite and biotite ; (iii) possible liberation of some iron in the form of magnetite. The new ferro-magnesian silicates which satisfy the conditions are rhombic pyroxene and a garnet of the almandine-pyrope-spessartine series, and the mineral usually produced is *hypersthene.* Although it has been recorded as a normal product of thermal metamorphism in the Oslo district, the Harz, the Comrie district, and elsewhere, it is not a mineral of very common occurrence at this stage. The reason is that sediments very rich in chlorite are usually more or less calcareous, and then give rise to hornblende instead of a rhombic pyroxene. The formation of hypersthene in high grade hornfelses has, as will be shown, a different significance. At its first appearance this mineral makes slender prisms or small crystal-grains and in no great abundance (Fig. 12, *A*). Its associates are cordierite, biotite, quartz, and magnetite, muscovite and orthoclase being obviously excluded. Andalusite, too, is excluded, in virtue of the relation :

$$2 \text{ Hypersthene} + 2 \text{ Andalusite} + \text{Quartz} = \text{Cordierite,}$$

[1] Hutchings, *cit.* Harker and Marr, *Quart. Journ. Geol. Soc.,* vol. xlvii (1891), p. 318 (Shap granite aureole).

or, if silica be in defect :

$$5 \text{ Hypersthene} + 5 \text{ Andalusite} = \text{Cordierite} + \text{Spinel}.$$

these reactions ensuring that hypersthene does not form until any andalusite present has become replaced by cordierite.

The place of hypersthene (or rather of hypersthene and cordierite) is sometimes taken by a red *garnet*, and the status of this mineral as a product of simple thermal metamorphism calls for some discussion. Although it is abundant in the inner aureoles of the Skiddaw, Foxdale,

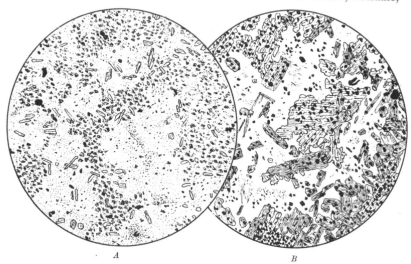

A   *B*

FIG. 12.—HYPERSTHENE-BIOTITE-CORDIERITE-HORNFELS, from the aureole of the Càrn Choìs diorite, near Comrie, Perthshire ; × 23.

A. Little prisms of hypersthene, minute scales of biotite, granules of magnetite, and fine dust probably of graphite are set in a matrix of cordierite-mosaic, without quartz.
B. Here, in a higher grade of metamorphism, the hypersthene and biotite make crystals of conspicuous size. The abundant little nearly opaque granules are octahedra of pleonaste, and mark this rock as one deficient in silica (see below).

New Galloway, and Leinster granites, and is recorded from some European areas, the occurrence of garnet (other than a lime-garnet) in such connexion is at first sight anomalous. We shall see that the common red garnet, approaching almandine in composition, is a highly characteristic mineral of regional metamorphism ; but, when the mica-schists of the Highlands enter the aureole of one of the ' Newer Granites ', the garnet is destroyed. This marks almandine as a mineral proper to metamorphism only under stress-conditions. It is important therefore to note that such analyses as we possess of garnets from thermally metamorphosed non-calcareous sediments indicate in general, not pure almandine, but a mixed garnet containing a notable proportion of spessartine. Sollas [1] found 18·55 per cent of manganous oxide in a garnet from Carrickmines, Co. Dublin. Tilley [2] found 14·88 per cent. in the New Galloway garnet and 6·02 in that of Grainsgill (Skiddaw granite aureole). We may infer with some confidence that the formation of garnet in thermal metamorphism is favoured by the presence of a certain amount of manganese in the rock, and that the garnet produced has then a composition between almandine and spessartine. It is perhaps significant that spessartine, unlike almandine and pyrope, can be reproduced in the laboratory from its constituent oxides. Goldschmidt's researches at Stavanger prove that rocks with a very small content of manganese may yield a richly manganiferous garnet.

It cannot, however, be ruled therefore that the presence of manganese (or lime) is absolutely essential. A garnet from Gadernheim in the Odenwald, in the aureole of an intrusion of hornblende-gabbro, was found by Klemm [3] to be a nearly pure almandine, with only 1·90 per cent of manganous oxide and 1·45 to 1·93 of lime. Here we have to do undoubtedly with a very deep-seated intrusion, and we may conjecture that under a sufficiently great pressure even almandine may form freely. Another instance is the well-known garnet of Botallock, in the aureole of the Land's End granite, Cornwall.[4]

The garnet makes idioblastic crystals with the usual dodecahedral habit. There may be a crowd of minute crystals enclosed in biotite or cordierite or quartz (Fig. 13, C) ; but more commonly garnet is from the first of conspicuous size (Fig. 3, B) and still more so in the highest grades of metamorphism (Fig. 17, B).

In argillaceous rocks originally poor in chlorite there is more white mica than can be consumed in the early reactions which yield biotite, etc., and the excess appears as recrystallized muscovite. It seems that under certain conditions this mineral may survive into an advanced grade, and Tilley [5] has suggested that this is the normal course in presence of a sufficiency of water. It should be remarked, however, that the abundant white mica often found in the neighbourhood of a granite-contact is certainly a pneumatolytic product. In the general case it appears that, with advancing metamorphism, muscovite dwindles and disappears, and *orthoclase* is formed in its place.

If we assume the mica to have the ideal muscovite composition,

[1] *Sci. Proc. Roy. Dubl. Soc.* (2), vol. vii (1891), p. 49.

[2] *Min. Mag.*, vol. xxi (1926), p. 49.

[3] *Notizbl. Ver. Erdk. Darmst.*, v. Folge, Heft 4 (1919), p. 20.

[4] Alderman, *Min. Mag.*, vol. xxiv (1935), pp. 42–3.

[5] *Min. Mag.*, vol. xxi (1926), p. 49.

and conceive it as derived originally from orthoclase, it is clear that in the process the felspar must have parted with two-thirds of its alkali and silica, besides taking up water. The alkali so removed being lost irretrievably, the mica can be reconverted to felspar only by the abstraction of two-thirds of its alumina. In presence of free silica the alumina so liberated goes to make andalusite :

$$H_2KAl_3(SiO_4)_3 + SiO_2 = KAlSi_3O_8 + Al_2SiO_5 + H_2O.$$

Andalusite thus produced as a by-product of the formation of felspar and thus a different significance from that generated in lower grades

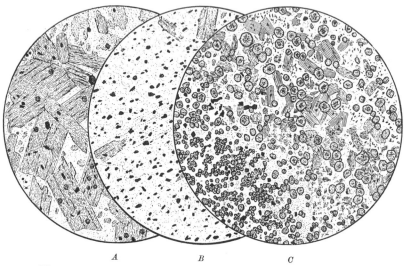

A B C

FIG. 13.—BIOTITE-CORDIERITE-HORNFELS, from the metamorphosed Skiddaw Slates of the Caldew Valley, Cumberland ; × 25.

A and B represent different bands in the same thin slice, one rich in biotite with the usual deeply coloured haloes, the other mainly of cordierite with many little crystals of magnetite.

C is a garnetiferous variety, also banded. One part shows a crowd of minute garnets set in quartz ; another part is mainly of cordierite and biotite with rather larger garnets.

of metamorphism. Since the white mica of these rocks constantly contains a certain proportion of soda, some albite is produced with the orthoclase. If, however, the ratio of soda to potash is not higher than 1 : 4, the albite may be concealed by solid solution in the orthoclase.

At its first appearance, in a medium grade of metamorphism, the alkali-felspar figures as minute granules, which rarely show twinning, and, when involved in a fine mosaic with quartz, may easily be overlooked. A useful criterion is that against cordierite quartz is idioblastic but orthoclase xenoblastic. Albite can often be identified by its low refractive index. When albite-oligoclase or oligoclase occurs instead, this is of course due to some small content of lime in the original sediment. With advancing metamorphism the felspar, like other minerals, is usually in larger elements, and then often shows the characteristic twinning. In the highest grade of metamorphism felspar sometimes becomes abundant. Much of it then is newly formed, and comes from the breaking up of biotite, a matter which calls for more particular notice.

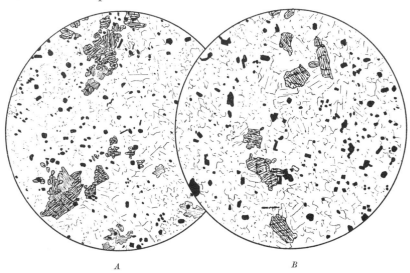

A   B

FIG. 14.—HYPERSTHENE-ORTHOCLASE-CORDIERITE-HORNFELS, from the inner aureole of the Càrn Chois diorite, near Comrie, Perthshire ; × 23.

A. Some biotite is present, but is corroded and in process of destruction.
B. Here biotite is wanting, and orthoclase is more abundant. A band down the middle of the field is composed almost wholly of hypersthene and orthoclase.

HIGHER GRADES OF METAMORPHISM IN ARGILLACEOUS SEDIMENTS

The dark mica more often than the white maintains its stability to the last, and it may be very abundant in highly metamorphosed rocks (Figs. 2, *A* ; 3, *B* ; 10 ; etc.). Especially is this found in the inner aureole of one of the more acid granites and in xenoliths enclosed in such a rock, the magma of which was capable of furnishing an ample supply of water and perhaps of other volatile bodies. Under other conditions it is often very evident that, as we pass to the highest grade of metamorphism, biotite dwindles and disappears. The potassic part goes to make orthoclase, with some aluminous silicate, while the ferro-magnesian part gives rise to hypersthene or possibly to garnet. This is the most usual origin of hypersthene in thermal metamorphism. When that mineral is produced at an earlier stage, as we have seen, its existence is incompatible with the presence of orthoclase. In the rocks now in question hypersthene and orthoclase are associated in abundance, together with cordierite. These and a little magnetite are often the only constituents. If biotite is present, it is seen to be in process of destruction. Characteristic examples occur near the diorite contact in the Comrie district (Fig. 14).

The weak mineral cordierite has, at these high temperatures, an enhanced force of crystallization. It shows idioblastic outline, though

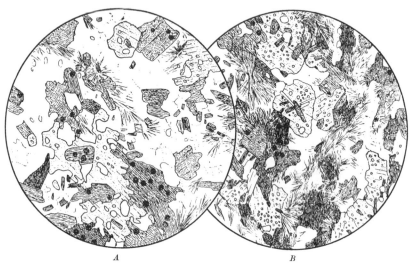

A   *B*

FIG. 15.—SILLIMANITE-CORDIERITE-GNEISS, M'Phatleles Location, Northern Transvaal ; × 23.

From the aureole of the Bushveld plutonic complex (Hall, *Tsch. Min. Pet. Mitt.*, vol. xxviii (1909), p. 135). The sillimanite, in closely packed bundles of fine needles, is forming at the expense of the biotite. Cordierite, enclosing quartz-granules, is conspicuous (over-emphasized in the drawing). Other constituents of the rock are orthoclase, quartz, some andalusite and muscovite, and minute zircons, causing dark haloes in the biotite.

seldom crystal-shape, against the still weaker orthoclase. The aluminous silicate which is formed together with orthoclase in this highest grade is not andalusite but *sillimanite*, which thus comes to have a critical significance. Its relation to the biotite from which it is derived is sometimes apparent (Fig. 15) ; but doubtless sillimanite may arise also in other ways, and in particular from the inversion of andalusite formed at some earlier stage. There is no evidence that such inversion takes effect at a determinate temperature, and indeed andalusite is sometimes found in company with the higher and doubtless stable form. Sillimanite occurs in slender prisms [1] of fine needles, often crowded together with a sub-parallel or sheaf-like arrangement (Figs. 15, 17, *B*).

We have seen that, in the reactions by which the aluminous silicate-minerals are built up, there is in almost every case a taking up of silica ; and we have hitherto assumed a sufficiency of silica, originally present in the form of detrital quartz, to satisfy all requirements. This condition, however, is not always realized ; and, when the free silica is exhausted before the end is reached, metamorphism in the

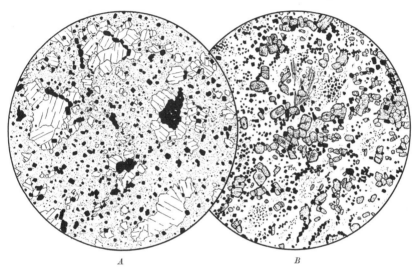

A   *B*

FIG. 16.—HIGH-GRADE TYPES OF HORNFELS POOR IN SILICA, from the aureole of the Càrn Chòis diorite, near Comrie, Perthshire ; × 23.

A. Pleonaste-Orthoclase-Cordierite Hornfels. The nearly opaque octahedra are of a spinel near pleonaste ; the clear crystals showing cleavage are of orthoclase ; the rest is a granular aggregate of cordierite.
B. Corundum-Cordierite Hornfels. The conspicuous crystals showing in relief are of corundum ; the opaque mineral is pleonaste ; and the other constituents are alkali-felspars and cordierite. This is the rock analysed by Tilley, *Quart. Journ. Geol. Soc.*, vol. lxxx (1924), p. 46.

highest grades must necessarily follow in part different lines. Owing to the deficiency of silica, cordierite comes to be represented in part by a spinel mineral, and andalusite and sillimanite give place partly or wholly to corundum. The spinellid occurs in very numerous little octahedra, of so deep a green as to be barely translucent in thin slices (Figs. 12 *B* ; 16). It may be set down as a *pleonaste* approaching hercynite in composition. It is probably more highly ferriferous than

[1] It is one of the few minerals of metamorphism in which the common habit does not conform with the principal cleavage, which is pinacoidal.

the cordierite which it represents and accompanies, and with this we may correlate the usual absence of magnetite in rocks of this type. The *corundum* is in crystals usually of no great size (Fig. 16, *B*), the common habit being that of imperfect hexagonal discs, though the ' barrel ' shape, with elongation along the vertical axis, is also found. A bluish tint is often perceptible in thin slices and sometimes a strong blue colour (sapphire), e.g., near Cape Cornwall.

Rocks carrying corundum are, of course, normally devoid of quartz, but it is easy to find apparent exceptions to the general rule.   We

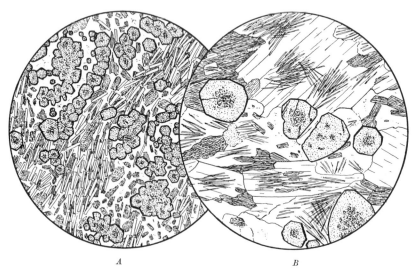

FIG. 17.—HIGHLY METAMORPHOSED ARGILLACEOUS SEDIMENTS of Silurian age bordering the Cairnsmore of Fleet granite at New Galloway ;   × 25.

A. Garnet-Gneiss. This has been a rather siliceous shale, and contains abundant quartz. The other chief constituents are garnet, muscovite, and biotite.
B. Sillimanite-Gneiss : a very coarse-grained rock consisting of garnet, two micas, quartz, and some felspar, with abundant needles of sillimanite, mostly enclosed in the muscovite.

may even see in a rock-slice a crystal of corundum or pleonaste very near to a grain of quartz, though not in contact with it.   The original sediment consisted of more siliceous and less siliceous seams, and the very narrow amplitude of diffusion possible in metamorphism has imposed a corresponding check upon chemical reactions.   In truth the ' system ' for which Goldschmidt's adaptation of the Phase Rule may hold good (p. 4) is not the rock as a whole, but merely any portion of it small enough to be comprised within the sphere of free diffusion.

The most highly metamorphosed rocks, quartz-bearing or not, take on often, though by no means universally, a decidedly coarse texture.   The principal constituent minerals—cordierite, micas, garnet, felspars, quartz, in different types—tend to relatively large dimensions, though some other minerals, such as sillimanite and the spinellids, are still in only small crystals.   While the general type of structure remains the same, the term ' hornfels ' is not appropriate to these coarse-grained rocks, and they are generally designated by the name (itself vague and unsatisfactory) of ' gnesis '.   To this is added some prefix, either of the most abundant mineral, as *cordierite-gneiss* (Fig. 27), or the mineral most distinctive and significant, as *garnet-gneiss* (Fig. 17, *A*), or *sillimanite-gneiss* (Fig. 17, *B*).   Very characteristic is a more or less pronounced banded structure, evident in the field or under the microscope.   This has been determined by original differences of composition, but accentuated by a certain amount of segregation, rendered possible by the enlarged amplitude of diffusion at the highest temperatures of metamorphism (p. 19).

In strong contrast with these coarse-textured types is another, rather exceptional, class of products, which equally represent a very high grade of metamorphism.   These are the *vitrified* shales and slates already referred to (p. 27).   They consist essentially of a brown glass ; but the solution of the original substance has not always been complete, and residual elements are often present, especially corroded granules of quartz.   In addition there has always been a certain amount of new crystallization, yielding minute but perfectly formed crystals of cordierite, sillimanite, pleonaste, magnetite, etc.   Locally fused phyllites showing these characters are found bordering basic and ultrabasic dykes in the Highlands of Argyllshire.[1]

[1] *Geology of Oban and Dalmally* (*Mem. Geol. Sur Scot.*, 1908), pp. 129–32, and *Geology of Colonsay, etc.* (1911), pp. 94–5.

CHAPTER V

THERMAL METAMORPHISM OF NON-CALCAREOUS SEDIMENTS (*continued*)

Successive Zones of Thermal Metamorphism—Thermal Metamorphism of Aluminous and Ferruginous Deposits—Constitution of Arenaceous Sediments—Thermal Metamorphism of Purer Arenaceous Sediments—Thermal Metamorphism of More Impure Arenaceous Sediments.

SUCCESSIVE ZONES OF THERMAL METAMORPHISM

ON an earlier page reference has been made to the possibility of dividing a metamorphic aureole into successive *zones*, marked by the coming in of different new minerals generated with the advance of metamorphism (p. 24). In the light of what we have now learnt, it is possible to give an answer to the question there posed. In any given aureole, sufficiently well exposed, careful study will enable us to recognize certain definite landmarks of the kind sought ; but these are likely to be different in different cases, and no complete scheme of general application can be laid down. Among what we roughly group together as ordinary argillaceous sediments there is in fact a diversity of composition which leads in thermal metamorphism to quite different mineralogical developments.

The most constant indices are, in a low grade, the production of biotite ; in a middle grade, the first formation of orthoclase at the expense of muscovite, or alternatively the appearance of hypersthene or garnet ; and, if a very high grade is reached, the coming in of sillimanite or corundum. Even here, as we have seen, the breaking up of muscovite to yield orthoclase may be indefinitely postponed, and this owing, not to anything in the initial composition of the rock, but to the proximity of a granite-contact. A like remark applies to the possible breaking up of biotite in a much more advanced grade. The most characteristic products of metamorphism, cordierite and andalusite, may make their first appearance at an earlier or later stage, and have therefore no precise value as indices.

THERMAL METAMORPHISM OF ALUMINOUS AND FERRUGINOUS DEPOSITS

We go on to consider briefly the effects of thermal metamorphism in those highly aluminous and ferruginous deposits which stand apart from more ordinary argillaceous sediments. A pure kaolin clay can yield only 'an aggregate of andalusite and quartz (p. 49), or in a very high grade sillimanite and quartz. If such rocks are to be found, they must be of rare occurrence. A deposit consisting simply of aluminium hydrate merely suffers dehydration when metamorphosed. Gibbsite (Al_2O_3 . $3H_2O$) and bauxite (Al_2O_3 . $2H_2O$) are thus reduced to corundum, giving the rock known as emery, though the best-known occurrences of emery are related to metamorphism of the regional type. Often there is some admixture of flaky diaspore

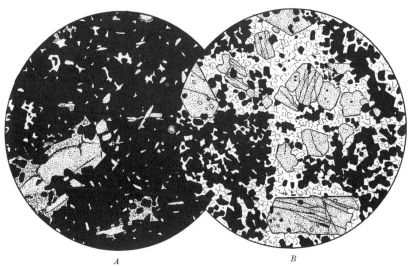

FIG. 18.—METAMORPHOSED BOLE, near Kilchoan, Ardnamurchan ;   × 23.

A. Essentially of hercynite and corundum, the latter encrusted in places with finely flaky diaspore. A little clear anorthite occurs interstitially.
B. Porphyroblasts of corundum and little octahedra of hercynite are embedded in a granular aggregate of anorthite.

(Al_2O_3 . H_2O), either from incomplete dehydration or by subsequent alteration of corundum. Bauxitic deposits, however, are usually more or less ferruginous, and give rise then to an iron-spinel (hercynite) in addition to corundum. If there was any calcareous admixture in the original deposit, this is represented by anorthite. Examples of these rather peculiar types of rocks have not often been recorded. Klemm [1] has described an interesting occurrence in the contact-aureole of a diorite intrusion in the Odenwald. The emery here consists of corundum, deep-green spinel, and iron-ore. It makes part of a banded series with various types of hornfels rich in cordierite

[1] *Notizbl. Ver. Erdk. Darmstadt*, v. Folge, Heft 1 (1916), pp. 1–41.

and sometimes containing corundum.   In Ardnamurchan [1] one of the
red bands of bole, due to destructive contemporaneous weathering of
the Tertiary basalt lavas, has been metamorphosed by a later intrusion
of gabbro.   It is converted to an aggregate of blue corundum and
nearly opaque hercynite with often abundant anorthite (Fig. 18).
At the same place another variety, which has been mainly ferruginous
in composition, is represented by a rock essentially of magnetite.

When a deposit of hydrated ferric oxide undergoes metamorphism,
there is a reduction first of limonite to haematite and then of haematite

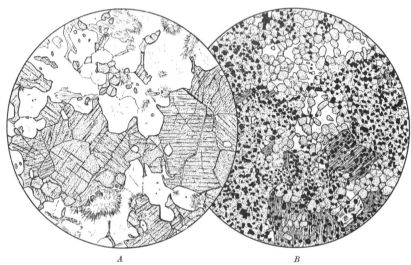

FIG. 19.—METAMORPHOSED IMPURE HAEMATITE ORES, Riekensglück,
Harz ;   × 23.

A.   Fayalite-Quartz-rock.   Reaction between the two minerals has produced in
places the iron-amphibole grünerite, partly in radiating fibres, partly in compact crystals.
B.   Fayalite-Pleonaste-rock, with garnet, biotite, and cordierite.

to magnetite.   The latter process may be incomplete, and so, besides
bedded magnetite-ores, we have others composed of magnetite and
haematite in varying proportions.   Aluminous and other impurities
may give rise to other minerals in addition.   In the contact-aureole
of the Rostrenan granite in Brittany, Barrois [2] records impure limonite,
converted to a mixture of magnetite and the iron-chlorite chamosite.
So too the presence of lime gives rise to andradite, as in some localities
in the Harz [3] and elsewhere.

[1] *The Geology of Ardnamurchan* (*Mem. Geol. Sur. Scot.*, 1930), pp. 233–4.
[2] *Ann. Soc. Géol. Nord* (2), vol. xii (1885), p. 78.
[3] Lossen, *Zeits. Deuts. Geol. Ges.*, vol. xxix (1877), p. 206.

In general any free silica which may be present recrystallizes side by side with the magnetite without any mutual reaction, but there are interesting exceptions. Ramdohr [1] has described the metamorphism of the haematite deposits of the Harz in the aureole of the Brocken granite. The original ores were in varying degree siliceous, and usually magnetite and quartz have crystallized together. Under some conditions, however, there has been a reaction, giving rise to fayalite, and sometimes a further reaction converting the orthosilicate to the metasilicate grunerite (Fig. 19, *A*). With aluminous and other impurities present in the ore, fayalite comes to be accompanied by pleonaste, cordierite, garnet, biotite, etc. (Fig. 19, *B*).

CONSTITUTION OF ARENACEOUS SEDIMENTS

The arenaceous sediments, representing the coarser detritus from the erosion of land-surfaces, are derived in the main from the mechanical disintegration, not chemical degradation, of the parent rock-masses. These may be conceived as crystalline rocks of various kinds, including crystalline schists. In so far as a sand may come from the breaking down of older sandstones, this merely introduces an intermediate step between the deposit and its ultimate source. Sand-grains consist accordingly of those minerals which make up the prevalent types of crystalline rocks within the drainage-basin, but *selected* with reference to their durability under atmospheric conditions.

First in importance stands quartz, which makes the bulk of most ordinary sands. With it, in much smaller quantity, muscovite has a wide distribution, being chemically indestructible under the conditions implied in erosion and transportation. Felspars, pyroxenes, hornblende, and biotite are in varying degree liable to suffer decomposition, and are consequently of less common occurrence in sands deposited under normal climatic conditions ; and the most easily destructible minerals, such as olivine and nepheline, are not found. The climatic conditions here contemplated are those which must always prevail over most parts of the globe, implying a sufficiency of moisture to facilitate the usual chemical changes. In a very arid climate, and also under arctic conditions, weathering is practically in abeyance ; so that a desert sand, for instance, may contain abundant fresh felspar. This we may regard as a special case.

The distinctive minerals of crystalline schists and other metamorphosed rocks make their contribution to sands. Tourmaline and common garnet, and in a less degree cyanite and staurolite, are

[1] *Centr. f. Min.*, 1923, pp. 289–97 ; *Neu. Jahrb.*, Beil. Bd. lv (1927), pp. 333–92.

sufficiently stable to persist : some other minerals, such as andalusite and cordierite, have usually perished. Of interest too are the minor accessory minerals which occur as minute crystals in igneous and other crystalline rocks : zircon, rutile, anatase, apatite, and the rest. Most of these are chemically stable, and so come to have a wide distribution, in very sparing amount, in detrital sediments. They are, as a class, heavy minerals, and for this reason, despite the small size of the crystals, they figure rather in sands than in clays. The densest minerals—magnetite and ilmenite, pyrites and pyrrhotite—tend to be concentrated in particular seams of the deposit.

In all arenaceous sediments other than incoherent sands there is, in addition to the detrital grains, some cementing material, which serves to bind the grains together. This may be furnished from the substance of the grains themselves, especially by recrystallization of a small part of the quartz. In many of our Lower Palaeozoic grits the interstitial cement is of quartz enclosing chlorite, sericite, or kaolin, these being derived from the decomposition in place of grains of felspar, hornblende, and other minerals. Other common cementing substances are iron-oxide (haematite or limonite) and calcite, the last probably often introduced in solution from an extraneous source It is to be recognized also that the ideal severance between coarser and finer detritus as laid down under water and the deposition of calcareous material on separate areas of the sea-floor are in fact only imperfectly realized. For these reasons most arenaceous sediments include, in addition to the clastic grains, some admixture of material having a different origin. Ferruginous and partly argillaceous sand-stones may be treated together with those of purer constitution ; but the presence of any notable amount of carbonate so modifies the course of metamorphism, that calcareous sandstones will be more conveniently considered in another place.

THERMAL METAMORPHISM OF PURER ARENACEOUS SEDIMENTS

It is in the adventitious element present in most arenaceous rocks that the earliest effects of metamorphism are commonly to be observed. In a pure quartzose sandstone or quartzite no change is to be expected, until a temperature is reached at which a general recrystallization sets in. The only exception is the dissipation of fluid-inclusions in the quartz-grains, an effect first noted by Sorby.[1] A pure quartz-rock recrystallized in metamorphism presents a typical example of the simple mosaic structure, in which all traces of the former clastic nature are obliterated (Fig. 20, A). At the same time, any slight

[1] *Quart. Journ. Geol. Soc.*, vol. xxxvi (1880), Proc., p. 82.

original impurity betrays itself by the formation of granules or scales of some new mineral scattered through the quartz-mosaic. A chert, composed originally of cryptocrystalline silica, likewise recrystallizes to a quartzite in thermal metamorphism.

A feature of these metamorphic quartzites is their even-grained texture, connected, as we have seen, with the influence of surface-tension on solubility (p. 20). In a rock of such simple constitution size of grain may afford a rough index of the grade of metamorphism attained.

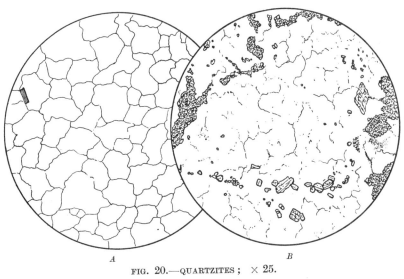

A   *B*

FIG. 20.—QUARTZITES ;   × 25.

A. Manx Grit, metamorphosed near the Foxdale granite, Isle of Man : the whole recrystallized to a simple mosaic of quartz.
B. Grit in Coniston Flags, metamorphosed near the Shap granite, Westmorland. In addition to quartz, there is some augite, formed by reaction between chlorite and calcite in the original sediment.

Increasing coarseness of texture is not, however, the only possible change in a purely quartzose rock which, already recrystallized, is subjected to further rise of temperature. The point 575° C., which marks the inversion between the two forms of quartz, must often be passed in metamorphism of an advanced grade, and the change is one which takes effect promptly in either direction. Doubtless there are among our metamorphic quartzites not a few in which the quartz has been of the highest form, and has inverted to the lower on cooling ; but the signs which mark this change in the coarser granites and pegmatites [1] will probably not be detected in ordinary quartzites.

[1] Wright and Larsen, *Amer. Journ. Sci.* (4), vol. xxvii (1909), pp. 421–47.

Another significant point on the ' geological thermometer ' is 870° C., the inversion-point between quartz and tridymite. It is probable that this temperature also is sometimes attained in metamorphism of the highest grade, as in the near vicinity of an ultrabasic intrusion. It is well known that the inversion is one which proceeds with extreme reluctance, and may be indefinitely suspended. There is also the possibility that tridymite may be formed at temperatures below the theoretical inversion-point.[1] Whether the production of tridymite in high-grade metamorphism leaves any permanent trace in the cooled

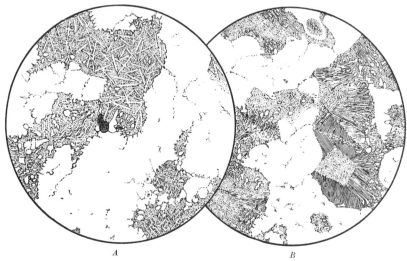

A   B

FIG. 21.—HIGHLY METAMORPHOSED FELSPATHIC SANDSTONE, in the Torridonian bordering ultrabasic intrusions, Isle of Rum ; × 25.

A. The rock is wholly recrystallized, and much of the quartz has given rise to flakes of tridymite. These, embedded in felspar, preserve their outlines, although now replaced by quartz.

B. Here much of the recrystallized quartz has entered into delicate micrographic intergrowth with the felspar.

rock must depend upon the circumstances of the case. With relatively rapid cooling it is perhaps possible that the tridymite itself may be preserved as a characteristic aggregate of minute flakes, while a somewhat slower rate of cooling may yield quartz in shapes pseudomorphic after tridymite. More often, it may be, sufficiently gradual cooling in presence of a solvent will permit of a rearrangement in which such residual structure is lost.

A case more favourable for study is that of a *felspathic sandstone*. Here the shapes of tridymite, formed in a high grade of metamorphism, are preserved by being embedded either in recrystal-

[1] *Cf.* Larsen, *Amer. Min.*, vol. xiv (1929), p. 87.

lized felspar or in glass. The Torridon Sandstone of the North-West Highlands is composed essentially of quartz and abundant fresh felspar, mostly a red microcline. At several places in the Isle of Rum it is highly metamorphosed near intrusions of eucrite and peridotite. As in other red sandstones, the first sign of change is the disappearance of the red colour, which results from the disseminated minute scales of haematite being reduced to magnetite. Ultimately quartz and felspar are alike recrystallized. In places where all was quartz the usual mosaic structure is seen ; but where

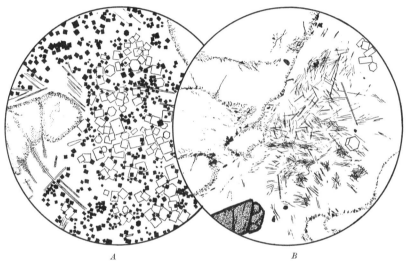

FIG. 22.—VITRIFIED SANDSTONES, showing corroded relics of quartz in a colourless glass ; × 125.

A. Isle of Soay, near Skye : with minute crystals of cordierite, magnetite, and pyroxene.
B. Corrary, Islay : showing cordierite and fine needles probably of mullite. A detrital crystal of zircon is fractured but not otherwise changed.

quartz was in contact with felspar, it shows a crenulated outline due to corrosion ; and there has been an abundant production of little tridymite flakes, either as a fringing growth or more widely dispersed (Fig. 21, A). These are now changed to quartz. Elsewhere recrystallizing quartz has entered into micrographic intergrowth with felspar, showing much variety of detail (Fig. 21, A, B). Indeed, except in the preponderance of quartz over felspar, some of these metamorphosed arkoses reproduce all the features of granophyres and spherulitic quartz-porphyries.

In some occurrences of metamorphosed sandstones, comparable with those near the peridotites of Rum, but where the cooling has been more rapid, more or less glass is produced, having the composition of a mixture of felspar and quartz ; and, embedded in this, pseudomorphs after tridymite may be preserved. A good example is the Old Red Sandstone at its contact with the Bartestree dyke, near Hereford.[1] The same thing is seen more frequently in partly fused xenoliths of sandstone enclosed in basic intrusions, such as those described by Thomas [2] from Mull.

It should be remarked that sandstones, no less than slates, may be more or less completely vitrified under favourable conditions, and

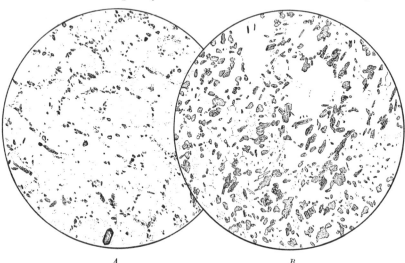

A   B

FIG. 23.—METAMORPHOSED ARENACEOUS ROCKS, near the Lausitz Granite, Saxony; × 25.

A. Quartzite with some biotite and epidote, the distribution of these outlining the grains of the original sandstone. At the bottom is a small crystal of zircon, unaffected by the metamorphism.
B. This belongs to a somewhat higher grade of metamorphism. Numerous little flakes of biotite are scattered through the mosaic of quartz (with some felspar), only indistinctly and in places marking the outlines of the original clastic grains.

that this is more likely to befall a felspathic sandstone than a purely quartzose one.   In the isle of Soay, near Skye, the Torridon Sandstone is vitrified at its contact with certain basic and ultrabasic sills of no great thickness.[3] It has yielded a clear glass, enclosing very numerous minute crystals of cordierite, magnetite, and sometimes tridymite, with corroded relics of quartz (Fig. 22, A).   Inclusions of sandstone

[1] Reynolds, *Quart. Journ. Geol. Soc.*, vol. lxiv (1908), plate lii, fig. 6.
[2] *Quart. Journ. Geol. Soc.*, vol. lxxviii (1922), pp. 239–40, and plate vii, fig. 5.
[3] *Tertiary Igneous Rocks of Skye* (*Mem. Geol. Sur. U.K.*, 1904), pp. 245–6, and plate xxi, fig. 3.

in basalt are often partly vitrified, and numerous occurrences have been described under the name ' buchite '. The pale or brown glass contains minute crystals of cordierite, mullite, and other minerals. Analyses [1] show from 3 to 5 per cent. of water in the inclusion as a whole, while for the glass the figure may be as high as 10 or 12. Clearly an abundant supply of solvent has been present, which has been fixed in the buchite, while it is lost from the enveloping basalt.

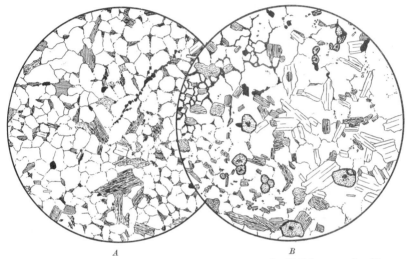

A   B

FIG. 24.—METAMORPHOSED GRIT, Skiddaw Grit, near the Skiddaw granite, Sinen Gill, Cumberland ;   × 25.

A. Biotite-Quartz-Hornfels : a mosaic of quartz and biotite, with some muscovite and magnetite.
B. A more advanced grade of metamorphism, in which garnet also figures. The rock shows banding, the right half of the field being of coarser grain and containing muscovite, while the left half is richer in biotite.

THERMAL METAMORPHISM OF MORE IMPURE ARENACEOUS SEDIMENTS

In a sandstone composed entirely of quartz, or of quartz and fresh felspar, no chemical reaction can be set up in metamorphism ; and most of the accessory minerals of detrital origin undergo no change other than recrystallization. This may take place sooner or later, according to the nature of the mineral. Rutile and tourmaline recrystallize at a very early stage ; magnetite, muscovite, etc., follow in their turn ; only zircon remains untouched in the highest grade of metamorphism (Fig. 22, *B*). It is otherwise with such substances as chlorite, sericite, kaolin, limonite, calcite, etc. These are decomposi-

[1] Lemberg, *Zeits. Deuts. Geol. Ges.*, vol. xxxv (1883), pp. 563–8.

tion-products, which whether formed before or after the deposition of the sediment, are essentially low-temperature minerals. They are very ready to enter into various reactions with one another and with quartz, and some of these reactions demand no great elevation of temperature. Sericite, chlorite, and limonite may give rise to biotite ; calcite and kaolin to epidote ; chlorite, calcite, and quartz to augite (Fig. 20, *B*) ; and so for other combinations. These various new minerals, at their first appearance, figure as numerous little flakes or granules in the interstices of the quartz-grains, and serve to indicate

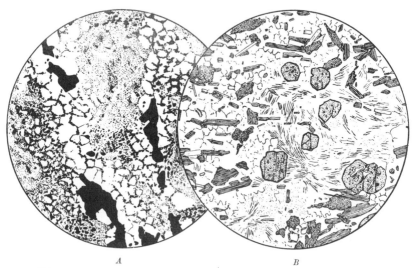

A *B*

FIG. 25.—HIGHLY METAMORPHOSED GRITS, in the Silurian, near the granite of New Galloway ; × 25.

A. This rock has had a cement of iron-oxide, now represented by a network of magnetite granules and minute octahedra. There is also some pyrrhotite.

B. The clear part is a mosaic of quartz with some felspar and cordierite. Through this are scattered biotite, muscovite, garnet, and swarms of minute sillimanite-needles.

the original clastic nature of the rock, even after the quartz itself is recrystallized to a new mosaic. With advancing metamorphism this appearance fades out in consequence of the enlarged latitude of diffusion (Fig. 23).

An analogous case is presented by radiolarian cherts in the South of Scotland, where they come within the aureoles of the Galloway granites.[1] At first minute flakes of biotite appear scattered through the rock, except in clear oval spaces which represent radiolarian tests. Later the outlines of these are obliterated, while the quartz-mosaic

[1] Horne, *Rep. Brit. Assoc.* for 1892 (1893), p. 712 ; *Ann. Rep. Geol. Sur.* for 1896 (1897), pp. 46-7.

acquires a coarser texture, and the biotite-flakes become fewer and larger.

Somewhat similar is the behaviour in metamorphism of sandstones with a ferruginous cement. Whether limonite or haematite, this is speedily transformed to magnetite with interstitial occurrence. In this case, however, this original distribution may persist into a high grade of metamorphism (Fig. 25, *A*). The magnetite is either in granules or in strings of little octahedra.

The non-detrital material in sandstones, when it is not ferruginous

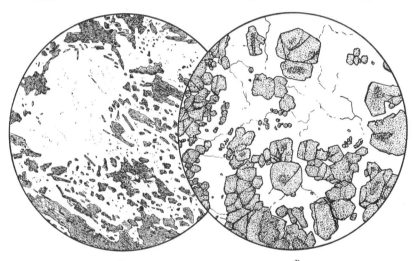

A   *B*

FIG. 26.—HIGHLY METAMORPHOSED GRITS, in the Silurian, near the granite of New Galloway ; × 25.

A. Originally an impure pebbly sandstone : now composed of biotite and quartz with some cordierite. The outlines of small pebbles are still indicated, though no longer sharp.
B. From a lenticular streak consisting entirely of red garnet and quartz.

or calcareous, is broadly comparable with the substance of ordinary argillaceous sediments. The new minerals produced in metamorphism are therefore in general the same that we have already met with. But while a slate and an argillaceous sandstone may be represented, at a given stage of metamorphism, by like mineral-associations, the relative proportions of the several minerals will be by no means the same. Rocks resulting from the metamorphism of impure sandstones come therefore to have characters sufficiently distinctive. Quartz is here much more abundant, while the more aluminous silicates, andalusite and cordierite, figure much less prominently. Orthoclase and plagioclase (usually oligoclase) are much more widely distributed than in metamorphosed slates, and are often plentiful : they come then from the recrystallization of detrital felspar. Corundum and the spinels—minerals which do not form in presence of free silica—are in general absent.

Quartz and felspar, with cordierite if present, are the minerals with the lowest force of crystallization,. and build a mosaic which constitutes the main bulk of the rock. In this as matrix the other minerals are embedded, with idioblastic habit. Since these other minerals have also higher refractive indices, and some of them are coloured, they are unduly conspicuous, and the rock, though still

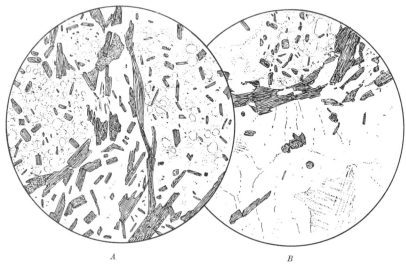

A   B

FIG. 27.—CORDIERITE-GNEISS, a highly metamorphosed Devonian greywacke near the Lausitz granite, Saxony; × 25.

The cordierite makes large ovoid and lenticular grains, with numerous inclusions. The rest is a coarse aggregate of biotite, quartz, and (in the second figure) felspar. See Beck, *Tsch. Min. Pet. Mitt.*, vol. xiv (1893), pp. 332-7.

with a large preponderance of quartz, ceases to have any superficial likeness to a simple quartzite.

In the progressive metamorphism of an argillaceous sandstone there is in general the same sequence of mineralogical transformations that we have traced in the purely argillaceous sediments. An early landmark is the appearance of biotite. The little flakes, at first interstitial between the quartz-grains, take on idioblastic shape as the quartz recrystallizes freely (Fig. 23, compare *A* and *B*), and biotite continues to be prominent, usually in larger flakes, in the higher grades of metamorphism. Garnet marks a more advanced stage, but this mineral, as we have seen, demands special conditions for its formation (p. 54). Good examples of garnetiferous rocks may be studied on Knocknairling Hill, near New Galloway, in the aureole of the Cairnsmore of Fleet granite (Figs. 24, *B* ; 25, *B*). The crystals, in addition to the common dodecahedron, show sometimes the trapezohedral form, which is known to be characteristic of the manganese-bearing garnets.

Of the more aluminous silicates, cordierite is the most common, but in the highest grade of metamorphism sillimanite becomes a characteristic constituent, occurring as usual in very numerous fine needles (Fig. 25, *B*). The rocks then may acquire a notably coarse texture ; so that the principal minerals, often including felspars, can be readily identified on a specimen. Owing to the enlarged latitude of diffusion, even quartz-pebbles may be reduced to rather vague patches with ill-defined outline (Fig. 26, *A*). There is for the same reason, in this highest grade of metamorphism, a strong tendency to the segregation of particular minerals in lenticles and inconstant bands. On Knocknairling Hill there are streaks, an inch or two in width, composed wholly of garnet and quartz (Fig. 26, *B*). Even cordierite, when abundant, may form ovoid and lenticular patches of some size, crowded, as is usual in this mineral, with inclusions of biotite and quartz (Fig. 27). In this way, owing to slight original differences between successive seams, exaggerated by a process of segregation, there arises a pronounced gneissic banding. The rocks may be styled *sillimanite-* or *garnet-* or *cordierite-gneisses*, according to the most distinctive mineral which they present.

CHAPTER VI

THERMAL METAMORPHISM OF CALCAREOUS SEDIMENTS

Pure Carbonate-rocks—Special Features of Semi-calcareous Rocks—Thermal Metamorphism of Impure Non-Magnesian Limestones—Thermal Metamorphism of Impure Magnesian Limestones.

PURE CARBONATE-ROCKS

UNDER the head of calcareous rocks are comprised in the first place those which consist wholly or mainly of carbonates, viz. the carbonates of calcium and magnesium and in less abundance those of iron and manganese. The non-calcareous element in such rocks may be of argillaceous or arenaceous nature, or may be material of direct volcanic origin. The behaviour of calcareous and semi-calcareous rocks in thermal metamorphism presents some features which differentiate these deposits fundamentally from non-calcareous sediments, and the new minerals produced are also in great part different. Moreover, a very moderate content of carbonate in a sediment is enough to determine its metamorphism along these special lines. For this reason we shall group together here, not only limestones and dolomites, pure and impure, but also rocks which the field-geologist would name calcareous shales, sandstones, or tuffs.

We will consider first a pure carbonate-rock. An ordinary limestone, as laid down, is essentially of calcic carbonate with only a small proportion of magnesian. It may, without any change in total composition, suffer a recrystallization by aqueous agency at ordinary temperature ; aragonite, if present, being transformed to calcite in the process. Again, it may undergo a more radical change, involving metasomatism, one half of the lime being replaced by magnesia to give dolomite ; a process sometimes almost contemporaneous with deposition, sometimes long subsequent. Another metasomatic change, often associated with dolomitization, is the replacement of calcic by ferrous carbonate, yielding chalybite. All these changes, not implying metamorphism in our acceptation of the term, involve nevertheless a total alteration in the fabric of the rock, with obliteration of organic and other original structures except those of the larger order.

Since a limestone may be partially or wholly dolomitized, the molecular ratio MgO : CaO for the rock in bulk may have any value up to unity, but not higher. Magnesite is not found in this association, but belongs to altered igneous rocks of the ultrabasic group.

The metamorphism of a pure carbonate-rock is necessarily a simple process and without gradations. The dissociation of a carbonate, since it involves liberation of a gas, is a reaction resisted by pressure. Under the normal conditions of thermal metamorphism the pressure is always sufficient to prevent the dissociation of calcite in a pure carbonate-rock, and a simple non-magnesian limestone merely recrystallizes, yielding an even-grained *marble*. Such a marble differs in no wise from one recrystallized at ordinary temperatures. As seen in section, the grains are of irregular shape, meeting one another in sinuous or zigzag boundaries.

The behaviour of dolomite in thermal metamorphism is not always the same. In some cases it merely recrystallizes, like calcite. A *dolomite-marble*, however, is usually of finer grain than one of calcite, and the structure is of a simpler type. Sometimes indeed the individual grains offer some suggestion of crystal-shape, always that of the primitive rhombohedron. Under different conditions, viz. with a lower pressure, dolomite dissociates, but only as regards its magnesian part,[1] so that the reaction implies dedolomitization :

$$CaMg(CO_3)_2 = CaCO_3 + MgO + CO_2.$$

The resulting rock is accordingly a *periclase-marble*, composed of periclase and calcite. The periclase, a mineral of pronounced idioblastic habit, appears as little octahedra embedded in the calcite matrix. It has, however, almost always suffered change by hydration, and is then represented by flaky pseudomorphs of brucite, $Mg(OH)_2$ (Fig. 28, *A*). The rocks known in the Tyrol and elsewhere by the names *pencatite* and *predazzite* are of this nature, and good examples are found in Skye and in the Assynt district of Sutherland. Pencatite, formed from a pure dolomite-rock, has calcite and brucite in equal molecular proportions. It is a close-grained white rock with the very low specific gravity 2·57. Predazzite, derived from a partly dolomitized limestone, has a larger proportion of calcite.

Ferrous carbonate, like that of magnesium and probably more easily, suffers dissociation when heated under natural conditions, the resulting product being magnetite. In Sweden and elsewhere there are bedded magnetite-ores which have been attributed to the meta-

[1] The dissociation-temperature at a pressure of one atmosphere is for calcic carbonate 898° C., but for magnesian carbonate only 402°.

morphism of chalybite, but these are intercalated among crystalline schists. A like transformation may sometimes be observed on a small scale even against a basalt dyke.[1] When sulphides are involved as well as carbonates, the reactions are more complex. Schneiderhöhn [2] has described the thermal metamorphism of some Westphalian ore-deposits, in which the characteristic reaction is the replacement of chalybite and chalcopyrite by haematite and bornite, the latter mineral reverting on cooling to chalcopyrite and chalcocite.

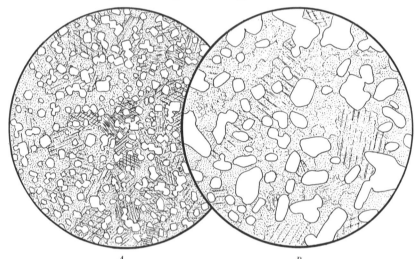

A   B
FIG. 28.—METAMORPHOSED DOLOMITIC ROCKS ;   × 25.
A.   Pencatite (brucite and calcite), Kilchrist, Skye.
B.   Ophicalcite (serpentine and calcite), Ledbeg, Sutherland.

Manganese and zinc carbonates too are easily decarbonated by heat. At Franklin Furnace, in New Jersey, thermal metamorphism of dialogite has given rise to hausmannite and franklinite. Of these the former, a tetragonal mineral, is perhaps Mn_2MnO_4, while the latter is of the magnetite type, and has the approximate composition $MnFe_2O_4$, with some replacement of manganese by zinc.

SPECIAL FEATURES OF SEMICALCAREOUS ROCKS

We turn now to rocks which, in addition to carbonates, contain a certain amount of non-calcareous material, which may be regarded as an impurity. If we set aside a possible admixture of volcanic ash, the extraneous material is in general siliceous or argillaceous, and the substances of most importance in the chemistry of the meta-

[1] Busz, *Centr. f. Min.*, 1901, pp. 489–94.
[2] *Zeits. Krist.*, vol. lviii (1923), pp. 309–29.

morphism are silica and alumina. We have to do with a large number of possible reactions between the calcareous and the non-calcareous, constituents of the rock, the simplest type being :

$$CaCO_3 + SiO_2 = CaSiO_3 + CO_2.$$

Calcite, which by itself would recrystallize without change, is, under any but the greatest pressures in thermal metamorphism, readily decomposed when silica is present to take the place of the expelled carbon dioxide. Dolomite is *à fortiori* even more readily affected. The actual reactions are generally more complex than that set forth as a type, and the resulting products in different cases include a long list of minerals—silicates and aluminosilicates of magnesium and calcium.

Rocks which are partly carbonate, partly non-carbonate, whether they be impure limestones and dolomites or calcareous shales and slates, have, as already remarked, certain characteristics which differentiate their metamorphism from that of other classes of rocks. In their initial state they present an example of '*false equilibrium*'. The calcareous and non-calcareous elements are in an enforced association which has no relation to true chemical equilibrium ; and if no reaction takes place between them, it is merely because at atmospheric temperature the rate of any possible reaction is infinitely small. The rate being accelerated by rise of temperature, reactions come into play, and proceed with a rapidity limited only (since they are endothermic) by the supply of heat. In most other rocks the process is retarded by the very small quantity of solvent present, but here this is reinforced by a copious supply of carbon dioxide liberated by the reactions themselves. The peculiar characters thus inherent in the metamorphism of rocks of this class have important consequences, some of which may be noted in this place.

One result of the promptitude and rapidity of the reactions between calcareous and non-calcareous material is, that we cannot recognize here any such gradations as we can distinguish in the early stages of advancing metamorphism in simple argillaceous sediments. The reactions in question, not only begin, but are completed, at an early stage of metamorphism, as estimated by distance from an igneous contact or by comparison with neighbouring non-calcareous sediments. Where an ordinary shale shows only the beginning of change, a limestone had its impurities already converted to new minerals, and a calcareous shale is totally reconstituted, with loss of all its carbon dioxide. The most striking instances of ' selective metamorphism ' have arisen in this way.

Another consequence of the rapid formation of new minerals is the frequent failure to establish chemical equilibrium. In this way may be produced in particular instances various metastable forms, or again anomalous associations of minerals. The number of distinct minerals in some of these assemblages is in excess of that prescribed by Goldschmidt's adaptation of the Phase Rule (p. 4). Such aberrations are probably very general in the first rapid metamorphism, but, unless cooling also is rapid, they are likely to be obliterated by adjustment of equilibrium.

Another respect in which partly calcareous rocks are peculiar is that the new minerals formed depend from the first upon the total composition of the rock. More precisely, they depend upon the composition of the carbonate part, whether purely calcic or partly magnesian, and of the non-carbonate part, whether merely siliceous or also albuminous. The reason for stating the matter in this way will appear, if we examine the list of minerals actually found in this association. They are very numerous, and a given metamorphosed rock may embrace a considerable number of distinct mineral-species. We will enumerate in the first place only those which are of most common occurrence. The purely calcic minerals, beginning with that richest in lime, are wollastonite, lime-garnets, idocrase, prehnite (of doubtful status), sphene, zoisite and epidote, and anorthite (with other lime-bearing felspars). Of these wollastonite is the only simple silicate : the rest, except the silico-titanate sphene, are aluminosilicates, all of orthosilicate type. All, with the exception of sphene and anorthite, are foreign to igneous rocks. The magnesian and partly magnesian minerals, on the other hand, include both ortho- and meta-silicates, normally non-aluminous, and are comparable with common pyrogenetic minerals—olivines, pyroxenes, and amphiboles. Of the olivines, forsterite, with little or no iron, is the usual variety. The pyroxenes are of the monoclinic division, and are represented generally by diopside. Of the amphiboles, tremolite and allied actinolitic varieties are the most common, but various coloured hornblendes are also found. Fuller knowledge of their composition is a desideratum, more especially as regards their possible content of alumina.

THERMAL METAMORPHISM OF IMPURE NON-MAGNESIAN LIMESTONES

The simplest case is that of a non-magnesian limestone in which the only impurity is silica, either as quartz or as chert of organic origin. No great elevation of temperature is necessary to initiate a reaction, and the product is the metasilicate *wollastonite*. It is note-worthy that the orthosilicate, which forms readily from a melt of

appropriate composition, is unstable under the conditions of metamorphism, and is of extremely rare occurrence. The stability-
relations between the monoclinic and triclinic forms of wollastonite
have not been determined. Peacock[1] suggests that the former,
which he names parawollastonite, belongs to the lower temperatures.
The highest form pseudowollastonite is not to be expected in metamorphism of the ordinary kind, being stable only above 1150° C.[2]
Wollastonite appears as lustrous white crystals, often of considerable
size. With a sufficiency of silica, the limestone may be completely
decarbonated, yielding a simple wollastonite-rock (Fig. 2, *B*, above).
Since the ideal composition which we have supposed is not often
realized in fact, wollastonite is commonly accompanied by at least
small amounts of other silicates. Thus the presence of a little magnesia
is enough to produce some diopside (equivalent to wollastonite *plus*
enstatite), while any aluminous material present is likely to give rise
to grossularite (with the composition 3 wollastonite *plus* alumina).

Consider now a non-magnesian, or practically non-magnesian,
limestone containing a noteworthy quantity of argillaceous impurities.
Here is the material for a number of possible aluminosilicates. The
more important are all of the orthosilicate type, and have alumina
and lime in molecular proportions as follows:

$$\text{Idocrase,}^3 \ Ca_{19}MgAl_{12}(SiO_4)_{18}(OH)_4 \qquad 6:19,$$
$$\text{Grossularite,} \ Ca_3Al_2(SiO_4))_3 \qquad 1:3,$$
$$\text{Zoisite,} \ Ca_2(AlOH)Al_2(SiO_4)_3 \qquad 3:4,$$
$$\text{Anorthite,} \ CaAl_2(SiO_4)_2 \qquad 1:1.$$

They fall accordingly into two pairs, the one, relatively poor in
alumina, being characteristic of metamorphosed limestones, the other,
rich in alumina, of calcareous shales.

The most common product of metamorphism in argillaceous limestones is a lime-garnet, which forms very readily. It is sometimes
produced in abundance even in the vicinity of a large dyke (Fig. 29).
Large crystals may thus be rapidly built up, crowded with foreign
matter, their growth being analogous to that of early crystals of
andalusite and cordierite in argillaceous rocks (p. 49). Since most
limestones are poor in iron, the garnet is usually a *grossularite*.

[1] *Amer. Journ. Sci.* (5), vol. xxx (1935), p. 525.

[2] It has been recorded by McLintock in marls metamorphosed by the com-
bustion of hydrocarbons; *Min. Mag.*, vol. xxiii (1932), pp. 207–26.

[3] This formula, equivalent to 6 grossularite + $CaMg(OH)_4$, was deduced
by Machatschki from a large number of analyses: *Cent. Min.*, 1930, A, p. 293.
Warren and Modell, from an investigation of the crystal-structure, give
$Ca_{10}Mg_2Al_4Si_9O_{34}(OH)_4$: *Zeits. Krist*, vol. lxxviii (1931), pp. 422–32.

Varieties rich in iron, often found in plutonic contact-belts in company with ferriferous pyroxenes, have a special manner of origin, to be discussed later. Its high force of crystallization causes garnet to be idioblastic against most other minerals, wollastonite usually excepted. The forms are the dodecahedral and less commonly the trapezohedral. The well-known peculiarities of these lime-garnets are perhaps connected with their rapid development. They show distinct zones of growth ; they are more or less decidedly birefringent, this property varying in successive zones ; and they exhibit polysynthetic twinning,

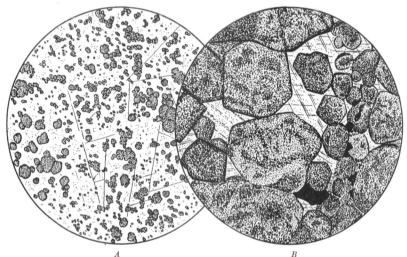

A   B

FIG. 29.—LIME-GARNET (GROSSULARITE) EMBEDDED IN RECRYSTALLIZED CALCITE, from calcareous rocks metamorphosed near dolerite dykes ; × 25.

A. Jurassic Limestone, Camasunary, Skye: showing a crowd of small crystals.
B. Carboniferous Limestone, Plas Newydd, Menai Straits. The garnet is birefringent and with polysynthetic twinning. Rapidly formed, the crystals enclose a large amount of foreign matter. The opaque mineral is pyrrhotite.

made evident in virtue of the birefringence. Grossularite is very often found in company with wollastonite.

Another common associate of lime-garnet is *idocrase* (vesuvianite), which has a very similar composition. Since, however, hydroxyl makes an essential part of its constitution, pressure must be one ruling condition of its formation. As a normal product of thermal metamorphism, idocrase is xenoblastic against wollastonite and grossularite but idioblastic against quartz ; usually also against calcite, though here the relations are rather variable. It should be added that idocrase is often found also partially or totally replacing grossularite as a result of later change (retrograde metamorphism).

The place of *zoisite* is often taken by *clinozoisite* or a slightly ferriferous epidote.   These allied minerals form readily in a calcareous sediment which is also sufficiently aluminous, and, like grossularite, may be produced in abundance in proportion as the bulk-composition of the rock when decarbonated approximates to that of the mineral in question.   Zoisite and epidote are strong minerals, but weaker than wollastonite and grossularite.

Another lime-silicate of like associations is *anorthite*.   Its composition is equivalent to that of wollastonite *plus* andalusite, and to produce it the non-carbonate part of the rock must be rich enough in alumina to yield andalusite.   Anorthite, however, belongs to a higher temperature-grade than the zoisite-epidote-minerals, and these may be regarded as a stage in the formation of the lime-felspar in advancing metamorphism.   The felspar is not always anorthite.   A mixed sediment may carry more or less detrital albite, or again albite may be generated at the expense of white mica by reactions already discussed (p. 56) ; and between this and the anorthite there arise accordingly various felspars of intermediate composition.   Often, too, a certain amount of *orthoclase or microcline* figures in the more highly metamorphosed sediments of this class.   The production of potash-felspar from muscovite and of lime-felspar from the zoisite-epidote minerals may be parts of the same reaction :

$$H_2KAl_3(SiO_4)_3 + 2Ca_2(AlOH)Al_2(SiO_4)_3 + 2SiO_2$$
$$= KAlSi_3O_8 + 4CaAl_2(SiO_4)_2 + 4H_2O.$$

The felspars are low in the crystalloblastic series, and the potash-felspars in particular constantly show a xenoblastic habit.

The zonary arrangement of different varieties of plagioclase in the same crystal, which is a conspicuous feature in so many igneous rocks, is much less in evidence in metamorphic felspars ; but it is found, and presents some points of interest.   In igneous rocks, allowing for some anomalies connected with supersaturation, the rule is that the core of the crystal is more calcic and the margin more sodic.   In the crystalline schists, as remarked by Becke, the reverse is found, though not without exceptions.   According to Goldschmidt,[1] the zoned felspars of thermally metamorphosed rocks present an intermediate case.   In all the more calcic felspars the crystal is most calcic in its core, and grows progressively more sodic, tending to a composition Ab_3An_1 at the margin.   Varieties with an average content of An of 26 to 20 per cent. show little or no zoning, and the more sodic felspars are most albitic at the centre.

The mineral *prehnite* is of not infrequent occurrence in meta-

[1] *Vidensk. Skr.* (1911), No. 11, pp. 292–301.

morphosed calcareous rocks, and is regarded by some petrologists as a normal product of thermal metamorphism ; but its status as such is not beyond question. In many occurrences it has certainly been produced by later reactions at the expense of other lime-silicates, and this is often manifest from its mode of occurrence.

A common product of metamorphism in argillaceous limestones is *sphene*. Its composition is equivalent to that of wollastonite *plus* rutile, and it owes its origin to the minute rutile needles which are so widely distributed in argillaceous sediments. A direct reaction between rutile and calcite would yield perovskite ; but this mineral is not known as a simply thermo-metamorphic product away from igneous contacts.

THERMAL METAMORPHISM OF IMPURE MAGNESIAN LIMESTONES

The metamorphism of dolomites and partly dolomitic limestones carrying various impurities presents some features of special interest. The salient fact that emerges is that *silica reacts with the magnesian in preference to the calcic carbonate*. It follows that, unless disposable silica is present in amount sufficient for complete decarbonation of the rock, one incident of the metamorphism is *dedolomitization*.[1] This is an effect which we have already observed in some pure carbonate-rocks as a consequence of the fact that the magnesian carbonate by itself is more easily dissociated than the calcic (p. 77).

Consider first a dolomitic rock containing silica as its only impurity. Here are the materials for making a number of compounds : the simple lime metasilicate wollastonite and orthosilicate larnite ; the double silicates diopside, tremolite, and monticellite ; and the purely magnesian silicates enstatite, anthophyllite, and forsterite. We find in fact that the first mineral to form, and with a limited supply of silica the only mineral, is the magnesian *forsterite*. This moreover, unlike wollastonite, is an orthosilicate, taking up therefore a double quantity of magnesia. The two other orthosilicates, monticellite and larnite, are unstable forms, and the magnesian metasilicates, enstatite and anthophyllite, are not found in this connexion. The commonest resulting rock is thus a forsterite-marble (Fig. 30, *A*). Idioblastic crystals of forsterite are set in a matrix which is of calcite or of calcite and dolomite, according as more or less silica was originally present. The forsterite has often been replaced by a pale serpentine, giving the rock known as *ophicalcite* (Fig. 28, *B*). There are sometimes interesting special structures, which are well exhibited in the metamorphosed cherty dolomites of Skye. Here much of the contained silica was

[1] Teall, *Geol. Mag.*, 1903, pp. 513–14.

originally in the form of sponges, and from these, at a time anterior to the metamorphism, had become diffused into the surrounding rock. Its distribution was of that rhythmical kind studied by Liesegang,[1] resulting in numerous thin concentric shells alternately rich and poor in silica, and these are now represented by alternations of serpentine and calcite.   The details of the structure at the same time illustrate those pseudo-organic appearances which once passed under the name of ' Eozoon '.[2]

If the original dolomitic rock contained more silica than would

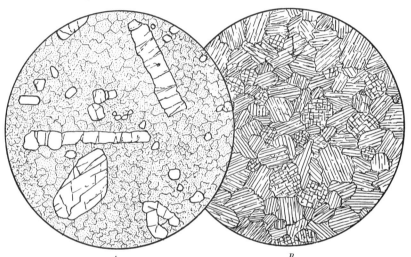

A   B

FIG. 30.—METAMORPHOSED CHERTY DOLOMITES in the Cambrian of Skye ;   × 25.

  A.   Forsterite-Marble, near the granite of Kilchrist : idioblastic crystals of forsterite set in a matrix of calcite.
  B.   Diopside-rock, near the gabbro of Broadford.   This represents a richly cherty band, now completely converted to diopside.   Note the decussate structure.

suffice to convert all the magnesia to forsterite, a lime-bearing silicate makes its appearance.   This, however, is not wollastonite but the double silicate *diopside*, which is then found accompanying or replacing the forsterite.   In Skye it takes the place of that mineral in certain beds or sometimes in patches which represent the vanished sponges, and there are even cherty seams converted to solid diopside-rock (Fig. 30, *B*).   It should be observed that the formation of diopside does not in itself import dedolomitization ; but, if the mineral is subsequently converted to serpentine and calcite, the same result is

---

 [1] *Geologische Diffusionen* (1913).
 [2] King and Rowney, *Proc. Roy. Ir. Acad.* (2), vol. i (1871), pp. 132–9 ; *An Old Chapter of the Geological Record* (1881).

reached indirectly. Adams and Barlow [1] have explained in this way the origin of some serpentinous marbles in Ontario, but doubtless the more usual derivation of serpentine in such rocks is from forsterite.

A colourless *tremolite*, in prismatic crystals or thickly felted needles, is a mineral of less general distribution in this connexion (Fig. 31, *A*). In Skye it is found especially as a skin investing the forms which represent siliceous sponges. Its situation here, with diopside inside and forsterite outside, accords with the intermediate composition of the amphibole. Tremolite does not, however, enter the inner ring

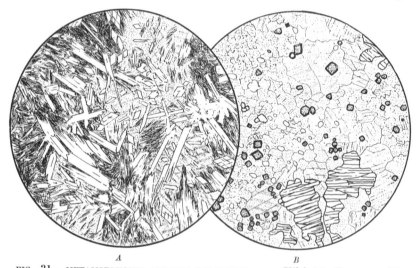

FIG. 31.—METAMORPHOSED CHERTY DOLOMITES, near Kilchrist, Skye ; × 25.

A. Tremolite-Marble : a dense aggregate of little prisms and fine needles of tremolite embedded in calcite.
B. Diopside-Marble with small octahedra of spinel.

of the aureoles, and it is presumably unstable at the highest temperatures of metamorphism. It is to be noted that the now generally accepted formula of this mineral is $Ca_2Mg_5Si_8O_{22}(OH)_2$, involving constitutional hydroxyl.

The ferrous and manganous carbonates, like the magnesian and in preference to the calcic, enter readily into reaction with silica. In a merely ferruginous limestone or dolomite the iron goes into such compounds as andradite and hedenbergite. The metamorphism of an impure chalybite-rock may give rise to ilvaite, with the formula $CaFe_2(FeOH)(SiO_4)_2$. From impure carbonate-rocks rich in manganese comes rhodonite with its zinc-bearing variety fowlerite.

[1] *Geology of the Haliburton and Bancroft Areas* (Mem. Geol. Sur. Can., 1910), p. 214.

Consider next the metamorphism of a magnesian limestone containing aluminous as well as siliceous impurities. If only a small quantity of alumina be present in a disposable form, it may be taken up into an amphibole, some member of the edenite-pargasite series being formed instead of tremolite.[1] In a higher grade, where the amphiboles cease to be stable, the alumina does not go into the pyroxene, but makes *spinel*, which is a very characteristic accessory mineral in forsterite- and diopside-marbles. It appears in little octahedra, colourless or of violet tint (Fig. 31, *B*).

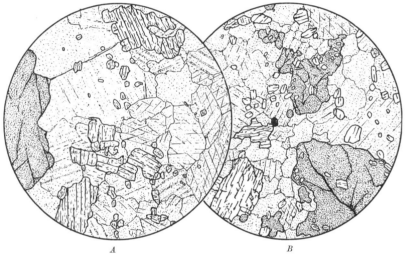

A   *B*

FIG. 32.—CRYSTALLINE LIMESTONES WITH GROSSULARITE AND DIOPSIDE, Crathie, Aberdeenshire ;   × 23.

The examples selected are of coarser grain than most rocks of this type. In *B* the irregularly shaped garnet, enclosing grains of diopside, suggests a hasty crystallization.

With a larger content of alumina the conditions are quite changed. Since no magnesian alumino-silicate figures in our list, the alumina goes now to make grossularite or idocrase. Diopside is found in company with one or both of these minerals, but forsterite is no longer formed (Fig. 32). In rocks not too rich in alumina, or too highly magnesian, wollastonite figures in addition. The association grossularite-diopside-wollastonite or idocrase-diopside-wollastonite is characteristic of a widespread class of metamorphosed calcareous rocks, the relative proportions of the minerals depending on the ratios $Al_2O_3 : MgO : CaO$ in the bulk-analysis of the rock. If the disposable alumina present passes a certain limit, the simple silicate wollastonite is ruled out, and the anorthite molecule takes its place.

[1] Tilley, *Geol. Mag.*, vol. lvii (1920), p. 454.

This, however, is a simplified view of the actual case. In an ordinary argillaceous limestone other components are present besides alumina and silica, the most important being alkalies, iron-oxides, titania, and perhaps sulphur. Some magnesia also enters in the form of detrital chlorite. The possibilities in respect of new minerals are thereby considerably enlarged. Note in the first place that some of the non-carbonate minerals in such a mixed sediment may merely recrystallize without change. This is true, for example, of *graphite* of detrital *tourmaline*, and in the lower grades of metamorphism of *muscovite*. A special case, which need not be discussed in detail, is that of a calcareous tuff, in which such minerals as felspars and pyroxenes may suffer no change beyond recrystallization. The finely divided *albite*, which is present in many argillaceous limestones, at first recrystallizes without other change. In a higher grade it is likely to become associated with new-formed anorthite, and we find accordingly *intermediate felspars* ranging from oligoclase to bytownite. In a high grade, too, we often see some potash-felspar, either *orthoclase* or *microcline*, and this can come only from the dissociation of white mica, present as sericite in the original sediment. Any original limonite is, as a first step, reduced to *magnetite* ; but at least part of the iron-oxides is ultimately taken up into the silicate-compounds, viz. as the hedenbergite molecule in diopside and perhaps as andradite in grossularite. Titanic acid, originally present as rutile, goes to make *sphene* : the titaniferous garnets, melanite and schorlomite, found in alkaline igneous rocks and in some contact-belts, have no place among normal products of metamorphism. There is usually sufficient silica present to convert all the alumina to silicates, but, if silica is deficient, a mineral of the spinel group is formed, often a *pleonaste*. Among the various types of metamorphosed limestones in the Carlingford district Osborne [1] has noted one composed essentially of diopside, grossularite, spinel, and calcite ; and in the same district calcareous rocks which have undergone pneumatolysis as well as metamorphism contain pleonaste as an abundant constituent (Fig. 53, *B*). Finally it may be remarked that many calcareous rocks have a certain content of sulphides. In a high grade of metamorphism pyrites is converted to *pyrrhotite* by the loss of part of its sulphur. The change takes place at some temperature above 500°, depending of course upon the pressure.[2] Near an igneous contact, however, pyrrhotite has often a pneumatolytic origin.

[1] *Geol. Mag.*, vol. lxix (1932), p. 224.

[2] Allen, Crenshaw, and Johnston, *Amer. J. Sci.* (4), vol. xxxiii (1912).

CHAPTER VII

THERMAL METAMORPHISM OF CALCAREOUS SEDIMENTS
(*continued*)

Lime-silicate-rocks—Stable and Metastable Associations—Goldschmidt's Classification of Types of Hornfels.

LIME-SILICATE-ROCKS

THE general course of metamorphism in impure limestones and dolomites is simple. By such reactions as we have specified, with rising temperature, the carbonates (and first the magnesian carbonate) are replaced by new minerals, of which the most characteristic are magnesia- and lime-bearing silicates, carbon dioxide being concurrently expelled. The reduced amount of residual (recrystallized) calcite finally associated with the new minerals gives a rough measure of the degree of impurity of the original sediment. If the non-calcareous part was initially in such quantity and of such a nature as to react with the whole of the carbonates, no calcite will remain, and we have what is conveniently styled a *lime-silicate-rock*. Mineralogically the name covers a rather wide range of diversity. In addition to the mineral-associations already noticed, a certain amount of quartz may enter, when the original rock was of a gritty nature—a mineral which would not be stable in company with calcite under the conditions of thermal metamorphism. In respect of grain-size, too, a wide range of difference may be observed. The coarser types are generally those made up mainly of one mineral, or sometimes an intergrowth of two minerals. The more common types, which are of composite nature, tend to illustrate the other extreme. Here the development of crystals of several different minerals from centres in close proximity has often given rise to a very fine-grained aggregate. For such close-grained lime-silicate-rocks, as developed among the metamorphosed Devonian strata round the Cornish granites, Barrow has used the term *calc-flintas*. It is a consequence of the large number of minerals which may possibly enter, that slight differences of initial composition in successive seams may result in very different mineral-associations, and accordingly a finely-banded arrangement is a characteristic feature of the calc-flintas (Fig. 33). In other cases a concretionary structure has given rise to a concentric arrangement of the new minerals (Fig. 52, *A*, below).

As regards the more important minerals formed in any given case, the controlling factor is to be sought, as we have seen, in the relative proportions of lime, magnesia, and alumina in the original sediment. One particular case is worthy of note. While in a pure carbonate-rock the magnesia (reckoned in molecules) can never be in excess of the lime, this relation is emphatically reversed in any ordinary type of argillaceous sediment. If then a dolomitic limestone

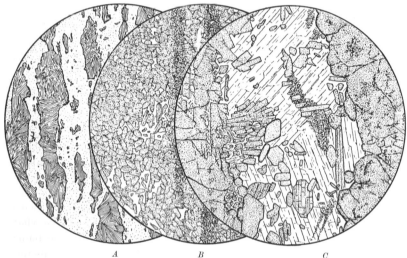

A   *B*   *C*

FIG. 33.—LIME-SILICATE-ROCKS (CALC-FLINTAS) from the Devonian of Cornwall and Devonshire ; × 25.

A. Gaverigan, near the St. Austell granite. Green hornblende makes discontinuous bands following the original lamination : the rest is a finely granular aggregate of felspar and quartz.

B. Ivybridge, near the Dartmoor granite : composed of little crystals of diopside with interstitial plagioclase.

C. Camelford, near the Bodmin Moor granite. This shows bands of zoisite on the left and garnet on the right : in the middle colourless tremolite enclosing crystals of zoisite.

contains an abundance of foreign material which is largely chloritic in composition, the ratio $CaO : MgO(+ FeO)$ in the mixed rock may well be too low to permit the formation of such silicates as garnet, idocrase, and wollastonite. The principal minerals produced are then pyroxenes, amphiboles, plagioclase felspars, and sphene, giving a well-characterized type of rock. It will be observed that these are all familiar pyrogenetic minerals. Indeed some sediments of the kind in question, apart from their content of carbon dioxide, do not differ much in composition from some basic igneous rocks ; and it is natural to find that, in a high grade of metamorphism, this resemblance should express itself in a like mineralogical constitution. A good example is a rock in the Ducktown district of Tennessee, originally described as a quartz-diorite.[1] Various types are illustrated on a small scale in the Cornish calc-flintas or in particular bands in that group.

In many partly calcareous sediments the non-carbonate element, instead of being an adventitious admixture, makes up the chief bulk. Such are the deposits styled *calcareous shales, slates, and sandstones.* Here the non-calcareous material is much in excess of what is requisite for the complete decarbonation of all the carbonate present, and the course of metamorphism is in consequence somewhat more complicated. In addition to the class of reactions which we have been discussing, there now come into play others of the kind formerly studied in the metamorphism of simple argillaceous sediments. Nor is it sufficient to picture the two sets of reactions as proceeding independently side by side or successively, for they exercise to some extent a modifying influence upon one another. One way in which this interaction makes itself felt is in ruling out what we may regard as the more extreme products on both sides, i.e. the more highly calcic and the more richly aluminous. The lime-silicates formed in the metamorphism of such a moderately calcareous sediment are characteristically the epidote minerals in a low grade and lime-felspars in a higher. Grossularite and idocrase are found only in bands which were rather more richly calcareous, and wollastonite is absent. Of magnesian minerals there occurs a pyroxene, probably near diopside, or a green hornblende, but not forsterite. On the argillaceous side, biotite often forms freely, and cordierite may be produced in the least calcareous bands, but not andalusite. By considering the composition of the several minerals, it is easy to see what associations are to be expected ; and the facts as observed are generally in accord with Goldschmidt's diagram, given on page 99 (Fig. 38). For instance, grossularite and idocrase are not found in company with either biotite or cordierite, while diopside may occur with biotite but not with cordierite (Fig. 34, *A*).

It should be remarked that the reactions for which we have premised a peculiar promptitude and rapidity are only those which take place between carbonate and non-carbonate. The metamorphism of the argillaceous part of the rock, in so far as it can be regarded separately, is a graduated process, though the adjustment of equilibrium from grade to grade is presumably facilitated by the solvent action of such free carbon dioxide as is present. Again, certain reactions characteristic of an advanced grade of metamorphism will

---

[1] Keith, *Bull. Geol. Soc. Amer.*, vol. xxiv (1913), p. 684 ; Laney, *U.S. Geol. Sur. Prof. Pap. 139* (1926), pp. 19–21.

be promoted in the presence of lime. This is true in particular of the reactions by which alkali-felspars are produced at the expense of micas (p. 56). Here the anorthite molecule affords a ready way of disposing of the excess of alumina :

$$H_2KAl_3(SiO_4)_3 + CaCO_3 + 2SiO_2$$
$$= KAlSi_3O_8 + CaAl_2(SiO_4)_2 + H_2O + CO_2.$$

The sodic element in the mica gives rise to albite. Orthoclase and plagioclase thus come often to be associated ; but there is no relation between their respective amounts in a rock, for alumina to make

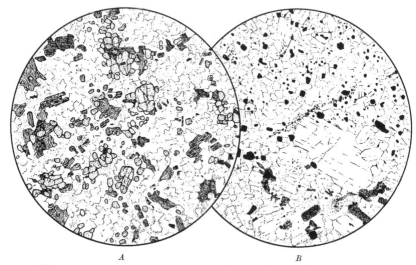

A   B

FIG. 34.—METAMORPHOSED CALCAREOUS SLATES, in the aureole of the Càrn Choìs diorite, near Comrie, Perthshire ;   × 23.

A. Diopside-Biotite-Hornfels ; composed of diopside, biotite, quartz, orthoclase, and some labradorite.
B. Plagioclase-Cordierite-Hornfels. The most abundant mineral is labradorite, in small crystals and some of larger size ; with this are cordierite, biotite, magnetite, and a little pleonaste. The sediment was one poor in silica.

anorthite may come from other sources, and albite has often been present in the original sediment. In this way arise rocks somewhat rich in plagioclase, usually of some intermediate variety—andesine or labradorite—and containing in addition orthoclase and biotite. The other minerals which may enter depend upon the proportion of carbonate in the original sediment. The more calcareous shales and slates yield diopside (Fig. 34, *A*) or in a lower grade sometimes a green hornblende ; with a lower lime-content hypersthene may take the place of diopside, or biotite may be the only coloured mineral ; and in only slightly calcareous rocks cordierite comes in (Fig. 34, *B*).

STABLE AND METASTABLE ASSOCIATIONS

We have already been led to the conviction that under the conditions realized in thermal metamorphism chemical equilibrium is in general quite promptly established. For reasons given, this is especially true when part of the rock metamorphosed is composed of carbonates. With continually rising temperature, various reactions are successively brought into play ; and, by observing the results as arrested at different stages, we are able to recognize *successive grades of metamorphism*. To present a schematic view of this succession, defined by particular index-minerals, is a less easy task, in view of the wide range of bulk-composition met with in this class of rocks, which expresses itself in different mineral-associations. Moreover, it is especially in this class that the influence of pressure as an independent controlling factor cannot be disregarded.

It is only at the lowest temperatures (in purely thermal metamorphism) that carbonates and free silica can coexist without mutual reaction ; and the first formation of wollastonite or (in a more argillaceous limestone) of grossularite or idocrase must be placed at a very early stage. Since, moreover, wollastonite and probably grossularite have no upper limit of stability in thermal metamorphism, they can be of no service as index-minerals. Idocrase too may persist to the highest grades, and its dependence on pressure and water-content complicates its stability-conditions. In more richly argillaceous sediments, such as calcareous shales, the first lime-silicates to form are minerals of the zoisite-epidote group, and these give place in a higher grade to lime-felspars. The production of anorthite as a distinct mineral marks always a high grade of metamorphism ; but it appears that in presence of albite the zoisite-anorthite reaction is initiated at much lower temperatures and carried on progressively. Any detrital albite contained in the original sediment recrystallizes, as we have seen, at a very early stage in the form of minute granules. When, at a somewhat later stage, the little grains of clear felspar are more developed, it can often be verified that they are no longer pure albite but rather an albite-oligoclase. Comparative study goes to show that oligoclase forms at a higher temperature and andesine and labradorite in turn only in a distinctly high grade of metamorphism. Whether these more calcic felspars can form at all, depends obviously upon the relative proportions of zoisite and albite. If then increased knowledge should establish the different varieties of plagioclase as known index-minerals, they must still be used only with due caution. In calcareous shales as in purely argillaceous rocks, the appearance of a potash-felspar marks a fairly advanced grade of metamorphism, though it is not to be assumed that the temperature indicated is necessarily the same in both cases (p. 92).

In magnesian limestones and in many of the more impure calcareous sediments some member of the amphibole group may figure among the products of metamorphism ; and, since these minerals are found in the lower and medium grades but disappear in a high grade, they are of significance in the present connexion. If such disappearance takes effect (for a given variety of amphibole) at a definite temperature-limit, they will evidently afford valuable indications. This may still be true, within reasonable limits, if these minerals are indeed merely metastable forms, as is perhaps suggested by their often inconstant or sporadic occurrence.

We have hitherto seen little reason to question the thesis laid down at the outset, that in thermal metamorphism chemical equilibrium is in general promptly made good. Some apparent exceptions do not carry conviction, especially those which are cited as showing too many different minerals in association, as judged by the standard of the Phase Rule. Here we must remember the very narrow limits of diffusion and also in some cases the difficulty of determining the exact number of components involved. We do, however, meet with certain mineral-associations which are on the face anomalous, and one of these is worthy of notice.

In the metamorphism of impure calcareous rocks we find grossularite and diopside associated either with wollastonite or with anorthite, but the two last-named minerals seem to be incompatible with one another. The presumption is then that the wollastonite and anorthite molecules combine to make grossularite :

Anorthite + 2 Wollastonite = Grossularite + Quartz.

Nevertheless, in the Carlingford district,[1] in Deeside,[2] and else-where, there are occurrences in which wollastonite and a lime-bearing felspar are found in close association, with or without grossularite (Fig. 35). It is no doubt conceivable that the association shown on the right side of the equation is stable below and that on the left above a certain temperature, and that the temperature-range of metamorphism embraces both cases ; but in fact the normal and the anomalous may be found near together with nothing to suggest any significant difference of temperature. Important variation of pressure

[1] Osborne, *Geol. Mag.*, vol. lxix (1932), pp. 223–4.

[2] Hutchison, *Trans. Roy. Soc. Edin.*, vol. lvii (1933), p. 574.

is equally ruled out,[1] and it might seem that the close association of wollastonite and a calcic felspar is to be explained as an instance of failure to adjust chemical equilibrium. It does not appear, however, that there are actual records of the occurrence of wollastonite with anorthite itself, but only with intermediate varieties of plagioclase ; and Osborne has suggested that the explanation may be found in the presence of the albite molecule.

The most indubitable examples of the non-adjustment of equilibrium occur in connexion with special geological conditions. They

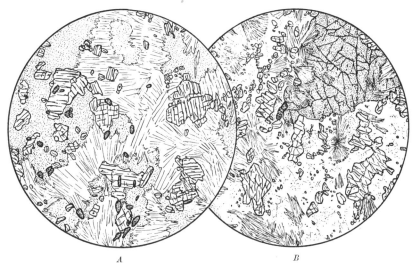

A   *B*

FIG. 35.—ANOMALOUS ASSOCIATION OF WOLLASTONITE WITH CALCIC FELSPARS, Pollagach Burn, near Cambus o' May, Aberdeenshire ; × 23.

A. Wollastonite, with fibrous habit and a tendency to radiate arrangement, is seen in contact with bytownite (dull from incipient change). The other minerals are diopside and sphene.
B. Here grossularite occurs in addition. The radiating needles of wollastonite are enclosed both in the clear felspar (andesine) and in the garnet.

are found, not within a regular aureole, but near contact with some minor intrusive mass ; that is, in a place where a high temperature was attained, but cooling was relatively rapid. The conditions, in short, were such as have elsewhere given rise to vitrification in ordinary shales and sandstones (p. 27). In such circumstances the metamorphism of impure calcareous rocks may yield, not only anomalous mineral-associations, but particular *rare mineral-species*, which may be confidently set down as merely metastable forms. Some of these minerals are known only from one or two localities, where perhaps several of them occur together.

[1] Higher pressure would favour the normal (garnet-bearing) association.

Among the minerals having this special manner of occurrence must be reckoned the melilites, including gehlenite and other varieties. A well-known locality for melilite [1] is in the metamorphosed Triassic dolomites of Monzoni (Fig. 36, *A*). More remarkable are the rare orthosilicates of calcium and magnesium :

Monticellite, $CaMgSiO_4$,
Merwinite,[2] $Ca_3Mg(SiO_4)_2$,
Larnite,[3] Ca_2SiO_4,
Spurrite,[4] $2Ca_2SiO_4 \cdot CaCO_3$,

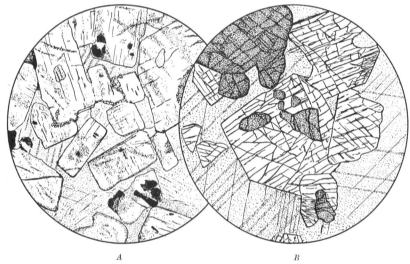

A   *B*

FIG. 36.—RARE MINERALS IN METAMORPHOSED DOLOMITE, Monzoni, Tirol ; × 23.
A. Melilite crystals set in calcite.
B. Fassaite and pleonaste in calcite.

the last being a compound of silicate and carbonate. These minerals are found in close association with one another. An interesting occurrence is that described by Tilley from the Chalk metamorphosed by a dolerite intrusion at Scawt Hill, near Larne, Co. Antrim. Here are found spurrite, larnite, melilite, merwinite, and pleonaste (Fig. 37).

Associated with some of these rare species are found also other minerals having an anomalous composition of the kind suggestive of

[1] Often named gehlenite, but see Buddington, *Amer. J. Sci.* (5), vol. iii (1922), pp. 71, 74.

[2] Larsen and Foshag, *Amer. Min.*, vol. vi (1921), pp. 143–8 (Crestmore, California).

[3] Tilley, *Min. Mag.*, vol. xxii (1929), pp. 77–86.

[4] Wright, *Amer. J. Sci.* (4), vol. xxvi (1908), pp. 545–54 (Velardeña, Mexico).

constrained or metastable solid solution. The pleonaste of Scawt Hill contains 17 per cent. of the magnetite molecule, a much larger proportion than is normally held in any spinel mineral.[1] A comparable case is presented by certain highly aluminous augites, such as the fassaite of Monzoni, found in company with melilite and monticellite (Fig. 36, *B*). An alumina-percentage 10–12 or more distinguishes fassaite sharply from the ordinary pyroxenes of thermal metamorphism, which are commonly referable to the diopside type, though here more actual analyses are a desideratum.[2]

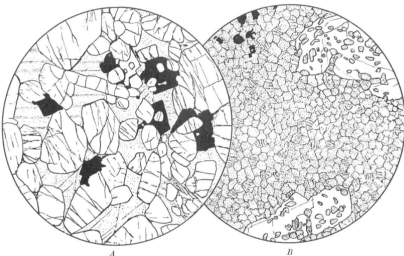

A   *B*

FIG. 37.—RARE MINERALS IN METAMORPHOSED CHALK at contact with dolerite, near Larne, Antrim ;   × 23.

A.   Spurrite with nearly opaque pleonaste and interstitial calcite.
B.   Porphyroblasts of spurrite set in an aggregate of larnite and enclosing grains of the same ; a little pleonaste.

GOLDSCHMIDT'S CLASSIFICATION OF TYPES OF HORNFELS

As an appendix to our discussion of metamorphism in sediments of various kinds, it will be useful to summarize very briefly Goldschmidt's classification of different types of ' hornfels ' (p. 4), as since amplified by Tilley.[3] It should be clearly understood, however, that such a scheme, concerned only with totally reconstructed rocks, which presumably represent final equilibrium associations of minerals,

[1] Vogt, *Vidensk. Skr.*, 1910, No. 5, p. 9.

[2] In regional metamorphism pyroxenes rich in alumina may be stable in association with spinelled minerals and calcite, but not in presence of free silica ; Tilley, *Geol. Mag.*, vol. lxxv (1938), pp. 81–5.

[3] *Quart. Journ. Geol. Soc.*, vol. lxxx (1924), pp. 32–56 ; *Geol. Mag.*, vol. lx (1923), pp. 101–7, 410–18, and vol. lxii (1925), pp. 363–7.

throws no light on the course of metamorphism. The equations written down by Goldschmidt are not to be taken as representing actual reactions, but merely as expressing the relations between certain minerals in respect of composition.

Goldschmidt inquires what different associations of minerals, from a selected list, are possible in accordance with the Phase Rule. This is done by discussing relations (or conceivable balanced reactions) such as:

$$A + B = C + D,$$

where the four letters stand for different mineral compounds. Here the possible associations are:

either AB, ABC, ABD,
or CD, ACD, BCD.

Which of the two sets of associations is to be adopted cannot be inferred theoretically, but must be determined by actual occurrences. For our purposes it will be sufficient to appeal directly to the petrographical evidence without following the steps of the argument, and the whole scheme can be conveniently presented in the form of a diagram.

It includes the various types of hornfels representing argillaceous and calcareo-argillaceous sediments ; and the selected list of minerals is : andalusite (with sillimanite), cordierite, enstatite (and hypersthene [1]), anorthite (with varieties of plagioclase), diopside, grossularite, and wollastonite. Quartz may also be supposed present, those rocks which are deficient in silica being reserved for later treatment. The inclusion of albite also makes no difficulty ; but that orthoclase and the micas find no place is a serious departure of the ideal scheme from realities. Biotite is in fact present in all types except those rich in lime, and orthoclase often enters in addition.

The diagram (Fig. 38) shows the relations of the several types which contain free silica, the numbers corresponding with the different classes distinguished by Goldschmidt, as given in the list which follows. The inset numbers indicate limiting cases, implying some particular adjustment of the total chemical composition. The strictly non-calcareous rocks are covered by the first three types, two of which have been supplied by Tilley.

(1) Andalusite, Cordierite.
(1A) Cordierite.
(1B) Cordierite, Enstatite.
(2) Andalusite, Cordierite, Anorthite.
(3) Cordierite, Anorthite.

[1] In the diagram ferrous are understood to be included with magnesian compounds.

(4) Cordierite, Anorthite, Enstatite.
(5) Anorthite, Enstatite.
(6) Anorthite, Enstatite, Diopside.
(7) Anorthite, Diopside.
(8) Anorthite, Diopside, Grossularite.
(9) Diopside, Grossularite.
(10) Diopside, Grossularite, Wollastonite.

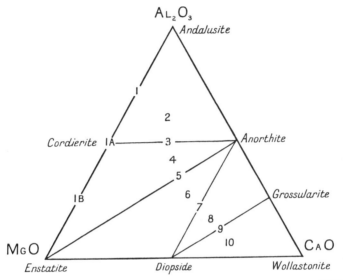

FIG. 38.—DIAGRAM SHOWING THE RELATIONS OF THE CLASSES OF HORNFELS DISTINGUISHED BY GOLDSCHMIDT.

In so far as the scheme is valid, the diagram exhibits clearly the manner in which the mineralogical constitution is determined by the total chemical composition. If we denote by A, M, C the relative proportions (in molecules) of alumina, magnesia, and lime, the conditions for the formation of the several minerals are seen to be as follows :

for andalusite,	$A > M + C$,
for cordierite,	$A > C$,
for diopside,	$A < C$,
for enstatite,	$A + M > C$,
and	$M + C > A$,
for anorthite,	$C < 3A + M$,
for wollastonite,	$C > 3A + M$,
for grossularite,	$C > A + M$.

From these relations it would appear, e.g., that cordierite and diopside are necessary alternatives, in the sense that any hornfels of the kind in question must contain one of these minerals, and cannot contain both ; and again (in lime-bearing hornfelses) wollastonite and anorthite (or some variety of plagioclase) figure as necessary alternatives. All such inferences, however, are subject to correction, having regard to the minerals biotite, orthoclase, and others, which are here ignored.

Coming now to the quartzless types, it is evident in the first place that, in any of the foregoing types, the amount of quartz may be supposed to dwindle to zero without disturbing equilibrium : this gives twelve limiting cases. If now we continue in imagination to abstract silica, it can come only from the breaking down of some silicate-mineral present, such as andalusite or cordierite. For these two minerals we have the relation :

$$5 \text{ Andalusite} + \text{Spinel} = \text{Cordierite} + 5 \text{ Corundum},$$

and it appears that the stable associations are those which include cordierite and corundum.[1] So Tilley gives the following types, here numbered consecutively for convenience :

> (11) Andalusite, Cordierite, Corundum.
> (12) Cordierite, Corundum.
> (13) Cordierite, Corundum, Spinel.
> (14) Corundum, Spinel.

These may be regarded as the non-quartzose representatives of (1) Corresponding with (1A) and (1B) we have in like manner :

> (15) Cordierite, Spinel.
> (16) Cordierite, Enstatite, Spinel.

Other types may be derived from (2) and (3) :

> (17) Andalusite, Cordierite, Corundum, Anorthite.
> (18) Andalusite, Corundum, Anorthite.
> (19) Cordierite, Spinel, Anorthite.

To secure an artificial simplicity by disregarding minerals of inconvenient composition is a device already practised, in regard to igneous rocks, by the authors of the Quantitative Classification. Though patently a source of error, it may be justified in the present instance by the undoubted utility of Goldschmidt's manner of treatment and the lack of any alternative scheme.

---

[1] There are, however, records of the association andalusite-cordierite-spinel and even of the four phases together.

CHAPTER VIII

THERMAL METAMORPHISM OF IGNEOUS ROCKS

Special Features of Igneous Rocks—Thermal Metamorphism of Basic Rocks —Thermal Metamorphism of Deeply Weathered Rocks—Thermal Metamorphism of Acid Rocks.

SPECIAL FEATURES OF IGNEOUS ROCKS

IN respect of their behaviour in thermal metamorphism igneous rocks have hitherto received much less general notice than those of sedimentary origin. They present nevertheless some features of more than common interest, and to bring out the significance of these special features, rather than a comprehensive treatment of the subject as a whole, will be the aim of what here follows. Since thermal metamorphism is, in its most general aspect, merely a readjustment of the constitution of a rock to more or less high-temperature conditions, it may perhaps appear, upon a hasty judgment, that igneous rocks, themselves of high-temperature origin, will be little susceptible to changes of this kind. Such inference, while containing a certain measure of truth, fails, however, to take account of some important considerations.

In the first place, the genesis of an igneous rock, starting from a fluid magma and ending normally in a crystalline aggregate, covers a wide range of declining temperature, and the several constituent minerals, as we now see them, belong to different stages of the prolonged process of cooling. In many rocks the latest-formed minerals have crystallized at temperatures which may be overtaken in metamorphism of quite moderate grade. Further, we know that the consolidation of an igneous rock cannot, in the most general case, be truly pictured as a simple separation of the several minerals in turn from the fluid magma. Later minerals may be derived in part at the expense of earlier ones which, crystallized at a higher temperature, cease to be stable at a lower temperature in contact with the changed magma, and are attacked by it. To such reactions most petrologists, following Bowen,[1] assign an important part in the normal course of petrogenesis.

---

[1] The Reaction Principle in Petrogenesis, *Journ. Geol.*, vol. xxx (1922), pp. 177–98 ; *The Evolution of the Igneous Rocks* (1928), chap. v.

Whenever the early crystals of important minerals have not been removed or in some way protected from contact with the magma, they will often be liable to partial or total resorption at a later stage ; but whether such reactions, demanded by chemical equilibrium, actually take effect or not will depend upon the conditions, and in particular upon the rate of cooling. In so far as any given igneous rock has actually passed through such changes with falling temperature, equilibrium being continually readjusted by the proper reactions, we may reasonably expect that the reactions will be reversed by rising temperature in thermal metamorphism.

If, on the other hand, owing to a too rapid rate of cooling or any other cause, these reactions making for equilibrium did *not* take effect as the magma cooled, some of the constituent minerals of the resulting rock must be in a metastable state. It cannot be doubted that a rock of such constitution will be eminently susceptible of thermal metamorphism. This is perhaps most clearly illustrated by considering an extreme case. If the rate of cooling be sufficiently rapid, crystallization is practically inoperative in the later stages, and the resulting rock consists partly of glass, which is essentially metastable. Now we know that, when such a glass is heated to a moderate temperature and so held for a time, devitrification is readily induced. Here crystallization, which is normally the result of cooling, appears as an effect of heating. It is an indirect effect, depending upon the fact that the higher temperature restores molecular mobility. Doubtless it also promotes atomic mobility ; and we may confidently infer that a mineralogical change in the direction of equilibrium with falling temperature, such as the uralitization of pyroxene, having failed of effect in the first instance, may be precipitated when the rock is again brought to a suitable temperature in the course of thermal metamorphism. We are also prepared to find that a suspended reaction of this kind, made effective in a moderate grade of metamorphism, will be reversed in a higher grade.

An igneous rock, then, even when freshly consolidated, is not likely to be immune from change when subjected to any notable elevation of temperature. Besides this, few rocks that we meet with are in a perfectly fresh state. Owing to secondary changes, whether correctly described as weathering or not, there has usually been at least some production of low-temperature minerals such as kaolin, sericite, chlorite, serpentine, calcite, iron-oxides, etc. These are in fact the same substances that we have met with as the constituents of argillaceous and calcareous sediments. Here, however, they have not been distributed into separate deposits, but remain for the most part in the rock in which they were generated. Some redistribution of the various secondary products within the rock is likely to be found, but it is as a rule narrowly limited. The process is a selective one, depending on relative solubility. Sericite and kaolin remain where they were produced, and so for the most part do serpentine and the iron-oxides. Epidote, chlorite, silica, and the zeolites are more liable to travel, and calcite, the most soluble of all, is also the most vagrant. These are the minerals commonly found in fissures, steam-vesicles, and other places of relief of pressure, the influence of pressure upon solubility being a controlling factor in the redistribution.

The earliest effects of thermal metamorphism in an igneous rock are shown by such low-temperature minerals, if any, as are present, including alteration-products of the kind just enumerated and sometimes the latest products of magmatic crystallization. Upon a moderate elevation of temperature these minerals readily undergo change, either individually or by reaction with one another. Incidentally there is an elimination of any oxygen or water or carbon dioxide taken up by the igneous rock in weathering or other destructive changes. Here we see in the successive reactions induced by rising temperature a *reversal* of those changes of the nature of degradation which affected the original rock with falling temperature (see p. 47). This principle, in which ' anamorphism ' appears as the opposite of ' catamorphism ', constituting the complementary part of a grand cycle of change, is applicable to thermal metamorphism in general, but it is in igneous rocks that it is most clearly exhibited. Here, since there has been no wide dispersal of the products of degradation, the new combinations which come from metamorphism are in general such as are familiar in pyrogenetic minerals, and the ultimate result is the restoration, as regards mineralogical constitution, of the original igneous rock. Some concrete examples will serve to set the matter in a clearer light.

THERMAL METAMORPHISM OF BASIC ROCKS

The amygdaloidal basalts of Tertiary age in Skye [1] and Mull [2] have in many places been metamorphosed by subsequent plutonic intrusions. In the non-metamorphosed basalts the contents of the amygdaloidal cavities may include chlorites, calcite, chalcedony, etc., but the principal and often the only minerals are lime- and soda-zeolites. Waiving for the moment the question of the precise mode of origin of these zeolites, we may conceive them as derived from plagioclase

[1] *The Tertiary Igneous Rocks of Skye* (*Mem. Geol. Sur.*, 1904), pp. 50-3.
[2] *Tertiary and Post-Tertiary Geology of Mull, etc.* (*Mem. Geol. Sur.*, 1924), pp. 151-5.

felspars by simple reactions involving hydration. We find accordingly that they are represented in the metamorphosed rocks by a crystalline aggregate of plagioclase felspar, with or without other minerals. The general correspondence between the plagioclase group and the lime- and soda-zeolites is sufficiently apparent. In both we see the ratios (in molecules) :

$$Na_2O + CaO : Al_2O_3 : SiO_2 = 1 : 1 : n,$$

where n ranges from 2 to 6. There is not indeed a correspondence term by term. It is to be remarked, however, that the zeolites seldom occur singly but in associations of two or three species together. It is easy to devise such equations [1] as :

$$Ab_2An_3 + 11H_2O = \text{Natrolite} + 3 \text{ Scolezite,}$$
$$Ab_2An_1 + 8H_2O = \text{Chabazite} + 2 \text{ Analcime,}$$
$$Ab_4An_1 + 9H_2O = \text{Heulandite} + 4 \text{ Analcime ;}$$

or, again, since free silica and other substances may accompany the zeolites :

$$\text{Albite} + H_2O = \text{Analcime} + \text{Quartz,}$$
$$2 \text{ Albite} + H_2O = \text{Natrolite} + 3 \text{ Quartz.}$$

Equations of this kind may well represent reversible reactions, which are driven towards the right with falling and towards the left with rising temperature.

This is, however, an incomplete view of the origin and metamorphism of the amygdales in these rocks, which present features of special interest. There is good evidence to show that in these basalts, and probably in many other amygdaloidal lavas, the minerals within the steam-vesicles are not secondary, but are the latest products of crystallization from a magma which had become rich in water and finally forced its way into the cavities. They are derived, not from the destruction of felspar crystals, but from anorthite and albite molecules becoming hydrolysed in the aqueous magma. Further, the several minerals so found in association do not all belong to the same stage in the process of cooling, and later minerals have often been formed at the expense of earlier ones. They constitute in that case a ' reaction-series ' as defined by Bowen. McLintock [2] has studied from this point of view the amygdaloidal basalts of the Ben More district of Mull and their metamorphism by subsequent intrusions

---

[1] I adopt here the text-book formulae for the various zeolites without inquiry concerning the significance of the contained water.

[2] *Trans. Roy. Soc. Edin.*, vol. lv (1915), pp. 1–33.

of granite. He makes it appear clearly that the reactions set up in metamorphism represent exactly the reversal of those which had taken place in the final stages of magmatic crystallization. The commonest zeolite in this district is scolezite, and this is the final term of a reaction-series which includes grossularite, epidote, and prehnite. In metamorphism the scolezite is first transformed to prehnite, this in turn to epidote, and so finally to grossularite.

Where basic rocks have suffered changes of the nature of weathering, calcite is a more or less abundant product ; and the redistribution

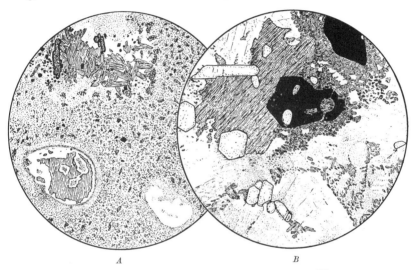

A   *B*

FIG. 39.—METAMORPHOSED IGNEOUS ROCKS ; × 25.

A. Amygdaloidal Pyroxene-Andesite, Wasdale Head, near the Shap granite, Westmorland. The only noticeable change in the body of the rock is the production of small flakes of biotite. The amygdales have a green hornblende instead, with some felspar and crystals of brown sphene : one at the bottom of the field is of quartz.
B. Gabbro metamorphosed by later acid intrusions, Caldbeck Fells, Cumberland. The gabbro is of a basic variety rich in apatite and iron-ore, which remain apparently intact. The augite is replaced by fibrous green hornblende with patches of brown biotite, the latter only in the neighbourhood of the iron-ore. The felspar has been cleared of its minute inclusions.

of this within the rocks, in virtue of its relatively free solubility, becomes an important factor influencing subsequent thermal metamorphism. The Ordovician lavas in the aureole of the Shap granite, Westmorland, afford very good illustrations.[1] The pyroxene-andesites on the west side of the granite were rocks not very rich in lime, and here calcite was mostly collected into the amygdales. In the general body of the andesite metamorphism has given rise to abundant

[1] Harker and Marr, *Quart. Journ. Geol. Soc.*, vol. xlvii (1891), pp. 292–301 ; vol. xlix (1893), pp. 359–65.

biotite, formed by reactions between chlorite, sericite, limonite, etc. ; but in the amygdales we find especially lime-bearing minerals, chiefly a green hornblende but also epidote, sphene, etc. (Fig. 39, *A*). To the north of the granite were basaltic lavas much richer in lime, and here calcite had been produced more plentifully. It was generally disseminated, as well as gathered in vesicles and fissures. Accordingly, in the metamorphosed rocks green hornblende is of general occurrence instead of biotite. There were large amygdales, lines with chlorite and chalcedony and filled in with calcite ; and here a number of

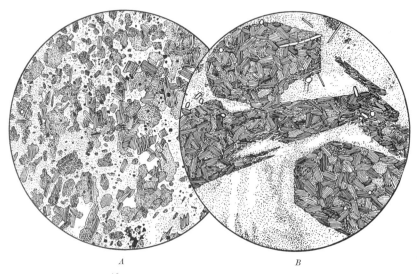

A   *B*

FIG. 40.—METAMORPHOSED IGNEOUS ROCKS ;　× 25.

A. Basic Tuff, metamorphosed near the Shap granite, Longfell Gill, Westmorland. The most conspicuous new minerals are flakes of brown biotite and little octahedra of magnetite.

B. Vogesite dyke near granite, Catacol, Arran. Well-shaped crystals of hornblende are replaced by aggregate of biotite. The felspar shows only the beginning of change, and the apatite is untouched.

lime-bearing silicates have been formed—epidote, green actinolitic hornblende, augite, sphene, and large crystals of grossularite. In the centre of the largest amygdales is calcite, recrystallized without decomposition, being too far from any source of silica to take part in chemical reactions.

Basic tuffs, owing to their original finely clastic state and consequent liability to weathering, are even more readily affected in thermal metamorphism than are lavas of like nature (Fig. 40, *A*).

So far we have discussed changes set up by metamorphism in the minor, and usually very late, minerals in igneous rocks. When, however, a medium grade has been reached, the principal constituent minerals of the rock begin to be affected in their turn, either by reactions producing definite new minerals or at least by what is in appearance a mere recrystallization. The fabric of the rock necessarily suffers change at the same time, though larger structures, such as the porphyritic and amygdaloidal, may still persist. Here again it is the basic rocks which furnish the readiest illustrations.

The most constant and noticeable change in any gabbro, dolerite, or basalt which has reached a certain grade of metamorphism is the

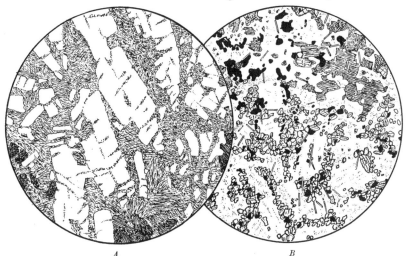

A   B

FIG. 41.—METAMORPHOSED DOLERITE DYKES, near the Granite of Beinn an Dubhaich, Skye; × 25.

A. Kilchrist. The ophitic augite is replaced by an aggregate of green hornblende, with a few little patches of biotite. The felspar is cleared of its inclusions, but not recrystallized. Narrow veinlets of hornblende traversing the clear crystals represent cracks which had been occupied by chlorite.

B. Torran. This dyke is enveloped in the granite, and shows a higher grade of metamorphism. Hornblende is seen in the upper part of the field, but has given place elsewhere to granules of new augite. The felspar and magnetite (but not the apatite needle) are recrystallized, and the original structure of the rock is only faintly suggested.

conversion of the augite to hornblende, commonly of a light green variety (Figs. 39, B; 41, A). As first formed, it makes distinct pseudomorphs. Hornblende may arise too from decomposition-products of augite, and this can often be distinguished by its manner of occurrence, e.g. as slender strings occupying cracks in the felspar. Rhombic as well as monoclinic pyroxene suffers uralitization, and even bastite pseudomorphs after hypersthene are converted to a pale green amphibole.[1] Closely associated with hornblende, as if taking its

[1] Harker, *Quart. Journ. Geol. Soc.*, vol. l (1894), p. 332 (Carrock Fell, Cumberland).

place, patches of deep brown biotite are not infrequently seen, and they usually occur about crystals of primary magnetite. What is here especially worthy of remark is the conversion of a higher mineral (pyroxene) to a lower (amphibole) as an incident of thermal metamorphism. This behaviour, at first sight anomalous, finds an explanation in considerations already noted (p. 102). It is a deferred or suspended reaction, which now takes effect when the appropriate temperature is realized.

The pyroxene and the replacing hornblende differing in composition, the transformation cannot be regarded as simple paramorphism. In general the reaction must involve also other minerals, though these may be merely minor constituents, such as magnetite, secondary chlorite, serpentine, etc. A case of special interest is that of eclogite enveloped in a granitic intrusion and thermally metamorphosed.[1] Here, as will be noted later, the pyroxene is of a peculiar kind, containing the elements of plagioclase felspar as part of its constitution. Both it and the associated garnet, rich in the pyrope molecule, are high-pressure minerals not stable under ordinary conditions; and amphibolization in this case is to be regarded as a reaction between the two minerals. It produces not only hornblende but plagioclase.

Any appreciable recrystallization of the felspar of an ordinary basic rock comes later than the uralitization of the pyroxene, but certain changes due to metamorphism may be developed at an earlier stage. The phenomena are not always the same. Often the dull felspar crystals are seen to become quite pellucid, an effect probably to be ascribed to the absorption of very minute inclusions of such minerals as zoisite and sericite (Figs. 41, 42). On the other hand there are numerous observations of relatively clear plagioclase, especially of the more calcic varieties, acquiring a peculiar cloudiness as a result of thermal metamorphism.[2] This is due to the development of a multitude of very minute opaque inclusions (Fig. 40, *B*). In some instances there has been a formation of magnetite from an original content of iron-oxide in the felspar, in other cases perhaps a development of secondary glass-inclusions. It seems that the subject is one which calls for further investigation. When recrystallization of the felspar sets in, the crystals may not at first lose their individuality in the process, and phenocrysts or the scattered crystals in a tuff may or may not be replaced by a granular mosaic. With advancing metamorphism, however, original outlines are lost, and the whole rock

[1] Eskola, *Vidensk. Skr.*, 1921, No. 8.

[2] MacGregor, *Min. Mag.*, vol. xxii (1931), pp. 524–38. See also Miss G. A. Joplin, *Proc. Linn. Soc.*, *N.S.W.*, vol. lviii (1933), pp. 152–6.

takes on a crystalloblastic type of structure. Since various chemical reactions are in progress, it is not to be assumed that recrystallization, whether of felspar or of other minerals, leaves their original composition unchanged.

Magnetite and ilmenite of primary origin have often furnished a certain amount of iron and titanium for the formation of biotite at a somewhat early stage, but a general recrystallization of the iron-ores belongs to a higher grade, following the stage of uralitization. It is still later, if at all, that olivine is affected. At a sufficiently high

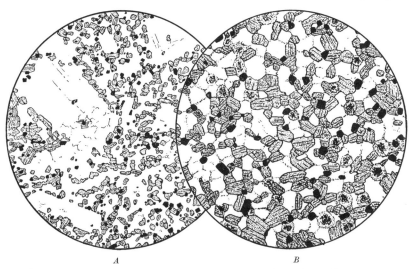

A   *B*

FIG. 42.—HIGHLY METAMORPHOSED BASALT LAVAS, enveloped by gabbro intrusion, Skye ; × 25.

A. Harta Coire. *B*. Druim an Eidhne. Metamorphism of a high grade has restored the original mineralogical composition (augite, plagioclase, magnetite) but with a crystalloblastic structure. The former figure still shows large porphyritic felspars, usually recrystallized as single individuals, but the one on the left partly replaced by a granular mosaic. *B* is the 'granulitic gabbro' of Geikie and Teall : see *Quart. Journ. Geol. Soc.*, vol. l (1894), p. 647, and Harker, *Tertiary Igneous Rocks of Skye* (1904), p. 115.

temperature it may become recrystallized ; but newly formed olivine may also be found with a manner of occurrence suggesting its formation from serpentine and other alteration-products.[1] Apatite often shows no sign of change even in very highly metamorphosed rocks (Fig. 39, *B*). Broadly speaking, the several constituents of the rock yield to metamorphism in an order the reverse of that in which they originally formed from the magma. Ilmenite has sometimes survived metamorphism of the most drastic kind, and the zircon, which is an occasional constituent of gabbroitic rocks, always remains intact.

[1] MacGregor, *Geol. Mag.*, vol. lxviii (1931), p. 508.

The highest grade of metamorphism in ordinary basic igneous rocks is marked by the total obliteration of all original structures except those of a large order such as the amygdaloidal. Mineralogically the most notable feature is the reappearance of augite, which now becomes the normal and stable ferro-magnesian constituent (Figs. 41, *B*; 42). The uralitization effected at an earlier stage is thus reversed, and hornblende formed by other reactions, e.g. in the interior of amygdales, likewise gives place now to augite (Fig. 43, *A*). The new augite has not necessarily the same composition as that of the

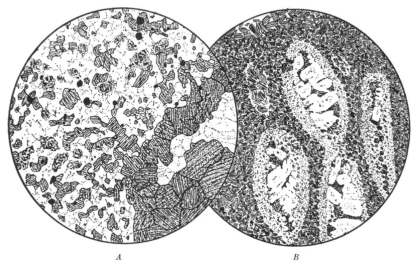

A   *B*

FIG. 43.—HIGHLY METAMORPHOSED AMYGDALOIDAL LAVAS ; × 25.

A. Basalt near the peridotite of An Sgùman, Skye. Totally recrystallized, showing augite, felspar, magnetite, and pseudomorphs after olivine. On the right is part of a large amygdale, occupied formerly by chlorite and zeolites, now by augite and felspar.
B. Pyrozene-Andesite, Wasdale Pike, near the Shap granite, Westmorland. The rock had suffered from weathering prior to metamorphism. Abundant flakes of biotite are the most prominent constituent, together with new felspar. The large crystals within the amygdales are of labradorite.

original rock, and it may be accompanied by hypersthene. Enclosed patches of basic rocks with ' granulitic ' structure are of frequent occurrence in the Tertiary gabbros and eucrites of Skye, Mull, and Ardnamurchan. Sometimes they represent basalt lavas (Fig. 42 *B*); but often they seem to come from the reconstruction of older (and more basic) intrusive rocks, belonging to the same series as the enclosing mass.[1] When an allivalite or felspathic eucrite has been

---

[1] *Tertiary and Post-Tertiary Geology of Mull* (*Mem. Geol. Sur. Scot.*, 1924), pp. 252–3 ; *Geol. of Ardnamurchan* (*Mem. Geol. Sur. Scot.* 1930), pp. 229–32, 308.

thus metamorphized, a dark green spinellid (pleonaste) is sometimes a conspicuous new product.[1]

THERMAL METAMORPHISM OF DEEPLY WEATHERED ROCKS

It has been shown that low-temperature alteration-products disseminated through an igneous rock are eminently susceptible to change in thermal metamorphism. More striking are the effects produced when igneous rocks which have been *deeply weathered throughout* are involved in a metamorphic aureole. Here again it is among basic and ultrabasic rocks that we find the most remarkable examples.

In serpentine we have an instance of a rock composed almost wholly of secondary minerals. Although the hydration and other changes which convert olivine and pyroxene to serpentine may not be correctly ascribed to atmospheric weathering, they are obviously of a kind which we may expect to be reversed by high temperature. It is found in fact that serpentine fused in a crucible recrystallizes as a mixture of olivine and enstatite :

$$H_4Mg_3Si_2O_9 = Mg_2SiO_4 + MgSiO_3 + 2H_2O,$$

but what elevation of temperature this reaction demands we are not informed. We cannot cite examples of peridotites which have certainly come from the metamorphism of serpentine-rocks, but some occurrences in the Eastern Alps possibly fall under this head.

The chief products from the destructive weathering of ordinary basic rocks are calcite and chlorite. If the abundant calcite remains in the rock, subsequent metamorphism follows the same general lines as in an impure limestone, giving rise to such minerals as grossularite, idocrase, and diopside. Often there has been some redistribution of the calcite within the rock, and this is especially true where there has been crushing and shearing, setting up a banded arrangement. Good examples are furnished by the Devonian spilitic lavas on the western border of the Dartmoor granite.[2] Some of these rocks are now composed mainly of green hornblende and lime-garnet, disposed in alternating parallel streaks. Other minerals which enter are epidote and zoisite, diopside, and some biotite.

Under other conditions basic igneous rocks may suffer weathering of a more drastic kind, the resulting calcite being more or less completely removed, while other constituents besides lime may also suffer reduction. The proportions of the more stable constituents are thus automatically raised, even when there is no actual accession of material.

[1] *Geol. of Ardnamurchan*, p. 317, with figure.

[2] *Geology of Dartmoor* (*Mem. Geol. Sur. Eng. and Wales*, 1912), pp. 20–3, plate II, fig. 4.

Quantitative estimates of the addition and subtraction of the various oxides involve an element of uncertainty ; for it is not safe to assume that a particular constituent, such as alumina, remains unchanged, and the alternative assumption of no change of total volume rests on no assured grounds.

The migration of dissolved material *within* a rock-mass subjected to destructive weathering may bring about some concentration if certain constituents in particular places ; but this process is probably much more effective as an incident of metamorphism in the vicinity of a plutonic intrusion, heated water furnished by the magma serving as solvent and carrier (see below, p. 115).

THERMAL METAMORPHISM OF ACID ROCKS

The basic igneous rocks have been treated at some length, because it is in these that the principles enunciated at the outset are most clearly illustrated. The *acid rocks* and those composed largely of alkali-felspars may be dismissed more summarily, except in so far as they introduce new points of interest. There are certain types consisting essentially of felspars and quartz, and it is evident that here no far-reaching chemical reactions are to be expected. At first there may be only mechanical effects, such as the shattering of the larger quartz-grains in a granite, a consequence of unequal expansion. In a sufficiently advanced grade felspar and quartz become, at least in part, recrystallized, with some tendency to a graphic intergrowth of the two minerals. The rhyolites or devitrified obsidians of Ordovician age in Westmorland are transformed near the Shap granite to a granular mosaic of felspar and quartz. A few scattered flakes of brown and white micas represent chloritic and sericitic material in the original lavas.

Among the ferro-magnesian minerals a very characteristic change is the replacement of hornblende by biotite, in an aggregate of flakes making a pseudomorph (Figs. 40, *B* ; 45, *A*). This is clearly a deferred reaction analogous to the conversion of augite to hornblende, for Bowen's ' reaction-series ' pyroxene-amphibole-biotite may be regarded as the normal course in magmas rich in potash. So also, to complete the parallel, a high grade of metamorphism reverses the reactions. Both hornblende and biotite are destroyed, giving rise to granular augite, accompanied in the case of biotite by much finely divided magnetite. These and other points are well illustrated by the granites and granite-gneisses of British Guiana,[1] where they are intersected by massive dykes of quartz-dolerite (Fig. 44).

[1] Harrison, *The Geology of the Goldfields of British Guiana* (1908), pp. 36–7, 126–7, etc.

In conclusion, something should be said of *xenoliths*—i.e. inclusions of relatively small dimensions—of one igneous rock in another later one.[1] These have naturally suffered thermal metamorphism after their kind. The *grade* of metamorphism, depending on the highest temperature attained, is determined by the nature of the enveloping rock, more basic magmas being intruded or extruded at higher temperatures than more acid ones. So, for example, a granite xenolith enclosed in a quartz-porphyry shows hornblende converted to biotite (Fig. 45); while a similar xenolith enclosed in a basalt has its horn-

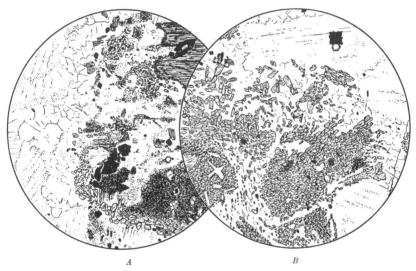

A   *B*

FIG. 44.—GRANITES METAMORPHOSED in the vicinity of great dykes of dolerite, British Guiana ; × 23.

A. Biotite-Granite, Tinamu Falls of Cuyuni River. Flakes of biotite are partly or wholly replaced by magnetite dust and granules of augite. The quartz has been shattered.
B. Hornblende-Granite, Great Falls of Demerara River. The hornblende is completely replaced by granular augite. Some part of the felspar and quartz has been recrystallized with a rude graphic intergrowth.

blende replaced by augite and its biotite by augite and magnetite. Often, and especially in a case like the latter, there is some mechanical breaking up of the xenolith and dispersal through the matrix. This facilitates further reactions, involving an actual interchange of substance between xenolith and matrix. The production in this way of heterogeneous and hybrid rocks makes an interesting study, but to pursue it would carry us beyond the limits of our present subject.

[1] There is an extensive literature of xenoliths. See especially Lacroix, *Les Enclaves des Roches Volcaniques* (Macon, 1893) and numerous later writings of the same author.

The circumstances may be such that, after a high temperature has been reached, cooling is comparatively rapid. This is especially likely to befall xenoliths in a volcanic rock, and accordingly these often show the effects of *local fusion and vitrification*. The case is comparable in a general sense with that of the ' spotted slates ' (p. 15) ; but the centres of local fusion, instead of being scattered fortuitously through the mass, are here determined by the particular minerals present. In a granite xenolith biotite is the mineral most easily affected. It yields a brown glass enclosing minute crystals of pleonaste,

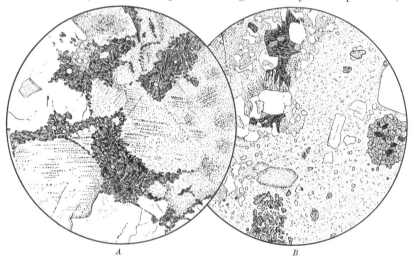

FIG. 45.—XENOLITH OF HORNBLENDE-GRANITE IN QUARTZ-PORPHYRY, Carrick Broad, Dundalk ; × 23.

A. Shows replacement of hornblende by biotite. Strings of chloritic alteration-products are replaced in the same way.
B. The same replacement is seen, with some small relics of hornblende ; but detached xenocrysts, enveloped by the quartz-porphyry magma, have given instead a new crystal-lization of hornblende.

magnetite, sillimanite, or sometimes hypersthene. The felspars in the same xenolith may be still intact, except that cleavage-cracks have been opened and secondary glass-inclusions developed. In the extreme case, however, a xenolith may be wholly fused, with the exception of such refractory minerals as zircon. This is a condition eminently favourable to intermingling of material with the surrounding magma.

We have tacitly disregarded the effects of *pressure* in the metamorphism of igneous rocks, although under deep-seated conditions this factor is certainly not negligible. Its influence is shown in promoting the formation of such minerals as pyroxenes, garnet, sphene, etc., in accordance with the Volume Law. This subject, however, will be more conveniently treated under the head of regional metamorphism.

CHAPTER IX

PNEUMATOLYSIS AND METASOMATISM IN THERMAL METAMORPHISM

Pneumatolysis Superposed on Metamorphism—Introduction of Borates— Introduction of Fluorides and Chlorides—Introduction of Sulphides and Iron Compounds—Introduction of Soda—Other Metasomatic Changes.

PNEUMATOLYSIS SUPERPOSED ON METAMORPHISM

IN the thermal metamorphism of various sedimentary and igneous rocks, as hitherto considered, the chemical reactions involved practically no material other than that furnished by the composition of the rocks themselves. From the igneous intrusion which was regarded as the cause of the metamorphism nothing was demanded beyond heat, and sometimes perhaps a modicum of water in supplement of that already present. In fact, however, it is often found that the rocks adjacent to an intrusion have been invaded by emanations from that source, which included in sensible quantity, not only water, but other volatile bodies chemically more active. These have entered into energetic reaction with the material of the rocks, as is often proved by the incorporation of one or other of these active substances in the final products. Chief among the bodies which play this part are borates, fluorides, and chlorides. At a sufficiently elevated temperature they are in the gaseous state, and this is implied in the term *pneumatolytic* as describing their chemical action.

It is not an arbitrary refinement that discriminates between the chemical action of water and the more restricted but more potent action of its associates. While the ubiquitous water officiates at every stage of metamorphism, the other volatile bodies, besides being restricted in range, become important only when a certain ' pneumatolytic phase ' is reached. This phase is often clearly indicated in the igneous rock itself, and is marked there as one of the latest episodes. The volatile constituents made part of the magma from the beginning, and assisted as fluxes and mineralizers throughout its crystallization. Only towards the close of that process, when the temperature had greatly declined, did the same substances begin to exercise a destructive effect upon minerals already crystallized. It was especially at this late stage that the volatile bodies, now liberated as gases, were able to escape into the surrounding rocks. Thermal metamorphism proper must clearly be assigned in the main to an earlier time, when the temperatures were higher, and we conclude that in general *pneumatolysis follows metamorphism, and is superposed upon it.* The geological evidence of this is convincing. It can often be verified that rocks have been jointed, brecciated, or faulted in the interval between metamorphism proper and pneumatolytic changes. Again, we see that characteristic structures of metamorphism, such as certain types of spotting in slates, were in existence before pneumatolysis supervened and wrought new changes. On the other hand, if this later action be of an energetic kind, all recognizable traces of the preceding metamorphism may be lost in a general reconstruction of the rock. The term *pneumatolytic metamorphism* is then a convenient one as embracing the final results of the two processes when they are not clearly separable.

Not all plutonic intrusions are attended by important emanations of gases. The generally accepted scheme of the evolution of different plutonic rocks, in an order of decreasing basicity and increasing alkalinity, involves also a progressive enrichment of the later derivatives in volatile constituents. It is in accordance with this that granites and nepheline-syenites are much more generally attended by important pneumatolytic effects than are rocks of more basic and calcic nature. The rule which would associate borates and fluorides especially with the former rocks and chlorides with the latter has no more than a loose and general validity.

While the minerals produced in simple thermal metamorphism draw their material solely from the substance of the rock metamorphosed, pneumatolysis introduces an extraneous element in addition. The composition of the new minerals, or of some of them, depends now upon two factors, which are quite independent. The second element, though it may be quantitatively much inferior, sets an unmistakable stamp on the whole, since without it the most distinctive minerals could not be formed. For this reason it will be convenient to arrange our observations, not primarily according to the original nature of the rocks affected, but with reference to the particular pneumatolytic agent involved.

INTRODUCTION OF BORATES

The most widespread type of pneumatolysis associated with granitic intrusions is due to *boric emanations*, and takes the form especially of tourmalinization. The characteristic mineral *tourmaline* is a borosilicate of the type $R_7B_3(Al_3Si_6O_{27})(O,OH)_4$ in which the principal base is alumina, while ferrous oxide, magnesia, soda, manganous oxide, lime, etc., are present in varying proportions.   Analyses show that the colourless and red varieties are highly aluminous and contain little iron or manganese ; the common yellow-brown kind has lower alumina and is rich in iron ; while blue colours are probably related to a note-worthy content of soda, and are much less common in metamorphosed rocks than in granites.   A mineral so rich in alumina (35 to 40 per cent.) is naturally associated especially with argillaceous sediments.

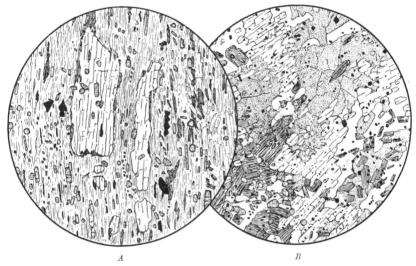

A B

FIG. 46.—TOURMALINIZATION OF METAMORPHOSED SLATES, Cornwall ;   × 23.

A.   Andalusite-Mica-schist, near the Bodmin Moor granite, Blisland.   Showing numerous little crystals of tourmaline.   These are mainly derived from biotite, but the andalusite is also beginning to be attacked.   The abundance of white mica is doubtless another effect of pneumatolysis.
B.   Xenolith of andalusite biotite-hornfels enclosed in the St. Austell granite.   Show-ing a more extensive production of tourmaline from both biotite and andalusite.   Part of the quartz is a by-product of the same transformation.

The Devonian slates near the granites of Cornwall and Dartmoor afford abundant material for study.

The rocks, being already metamorphosed and often in the state of dense compact hornfels, were not very freely permeable by gases. Tourmalinization may indeed be found at a considerable distance from a granite-contact, but only in proximity to tourmaline-quartz-veins, which mark the channels of supply.   The change commonly begins with the formation of little crystals of tourmaline, enclosed in those aluminous silicates which could furnish most of the material.

Biotite is first attacked, then cordierite and andalusite, and finally felspar if present.   As the process goes on, the gradual replacement by tourmaline of the various minerals of metamorphism is very evident (Fig. 46).   It does not, as a rule, yield sharply defined pseudomorphs, and it is clear that there is considerable freedom of diffusion within the rock—appreciably more than is possible in ordinary thermal metamorphism.   Yellow-brown tourmaline, undoubtedly an iron-bearing variety, is seen replacing andalusite as well as biotite.   The tourmaline, moreover, tends constantly to develop its proper crystal

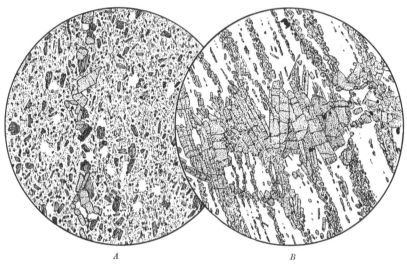

A   *B*

FIG. 47.—STAGES OF TOURMALINIZATION, in rocks bordering the St. Austell granite, Cornwall ;   × 25.

A.   The earliest stage : a muscovite-biotite-hornfels from Dennis, showing the formation of crystals of tourmaline at the expense of biotite.   They occur along a line, which marks the fissure by which the gases found entrance.
B.   The final stage : a tourmaline-quartz-schist from Roche.   The fissure of supply is still indicated by a belt of larger crystals crossing the bands.

shape, and sometimes shows a zonary distribution of colours, though this is much less common here than in tourmalinized granites.

Ultimately all the silicates are destroyed, and the final product of boric pneumatolysis in a slate, as in a granite, is a tourmaline-quartz-rock.   Here, however, it may be named a tourmaline-quartz-schist, for the little prisms of tourmaline have a common orientation, and are often crowded along particular bands (Figs. 47 *B* ;   48 *A*).   This preservation of old structures shows that the replacement has been effected wholly by molecular and atomic processes.   In this last stage, as in the earliest, the fissure by which the gases found access can often be identified (Fig. 47, *B*). Quartz is abundant in many of these rocks. It comes in part from the formation of tourmaline at the expense of less basic silicates, in part from the recrystallization of quartz contained in the hornfels ; but it is evident that there has often been also an introduction of new silica.

The chemical reactions here implied are clearly of a very drastic kind. Whether there has been any addition of substance, other than boric acid and silica, it is not easy to pronounce [1] ; but some removal, of potash at least, must be assumed. More knowledge is required for

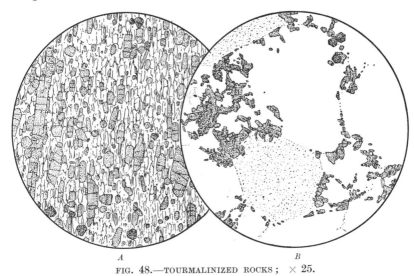

A   B
FIG. 48.—TOURMALINIZED ROCKS ; × 25.

A. Tourmaline-quartz-schist, enclosed in the Foxdale granite, Isle of Man. The parallel arrangement of the tourmaline crystals is well shown.
B. Tourmalinized grit, Cwm Dwythwc, near Snowdon. The clastic quartz is in part recrystallized, but the tourmaline is mainly confined to the interstices between the grains.

a full understanding of the chemistry of tourmalinization ; but it is probable that it is due only in part to pneumatolysis in the strict sense. The first formation of tourmaline in the rock may demand little more than an accession of boric acid or some volatile borate. The total reconstruction, which may or may not follow, is perhaps to be assigned to a somewhat later phase of igneous activity and to the agency of liquid solutions, to which are also attributable the associated tourmaline-quartz-veins. In confirmation of this we have the fact that tourmaline has sometimes been formed in a quartzose grit [2] or a

[1] Most analyses of tourmaline, however, show a small proportion of fluorine.
[2] Fearnsides, *Rep. Brit. Assoc.* for 1908 (1909), p. 704 ; Williams, *Quart. Journ. Geol. Soc.*, vol. lxxxiii (1927), pp. 354–6 (basal Ordovician grits of the Snowdon district).

limestone at a distance from any igneous intrusion (Fig. 48, *B*).   This, it would seem, is to be explained only by the bodily introduction of tourmaline, or at least of the sum-total of its constituents in some form.

The common boron-mineral in calcareous rocks is *axinite*.   This is essentially a borosilicate of calcium, $HCa_3Al_2B(SiO_4)_4$, though iron, manganese, and magnesium may also enter.   Like tourmaline, it occurs in evident relation with fissures in the rocks, but it has a more restricted distribution.   Good localities are Tregullan, on the northern

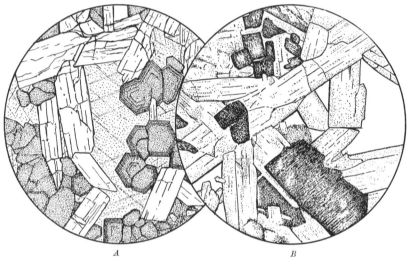

A   B

FIG. 49.—AXINITE-BEARING ROCKS, Tregullan, S. of Bodmin, Cornwall ;   × 25.

A.   Axinite-Andradite-Calcite-rock.
B.   Axinite-Hedenbergite-Quartz-rock.
That the garnet and pyroxene of these rocks are of varieties rich in iron is doubtless due to a metasomatic change of the kind described later as characteristic of many lime-stone-granite-contacts.

border of the St. Austell granite,[1] and Ivybridge and South Brent, on the fringe of Dartmoor.[2]   The common associates of axinite are andradite and hedenbergite (Fig. 49), sometimes also epidote and actinolite, and rarely some tourmaline.   Axinite is produced, not only in calcareous sediments, but often also in basic igneous rocks, which have been metamorphosed within a granite-aureole (compare Fig. 5, *A*, above).   Such rocks may have suffered alteration, with formation of calcite, prior to metamorphism.   In them, as well as in lime-silicate-rocks, it is not uncommon to find slender veins or strings of axinite, doubtless representing calcite veins.   A much rarer mineral, with

[1] Barrow and Thomas, *Min. Mag.*, vol. xv (1908), pp. 113–23.

[2] Busz, *Neu. Jb. Min.*, Beil. Bd. xiii (1900), pp. 125–32.

the same association as axinite, is datolite, $Ca(BOH)SiO_4$. It is recorded by Busz at South Brent in Devonshire.

In argillaceous rocks pneumatolytic metamorphism by the agency of *fluorides* is less prominently in evidence than that due to borates, but it plays a part by no means negligible.   In plutonic rocks themselves, such as the Cornish granites, greisenization represents an action no less energetic than extreme tourmalinization, but its operation is more restricted.   Topaz, so highly characteristic of the greisens, does not figure in metamorphosed sediments, except in injection-veins at an actual contact.   The more easily transported fluor has a wider distribution of the same kind, but its relation to any particular intrusion is not always evident.   The most usual repository of fluorides introduced into metamorphosed slate-rocks is to be seen in the *white micas* which have often been developed abundantly in the vicinity of a granite or greisen.   The production of white mica at the expense of such minerals as felspar, andalusite, and cordierite is a reversal of the processes by which those minerals were built up in thermal metamorphism, and points clearly to the intervention of a different factor. It is to be observed especially in the inner aureoles of muscovite-bearing granites, such as those of Cornwall, Dartmoor, Skiddaw, and Leinster.

It is to be regretted that we possess little knowledge concerning the nature of micas occurring in this manner, and more particularly their content of fluorine.   Muscovite is found, sometimes in great abundance (Figs. 46, *A*, 50, *C*) ; but more commonly the little flakes show the lower refractive index and weaker birefringence which are the properties of *lepidolite*, and may be provisionally distinguished by that name.   The lepidolite of the mineralogists contains 5 per cent. or more of fluorine, with 4 or 5 per cent. of lithia, and has silica nearly in the metasilicate ratio.   Whether our mineral can be identified with this must, however, remain a doubtful question.   Often the flakes are so woven into the texture of a hornfels that their formation must have been part of a total reconstruction of the rock (Fig. 50, *A*).   In other cases the structures of thermal metamorphism are only partly obliterated, and the mica—usually in very fine scales—can be seen replacing the crystals of cordierite, etc.   The magnesia of cordierite and biotite gives rise to chlorite, mingled with the mica (Fig. 50, *B*).

The not infrequent occurrence of abundant white mica in metamorphosed slates which are also partly tourmalinized may be taken to indicate the joint action of fluorides and borates.   Muscovite-tourmaline schists appear to be not uncommon in the Land's End district of Cornwall.[1]

It is, however, in those metamorphosed rocks which represent impure dolomites and magnesian limestones that the presence of fluorine most often reveals itself by the occurrence of distinctive minerals. The most widespread of these is the mica *phlogopite*. The colourless variety has, according to Clarke, the ideal composition $H_2KMg_3Al(SiO_4)_3$, with partial replacement of H by MgF, while yellow and brown colours indicate some content of iron.[2] The fluorine

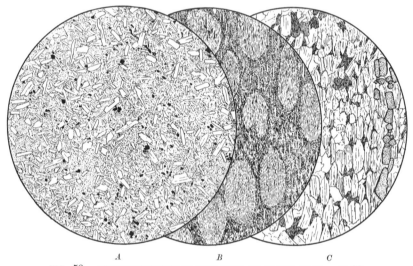

FIG. 50.—METAMORPHOSED SLATES RICH IN WHITE MICA ; × 23.

A. Lepidolite-Hornfels, close to the greisen of Grainsgill, Cumberland : composed of lepidolite and chlorite with some quartz and pyrites.

B. Pneumatolytically altered cordierite-mica-schist, near Bodmin, Cornwall : now consisting of lepidolite and chlorite, with a little quartz. The original schistose structure is not lost.

C. Muscovite-schist, twenty feet from granite, Warleggon, Cornwall : mainly of muscovite with biotite and quartz ; on the right a string of crystals of brown tourmaline.

may amount to as much as 4 or 5 per cent. Phlogopite is found in flakes scattered through a marble or as one among other minerals in a lime silicate rock (Figs. 51, *A*, 52). An interesting group of magnesian fluosilicates is that which includes *chondrodite*, humite, and clinohumite. Only the first of these is a common mineral, and the last is rare. Considering the composition of the three minerals :

$$\text{Chondrodite, } Mg_3(MgF)_2(SiO_4)_2 \text{ ;}$$
$$\text{Humite, } Mg_5(MgF)_2(SiO_4)_3 \text{ ;}$$
$$\text{Clinohumite, } Mg_7(MgF)_2(SiO_4)_4 \text{ ;}$$

[1] *Geol. of Land's End* (*Mem. Geol. Sur.*, 1907), p. 26.

[2] Structural formula $KMg_3AlSi_3O_{10}(OH)_2$, with replacement of OH by F.

we may regard them as arising from the union of one molecule of magnesium fluoride with two, three, and four molecules of forsterite, respectively. This does not, of course, assume that the fluorine entered the rock in the form of magnesium fluoride. A frequent associate both of phlogopite and of chondrodite is a blue or green *fluor-apatite*. The ideal compound $Ca_4(CaF)(PO_4)_3$ carries 3·8 per cent. of fluorine, but some part of this may be replaced by chlorine.

The calcium fluosilicates are rare minerals. Of these *cuspidine* has probably the formula $Ca_4Si_2O_7F_2$, while in *custerite* part of the

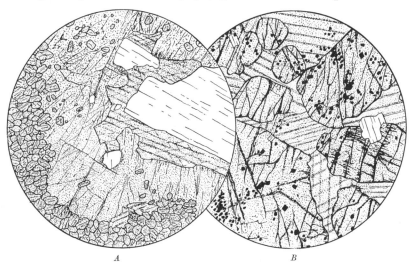

A   *B*

FIG. 51.—METAMORPHOSED AND PNEUMATOLYSED LIMESTONES, Grange Irish, near Carlingford, Co. Louth ;   × 23.

A. This shows a concentric arrangement of the new minerals, consequent upon an original nodular structure. A large flake of phlogopite is surrounded by calcite, then by idocrase, and finally by diopside.

B. Monticellite-Calcite-rock, with little octahedra of nearly opaque (deep green) pleonaste and a few small flakes of phlogopite. Some custerite and phlogopite occur elsewhere in the slice.

fluorine is replaced by hydroxyl. The latter mineral, first known from American localities,[1] occurs locally in some abundance in the Carlingford district of Ireland.[2] Here it is associated with phlogopite, idocrase, monticellite, calcite, pleonaste, and apatite (Fig. 52). Probably the custertite, like the monticellite (Fig. 51, *B*), is a metastable form indicative of failure to reach equilibrium ; and this is in accord with the varied diablastic intergrowths and poeciloblastic inclusions of the several minerals, pointing to a hasty crystallization of the whole.

[1] Umpleby, Schaller, and Larsen, *Amer. J. Sci.* (4), vol. xxxvi (1913), pp. 385–94 ; Tilley, *Geol. Mag.*, vol. lxv (1928), pp. 371–2.

[2] Osborne, *Geol. Mag.*, vol. lxix (1932), pp. 61–2, 219–20, 225–6.

The associations and relative proportions of the minerals vary greatly from point to point in the rock.

Fluorine enters exceptionally and in small amount into the composition of various common minerals of thermal metamorphism. Idocrase from Vesuvius contains about 1 per cent. and analyses of amphiboles of various kinds reveal traces of fluorine. Moreover, it is not impossible, even where no fluoride is now to be detected, that it may have played a part as a ' mineralizer ', perhaps determining the formation of an amphibole rather than a pyroxene [1] or idocrase rather than grossularite.

FIG. 52.—CUSTERITE-BEARING ROCKS, Grange Irish, near Carlingford, Co. Louth ;
× 23.

A. Idocrase-Phlogopite-Custerite-rock. The custerite is of xenoblastic habit, and sends out narrow strings between the idocrase crystals.

B. On the right are single large crystals of idocrase and phlogopite. The rest is an intergrowth of custerite and calcite. Both phlogopite and custerite enclose very abundant little crystals of pleonaste.

The pneumatolytic action of *chlorides* is shown principally in the production of *scapolites* in limestones. The minerals of this group constitute an isomorphous series parallel with that of the plagioclase felspars, and having as end-members :

$$\text{Marialite, } Na_4(AlCl)Al_2(Si_3O_8)_3 = Ab_3.NaCl.$$
$$\text{Meionite, } Ca_4(AlCO_3)Al_5(SiO_4)_6 = An_3.CaCO_3 ;$$

while sulphate may also enter in the compounds $Ab_3.Na_2SO_4$ and $An_3.CaSO_4.$ [2] The conversion of plagioclase to scapolite may often

---

[1] Von Eckermann, however, has described, from Mansjö in Sweden, a diopside containing 0·63 per cent. of fluorine : *Geol. För. Stock. Förh.*, vol. xliv (1922), pp. 356–8.

[2] Borgström, *Zeits. Kryst.*, vol. liv (1915), pp. 238–60.

be observed in progress (Fig. 53, *A*) ; but other lime-aluminosilicates, such as idocrase and grossularite, may also be scapolitized, so that the process sometimes affects the chief bulk of the rock (Fig. 53, *B*). The chloride enters into actual combination with the albitic constituent of the felspar only, but by its influence as a ' mineralizing agent ' the various lime-aluminosilicates become converted to meionite. The scapolities are idioblastic towards calcite and felspar, but xenoblastic towards the other lime-silicate minerals. Good illustrations of scapolitization are afforded by the Deeside limestone.[1]

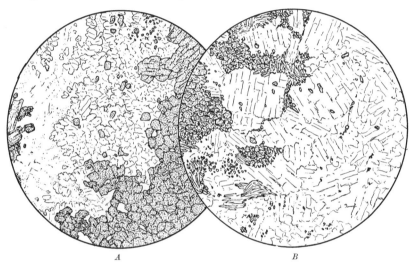

A   B

FIG. 53.—SCAPOLITIZED LIME-SILICATE-ROCKS, Etnich, near Deecastle, Aberdeenshire ; × 23.

A. Plagioclase felspar is seen on the left in process of replacement by scapolite, which occupies the centre of the field. The other minerals are grossularite and (above) idocrase.
B. Here scapolite fills most of the field, with diopside and some wollastonite. Elsewhere in the slice are grossularite, idocrase, and plagioclase.

A common associate of scapolite is *apatite*. In contradistinction to that which accompanies chondrodite and phlogopite, it is in the main a chlor-apatite. In both cases the mineral often occurs far too abundantly to be accounted for as part of the original substance of the rock, and we are led to the conclusion that there has been an introduction of *phosphate*, or rather of phosphorus in some volatile form. We may suppose this to react with water :

$$PCl_5 + 4H_2O = 5HCl + H_3PO_4,$$

the acids so produced then entering into reaction with some of the silicates present.

[1] Hutchison, *Trans. Roy. Soc. Edin.*, vol. lvii (1933), pp. 581–2.

INTRODUCTION OF SULPHIDES AND OF IRON COMPOUNDS

Various *sulphide* minerals have a pneumatolytic or post-pneumatolytic origin. An interesting special case is that of lapis lazuli, which Brögger and Bäckström [1] have shown to be a metamorphosed dolomitic limestone. The blue lasurite, which is its distinctive mineral, is one of the sodalite group, which the authors regard as alkali-garnets, and they assign to it the formula $Na_4(AlS_3Na)Al_2(SiO_4)_3$. Pyrites is one of the associated minerals.

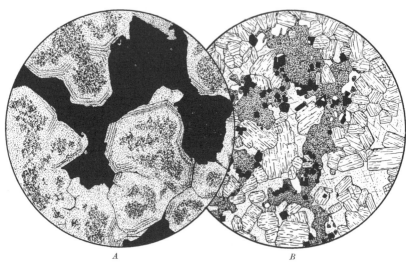

A   *B*

FIG. 54.—VARIETIES OF ' SKARN ', FROM the Oslo district ; × 25.

A. Andradite-Magnetite-Skarn, Gjellebaek.
B. Hedenbergite-Zinc-blende-Skarn. Nysaeter, near Grua. The opaque octahedra are of magnetite. Calcite and a little quartz occur interstitially.
These and other types have been described by Goldschmidt.

More important are the ore-deposits found in some districts at the contact of limestones with granite and other plutonic rocks. These are the *skarn* of Scandinavian geologists and the ' garnet-contact-zones ' of some American writers. Here sulphides of iron, zinc, lead, and copper, and in other occurrences magnetite, are associated with lime-silicates, chiefly garnet and pyroxene (Fig. 54). Kemp showed that the garnet is always of a variety rich in iron, containing up to 80 or 90 per cent. of the andradite molecule, and in like manner the pyroxene is near hedenbergite in composition. The iron in these silicates has clearly been derived from the igneous intrusion, which is equally the source of the ore-minerals. It is significant that these deposits are found in connexion with limestones, and that the ore-

[1] *Zeits. Kryst.*, vol. xviii (1890), pp. 231–75.

minerals have the appearance of replacing part of the substance of the limestone itself. The explanation of this is that an impure limestone suffers a very considerable diminution of volume when converted to an aggregate of garnet, pyroxene, wollastonite, epidote, etc. According to Barrell,[1] the shrinkage is as much as 30 or 40 per cent. of the volume. Allowing for the effect of pressure, it may still be believed that such a rock is sufficiently porous to give ready access to an invading fluid.

Whether the introduction of iron compounds into limestone rocks bordering a plutonic contact is truly a pneumatolytic process, is a question which has been debated. Goldschmidt [2] has suggested that the iron is introduced as a volatile chloride or fluoride, which reacts with calcite:

$$Fe_2Cl_6 + 3CaCO_3 = Fe_2O_3 + 3CaCl_2 + 3O_2.$$

The chloride or fluoride would then go to make scapolite or fluor ; minerals which are indeed found, but not generally or in abundance. Indeed it is clear from the descriptions of ores of this class in various countries that the presence of the ordinary pneumatolytic minerals is not usually a prominent feature. It is probable that the iron, etc., are introduced, not in gaseous form, but in liquid solutions at a somewhat later stage, and this is the view of most American geologists who have studied such occurrences.

The skarn type of metasomatism is found at numerous British localities, but usually as a narrow belt and with little of the impregnation with sulphides. Andradite-hedenbergite-rocks occur on the border of the Dartmoor granite [3] (Fig. 55). At Tregullan, near Bodmin, too, lime-silicate rocks, carrying zinc-blende, contain a yellow garnet of andradite composition.[4] In both cases borate minerals are also present (p. 120), and it is evident that the sequence of processes following the intrusion of the granite falls into three stages : (i) thermal metamorphism, (ii) pneumatolysis, (iii) invasion of iron-bearing solutions and introduction of sulphides. The occurrences described by Osborne [5] in the Carlingford district are especially instructive. Here the thermal metamorphism proper is due to an intrusion of eucrite, but the iron-bearing solutions were derived from a subsequently intruded granite magma. The conversion of grossularite to andradite and of diopside to hedenbergite begins at numerous isolated spots in

[1] *Amer. J. Sci.* (4), vol. xiii (1902), pp. 279–96.

[2] *Vidensk. Skr.*, 1911, No. 1, p. 214.

[3] Busz, *Neu. Jahrb. Min.*, Beil. Bd. xiii (1900), pp. 125–31, with analyses.

[4] Barrow and Thomas, *Min. Mag.*, vol. xv (1908), p. 118.

[5] *Geol. Mag.*, vol. lxix (1932), pp. 226–7.

the metamorphosed rocks at some little distance from the igneous contact. At the actual contact these changes are complete ; but the solutions have also reacted with residual calcite to produce the same lime-iron-silicates, and it is evident that there has been some accession of silica as well as of iron-oxides. Any wollastonite that may be present is left unchanged. Sulphides are represented only by a little pyrites.

INTRODUCTION OF SODA

Another type of metasomatism brought about by the agency of

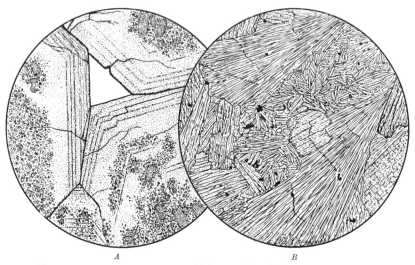

FIG. 55.—LIME-IRON-SILICATE-ROCKS, Aish, near South Brent, in the aureole of the Dartmoor granite ; × 23.

A. Andradite-rock ; composed mainly of large crystals of garnet, which in polarized light show rather strong birefringence and polysynthetic twinning. The marginal part of each crystal shows zonary growth, and the interior encloses crystals and grains of heden-bergite. The clear interstitial mineral is quartz.
B. Hedenbergite-rock, partly with a radiate grouping of slender prisms.

liquid solutions of magmatic origin is that which involves an accession of soda or of sodic compounds. A typical case is the albitization of the rocks bordering certain basic intrusions, whereby an argillaceous sediment is converted to an *adinole*. The igneous rocks responsible for this transformation are themselves rich in soda, and are generally interpreted as normal dolerites which, after their first consolidation, have been albitized by the action of ' juvenile ' liquid carrying sodic compounds. The same liquid solutions have invaded the adjacent rocks for a few feet from the contact, not merely along fissures but by intimate permeation, and have there brought about metasomatic changes of a radical kind

The classical examples of adinoles are found in the Harz,[1] but comparable phenomena are known in many other districts.[2] The best British adinoles are in the Devonian slates of North Cornwall at their contact with sills of albite-dolerite [3] (Fig. 56). The common type has the appearance of a chert, and in thin slices shows a very fine-grained texture ; but there are coarser varieties, including one made up chiefly of spherulitic growths of albite crystals. The Cornish adinoles which have been analysed consist almost wholly of albite. Those of the Harz, although derived from similar slates, are albite-

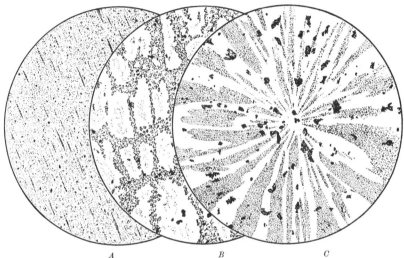

A   B   C

FIG. 56.—VARIETIES OF ADINOLE, from Dinas Head, near Padstow, Cornwall;
× 25.

A. Essentially a very fine-grained aggregate of albite : the original lamination of the slate is still indicated.
B. Containing a considerable amount of calcite in a network of veins. The albite here is in larger crystal-grains.
C. Showing a spherulite made by a radiate grouping of albite crystals. The opaque substance is white or yellow by reflected light, and is perhaps leucoxene : some chalybite is also present.

quartz-rocks with a variable proportion of other minerals, such as chlorite, epidote, actinolite, and iron-ore, sometimes with sphene or rutile. In a recent paper on the adinoles of Dinas Head in Cornwall, Agrell [4] distinguishes four main types : (i) Normal adinoles—grading

[1] Kayser, *Zeits. Deuts. Geol. Ges.*, vol. xxii (1870), pp. 103–78 ; Milch, *ibid.*, vol. lxix (1917), pp. 349–486.
[2] An interesting occurrence in Michigan is described by Morgan Clements, *Amer. J. Sci.* (4), vol. vii (1899), pp. 81–91.
[3] Howard Fox, *Geol. Mag.*, 1895, pp. 13–20 ; McMahon and Hutchings, *ibid.*, pp. 257–9.
[4] *Min. Mag.*, vol. xxv (1939), pp. 305–36.

into rocks composed essentially of dravite (magnesian tourmaline). (ii) Adinoles with pseudomorphs probably after andalusite. (iii) Adinoles with globular masses of ankerite, showing concentric structures. (iv) Polygonal and spherulitic adinoles.

The adinole transformation, like ordinary thermal metamorphism, begins at isolated points within the rock. If it is arrested at an early stage, there results the peculiar type of spotted slate known as *spilosite*.[1] The spots may be seen to consist of albite, chlorite, and quartz, while the surrounding matrix is still mainly sericitic. In a more advanced stage the spots are composed essentially of granular albite, while the matrix is of quartz, albite, chlorite, actinolite, etc. Sometimes the change begins, not in distinct spots, but along selected seams of the sediment, giving a finely banded structure (' desmoisite ').

The chemistry of the process presents a problem of some difficulty, but it is evident that a very radical replacement is implied. Besides addition of soda and silica, there has been a removal of magnesia, iron, and potash. Lime is probably removed also ; but albitization may be followed by carbonization, likewise due to magmatic emanations, and calcite, ankerite, or chalybite is conspicuous in some adinoles. The replacement of substance must have been effected molecule by molecule, for the original lamination is not obliterated. The most essential change is the accession of soda, which may be regarded as taking the place of potash. The simple constitution of the Cornish adinoles suggests an introduction of albite *per se*, while in the Harz a variable amount of silica has also been added. That albite may be introduced bodily into a rock seems to be indicated by its manner of occurrence in certain limestones.[2] A good example has been described by Lacroix[3] in the Pyrenees. The crystals show a variety of habit unlike anything seen in thermal metamorphism. They are associated with phlogopite, chlorite (leuchtenbergite), pyrites, quartz, and sphene (Fig. 57, *A*). The type of metasomatism of which adinole represents the extreme product, viz. an accession of soda which, now at least, is contained in albite, is a widespread effect at the contact of argillaceous rocks with dolerite sills. It is well shown near the Whin Sill in Teesdale,[4] and about Tremadoc and elsewhere in North Wales.[5] It is to be suspected whenever chemical analysis shows a marked preponde-

---

[1] Some writers have applied this name more generally to spotted slates, without reference to albitization.

[2] Spencer, *Min. Mag.*, vol. xx (1925), pp. 365–81 (Bengal).

[3] *Bull. Carte Géol. Fra.*, vol. vi, No. 42 (1895), pp. 85–6.

[4] Hutchings, *Geol. Mag.*, 1895, pp. 122–31, 163–9.

[5] Teall, *British Petrography* (1888), pp. 217–21.

rance of soda over potash.   Not only basic, but sometimes acid igneous rocks, themselves rich in soda, may be attended by like effects.   Goldschmidt has noted several instances at the contacts of nordmarkites and soda-granites in the Christiania district.

Metasomatism by the agency of soda-bearing solutions has also been invoked to account for the production of *paragonite*.   Such is the contention of Killig [1] in the case of the Ochsenkopf in the Saxon Granulitgebirge, but the evidence adduced is not wholly convincing. Here again the action premised is of a very drastic kind, not merely

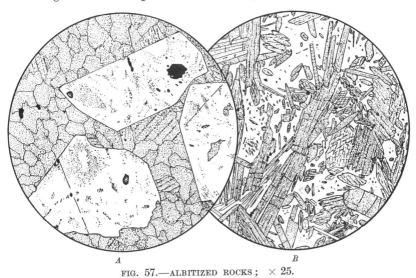

A   B

FIG. 57.—ALBITIZED ROCKS ;   × 25.

   A.   Albite crystals in metamorphosed Jurassic limestone near contact with lherzolite, Roc Tourné, Modane, Hautes Pyrénées.
   B.   Quartz-Glaucophane-schist, Ollard, New Caledonia.   Crystals of glaucophane, large and small, are embedded in a mosaic of albite and quartz.

potash-mica but a phyllite as a whole being converted to paragonite. Connected apparently with the same process, there has been also an introduction of sulphide-ores.

In some cases there has been an introduction of ferric iron, as well as of soda and silica.   Thus, in the sericite-phyllites of Winterburg in the Hunsrück [2] there has been an abundant production of a soda-amphibole (crossite).   There are indications that this preceded the injection with albite.   An introduction of soda-bearing amphiboles, and more rarely pyroxenes, is found also in some metamorphosed arenaceous rocks.   Goldschmidt [3] has observed this effect in a lenticle

---

[1] *Mitt. Nat. Ver. Greifswald*, 1912.

[2] Chudoba and Obenauer, *Neu. Jb. Min.*, Beil. Bd. lxiii, A (1931), pp. 77–80.

[3] *Neu. Jb. Min.*, Beil. Bd. xxxix (1914), pp. 193–224.

of sandstone enclosed in a pegmatite dyke in the Langesundsfjord. The percentage of soda has been raised from 0·89 to 3·47, and ferric oxide has probably been added also.   More remarkable are the *quartz-glaucophane-schists*.   Washington [1] has pointed out that glaucophane-schists in general fall into two main groups.   Those of basic composition represent igneous rocks modified by regional metamorphism. The rarer type with a high silica-percentage (75 to 80 or more) comes from metasomatic transformation of siliceous sediments.   In California not only sandstones but radiolarian cherts are proved to have

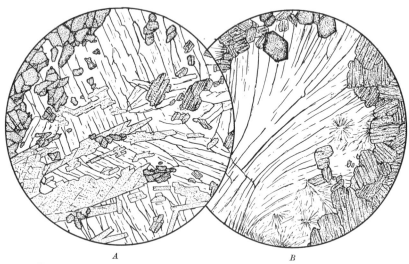

A   *B*

FIG. 58.—PREHNITIZED LIME-SILICATE ROCKS, Pollagach Burn, near Cambus o' May, Aberdeenshire ;   × 23.

A.   The coarsely crystalline prehnite, making more than half of the rock, comes in the main from lime-felspar and perhaps idocrase, but grossularite is also being attacked. Diopside remains untouched, but wollastonite is represented by pseudomorphs in calcite.
B.   This shows the common tendency of prehnite to sheaf-like and radiate-fibrous crystallization.   The other minerals are diopside and a few dodecahedra of grossularite.

suffered this change.   The rocks consist essentially of quartz and glaucophane with or without albite, and the percentage of soda may range as high as 6 (Fig. 57 *B*).

OTHER METASOMATIC CHANGES

Here should perhaps be included certain metasomatic transformations in which heated water, of magmatic origin, seems to have been the sole agent, or is at least the only one which has left direct evidence. Among processes which may be attributed primarily to this hydrolytic

[1] *Amer. J. Sci.* (4), vol. xi (1901), pp. 35–59.

action is the *prehnitization* [1] of various lime-aluminosilicates. In some localities, always in the near neighbourhood of a plutonic contact, this destructive action has been carried far, even in the extreme case to the reduction of the whole rock to an aggregate essentially of prehnite and quartz. Any lime-felspar present is first converted. Anorthite, however, has a lower ratio $CaO : Al_2O_3$ than prehnite, and accordingly, unless calcite be present, the more calcic minerals are attacked in their turn, viz. idocrase and then grossularite. The non-aluminous silicates, diopside and wollastonite, are exempt, but there may be a simultaneous replacement of wollastonite by calcite (Fig. 58).

It is well known that some geologists attribute far-reaching consequences to the agency of magmatic solutions as introducing, not only silica, soda, and iron-compounds, but alumina, magnesia and potash. The French school, and Lacroix in particular, have attached great importance to this metasomatic element as applicable to metamorphism on an extensive scale. It has also been invoked as an important factor by Adams and Barlow in Canada. [2] This is not the place to discuss a question which turns largely upon the interpretation of field-evidence, but one general consideration may be recalled. The normal course of crystallization in a cooling igneous rock-magma is now sufficiently well understood in its main lines. It enables us to account for the concentration in the final residual magma, in different cases, of soda or silica or iron compounds, as well as an enrichment in water and other volatile bodies. Magnesia, on the other hand, is selectively taken out in the earlier stages of crystallization to make olivine and pyroxenes and, it may be, later for amphibole and biotite (perhaps at the expense of the former minerals), with the result that the residual magma in the final stage is normally devoid of magnesia. This is not to be forgotten when magnesian solutions, of magmatic origin, are invoked to explain dolomitization or to account for the production of abundant biotite and hornblende in a supposed pure limestone.

Perhaps the strongest case for recognizing metasomatic changes of this kind is the abundant production of such magnesian minerals as cordierite and anthophyllite in some aureoles of metamorphism. A standard instance is that described by Eskola [3] in the Orijärvi

---

[1] Compare Hutchison, *Trans. Roy. Soc. Edin.*, vol. lvii (1933), pp. 575, 583–6 (Deeside Limestone).

[2] *Geology of the Haliburton and Bancroft Areas (Mem. No. 6, Geol. Sur. Can.,* 1910).

[3] Bull. No. 40, *Com. Géol. Finl.* (1914), pp. 252–63.

district of Southern Finland.   Here anthophyllite-cordierite-rocks and
allied types have been produced in the ' leptite ' formation bordering
an intrusion of oligoclase-granite.   There is evidence of the replacement of lime and alkalies by iron-oxides and magnesia, and this is
attributed to the action of emanations from the granite-magma.

It must be recognized, however, that, in other occurrences, drastic
changes of bulk-composition have been brought about in the near
vicinity of a plutonic intrusion by the agency of heated water alone.
An instructive example is that of certain greatly altered dolerites near

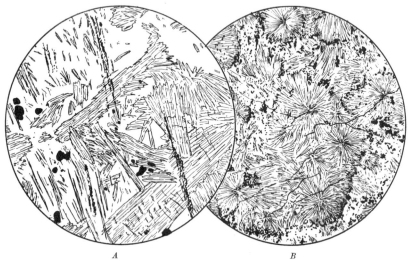

A   *B*

FIG. 59.—ANTHOPHYLLITE-CORDIERITE-ROCKS, Kenidjack, Cape Cornwall ;   × 23.

  A.   Columnar crystals and slender needles of anthophyllite are embedded in a mosaic
of cordierite.   The only other constituents are a few crystals of pleonaste and magnetite
and thin plates of ilmenite preserving their original position.
  B.   Here the anthophyllite has the radiate-fibrous habit.   Magnetite is rather
abundant, especially interposed between anthophyllite and cordierite.

the Land's End granite of Cornwall.[1]   These, prior to metamorphism,
had suffered shearing, and the metamorphosed rocks show in consequence a banded structure, with much variety in the different bands.
The simplest type consists essentially of a green aluminous hornblende
with plagioclase.   In other associated rocks, however, hornblende
gives place to one of the non-calcic amphiboles, cummingtonite and
anthophyllite, and plagioclase is represented by cordierite, a characteristic type being an *anthophyllite-cordierite-rock* (Fig. 59).   Other
significant minerals which may enter are pleonaste and diaspore.
Ilmenite and apatite occur as residual elements.

  [1] Tilley and Flett, *Sum. of Progr. Geol. Sur.* for 1929, Part II (1930), pp.
24–41 ; Tilley, *Min. Mag.*, vol. xxiv (1935), pp. 181–202.

The loss of lime, which is here the most evident chemical change, might conceivably be due to weathering of the dolerite, and this would automatically raise the proportion of alumina, magnesia, etc. Tilley has, however, made it clear that the more remarkable of these metamorphosed rocks have no equivalent among known products of atmospheric weathering. Analyses indicate, in different cases, in addition to and compensating the abstraction of lime, an accession of silica, iron-oxides, magnesia, and potash. These substances, nevertheless, were all contained in the original dolerite, and only the redistribution of them is ascribed to the agency of heated water from the granite magma.

PART II

DYNAMIC AND REGIONAL METAMORPHISM

CHAPTER X

STRESS AS A FACTOR IN METAMORPHISM

Analysis of Strain and Stress—Influence of Stress on Solubility—Influence of Stress on Chemical Reactions—Stress- and Anti-stress Minerals.

ANALYSIS OF STRAIN AND STRESS

WHEN rocks suffer metamorphism in response to rise of temperature under the relatively simple mechanical conditions hitherto assumed—viz. under no external force other than hydrostatic pressure—there is no deformation of the rock-masses affected. There may be a uniform compression, but, except in certain special cases (p. 127), the actual diminution of volume is not important. The internal stress experienced under such conditions is primarily a mere uniform pressure. It is true that in theory some shearing stresses must always be set up in addition, both by the compression of a heterogeneous mass (p. 6) and as a consequence of crystal-growth (p. 33) ; but such stresses are in great measure automatically relieved by the co-operation of causes already indicated.

In metamorphism of the most general kind, with which we are now concerned, these factors making for simplicity are no longer present. Rock-masses, at some lower or higher temperature, are subjected to external forces which are different in different directions, e.g. to a lateral thrust. They yield in greater or less degree according to the magnitude of the forces and the resistance of the rocks (at the given temperature). In so far as they yield, they suffer deformation. In so far as they resist, internal shearing stresses are set up ; and these stresses, being maintained or renewed so long as the external forces are operative, attain a magnitude far beyond anything that is possible in simple thermal metamorphism. The hydrostatic pressure may be very high at the same time, but is not necessarily so.

Since we meet here with conceptions, both geometrical and

mechanical, not hitherto encountered, it will be convenient to begin by recalling briefly some of the principles involved.

Ideally, deformation of a solid body may be continuous or discontinuous, the latter term implying finite slipping of one part against another, i.e. internal faulting on a small scale. The distinction which we shall actually make is indeed one of degree rather than of kind ; for *strictly* continuous deformation, such as is shown by india-rubber, is scarcely possible in a rock composed of discrete elements. Deformation of a plastic clay is, however, continuous to the eye, and we shall see that concurrent physical and chemical changes may confer a measure of effective plasticity even on a crystalline rock. We shall accordingly begin with a simplified discussion of deformation, or in mathematical phrase *strain*, of a continuous or quasi-continuous kind. The external forces which set up the strain and the internal stresses thereby called into play are not at present in question, the treatment being purely geometrical.[1]

In a large body strain is not necessarily uniform throughout ; but it is sensibly so for any small part of such body, and we shall confine our attention to this case. In general a strained body suffers both change of volume and change of shape, and these can be considered separately. There are accordingly two fundamental types of strain : (i) a uniform contraction or elongation in all directions (change of volume only) and (ii) a simple shear (change of shape only). The former needs no description. It may be a purely mechanical effect or an incident of chemical reactions in the rock. The diminution of volume is small, except where there is actual loss of material, as in the squeezing of water out of a clay or the expulsion of carbon dioxide in the metamorphism of an impure limestone.

The simple *shear* [2] calls for a more particular consideration. It is most easily pictured as the type of strain by which the cube of Fig. 60 is deformed into the oblique parallelepiped of Fig. 61. Here each horizontal plane in the body, without suffering distortion, is displaced, relatively to the base, in the direction DC through a distance proportional to its height above the base. The ratio DD' : AD, or tan θ, measures the ' amount of shear '. The character of any strain is, however, most completely and conveniently expressed in terms of the ' strain-ellipsoid ', i.e. the figure into which a sphere in the unstrained

---

[1] For a full analysis of strains and the correlated stresses see Thomson and Tait's *Natural Philosophy* or other work of reference.

[2] The term is here used in the mathematician's sense. As such it was introduced into geological literature by Fisher, but it has since come to be more loosely employed, so as to include deformation which is essentially discontinuous.

body becomes deformed. Let r be the radius of the sphere and a, b, c, the greatest, mean, and least semiaxes of the ellipsoid, a and c lying in the plane of the figure and b perpendicular to it. Then every line in the body parallel to a has been elongated in the ratio $a : r$, and every line parallel to c has suffered a corresponding contraction. The ratio $a : r = r : c = s$ is called the 'ratio of the shear', and the amount of shear is connected with this by the relation $\tan \theta = s - 1/s$.

Since all particles in the body are displaced in planes parallel to that of the figure (called 'planes of shearing'), any line perpendicular

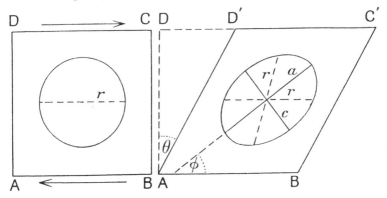

FIG. 60.—SECTION THROUGH A CUBE, PARALLEL TO ONE PAIR OF FACES.

FIG. 61.—THE SAME DEFORMED BY UNIFORM SHEAR PARALLEL TO THE BASE.

to these planes is unchanged in length, and the mean semiaxis $b = r$. The relation $ac = r^2$ or $abc = r^3$ expresses the fact that there is no change of volume.

The greatest axis is inclined to AB at an angle ϕ, such that $\cot \phi = s$. Now an ellipsoid with three unequal axes has two sets of plane circular sections, parallel to the mean axis and symmetrically inclined to the other two. One set is evidently given by horizontal planes, which are 'planes of no distortion', and the other set must be equally inclined to a on the opposite side. Consider first a shear of infinitesimal amount : then s differs only infinitesimally from unity, and $\phi = 45°$. It follows that everything is symmetrical about the diagonal AC ; the second set of circular sections is vertical ; and it is indifferent whether the infinitesimal displacement was made parallel to AB or to AD. In either case the effect of the shear is equivalent to a certain elongation parallel to the diagonal AC with a compensating contraction parallel to BD. If now we suppose shearing to continue, so as to bring about deformation to any finite extent, it is evident that this may be conducted upon more than one plan. We may distinguish two cases, both of which have their application to the deformation of rock-masses :

(i) First suppose AB held in a fixed position—say horizontal— and DC dragged continually to the right. As this goes on, the strain-ellipsoid is continually becoming more elongated and narrowed, and its greatest axis is rotated towards the horizontal ; i.e. s increases and the angle ϕ continually diminishes.

(ii) Next suppose instead that the axes of the strain-ellipsoid are fixed in direction—say with the greatest axis vertical (Fig. 62). We

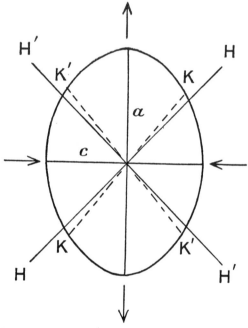

FIG. 62.—DIAGRAM TO ILLUSTRATE CONTINUED SHEARING BY LATERAL CON-
TRACTION AND VERTICAL ELONGATION

have seen that an infinitesimal elongation in the direction of a with a compensating contraction in the direction of c is equivalent to a certain shear, the planes of no distortion being those parallel to HH and H'H', making angles of 45° with the axes of the ellipsoid. Evi-dently continued elongation and contraction on the same lines will be equivalent to continued shearing. Planes parallel to HH and H'H', fixed in space but not fixed with reference to the body undergoing deformation, are at every stage planes of no instantaneous distortion, while the circular sections, parallel to KK and K'K', are planes of no resultant distortion.

The most general kind of strain that a body can undergo involves both change of volume and change of shape. Since the volume-change in metamorphism is in general a diminution, it is convenient to reckon contraction as positive and to regard expansion or elongation as negative contraction. Considering only the strain in a small part of the mass, so that it may be supposed uniform, we may still take the strain-ellipsoid as guide, and the voluminal contraction will be in the ratio $abc : r^3$. From the relation already pointed out between shearing and linear elongation and contraction, it is evident that the effect of the complex strain can be presented in various ways. It will be

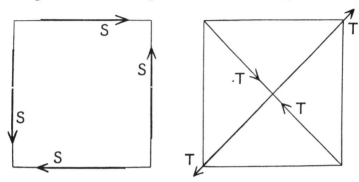

FIG. 63.—DIAGRAM TO ILLUSTRATE SHEARING STRESS

sufficient for our purpose, however, to consider, not the general problem, but a particular case, which is more or less closely realized in the ordinary types of crustal displacement; viz. that in which the rocks suffer neither elongation nor contraction in the direction of strike. This direction will therefore be taken as that of the b-axis, and the displacement of any particle is then in a plane parallel to that containing a and c. Accordingly $b = r$, and the volume-change in a strain of this type is in the ratio $ac : b^2$. We can then make alternative suppositions as before :

(i) Conceive that, while all lines in the body parallel to b remain unchanged in length, all lines perpendicular to b suffer contraction in a certain ratio, measured by $\sqrt{(ac)} : r$; and, following on this, let there be continued shearing in the manner of Fig. 61.

(ii) Suppose, as in Fig. 62, a linear contraction in the direction of c and elongation in the direction of a, but the elongation not of such magnitude as to compensate the contraction.

From strains we turn to the correlated stresses, but these can be treated briefly. Two fundamental types of stress are to be recognized :

(i) *Uniform pressure* (' hydrostatic pressure ') which is correlated with uniform compression. If a volume V be diminished by an amount dV, and the external pressure which causes this (bringing into play an internal stress equal and opposite) be P, then $P = k.dV/V$, where k is a specific constant, the 'modulus of compressibility'.

(ii) *Shearing stress.* We have seen how a cube is deformed into a parallelepiped by a simple shear (Fig. 61). The external force-distribution to cause or maintain this strain may be represented by an opposed couple acting in the horizontal direction, as S in Fig. 63 (bringing into play an internal stress equal and opposite). A similar

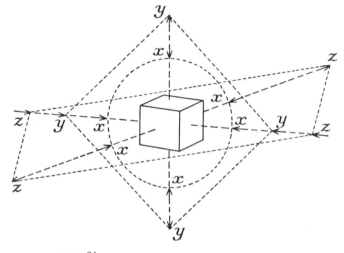

FIG. 64.—RESOLUTION OF A LINEAR PRESSURE

Each opposed pair of arrows represents a linear pressure (or tension) $\frac{1}{3}P$. The given pressure in a horizontal direction is thus divided into three equal parts. In each of the other two rectangular directions we are free to imagine a linear pressure with a tension compensating it. Now, grouping the arrows anew, we see that those marked x give a uniform pressure of amount $\frac{1}{3}P$, while the pressure and tension y are equivalent to a shearing stress in a vertical plane, and the pressure and tension z to a shearing stress in a horizontal plane.

couple acting in the vertical direction, as shown, would give the same deformation, and to eliminate bodily rotation we must suppose both couples to act simultaneously. The measure of the shearing stress is n times the amount of shear, where n is the 'modulus of rigidity', a specific constant.

The force-distribution shown in Fig. 63 is equivalent to a linear pressure in one diagonal direction with an equal tension (negative pressure) in the other diagonal (T in the diagram). It can also be shown that a single linear pressure is equivalent to a uniform pressure of one-third of its measure together with certain shearing stresses (Fig. 64).

INFLUENCE OF STRESS ON SOLUBILITY

Having, sufficiently for our purpose, considered stress in rocks in its purely mechanical aspect, we have next to inquire how this factor enters, jointly with temperature, as controlling physical and chemical processes.[1] More precisely, resolving stress into its two fundamental types, we have to ask how the mineralogical changes comprised in metamorphism are dependent upon the three governing conditions— temperature, hydrostatic pressure, and shearing stress.

Unfortunately it must be confessed at the outset that to the problem as thus stated no complete answer can at present be given. Concerning the distinctive properties of solids in a condition of shearing stress chemical theory has, as yet, little to tell us. Nor is more to be learnt from the experimental side. In ordinary laboratory practice the chemist studies the behaviour of systems under conditions which are sufficiently defined by temperature, hydrostatic pressure, and concentration. He may be aware that grinding in a mortar will sometimes bring about a reaction which no application of mere pressure could effect; but otherwise it is never brought to his notice that the chemistry of stressed solids is in any respect different from that of unstressed. Until research comes to be directed expressly to this question, there remains therefore a serious hiatus in our knowledge, and our inquiry is hampered accordingly.

Consider first the effect of stress upon fusibility and solubility, which may be included together, since melting is only a particular case of solution. The dependence of melting-point upon simple uniform pressure is expressed by Thomson's equation (p. 8):

$$\frac{dT}{dp} = (v_2 - v_1)\frac{T}{L}.$$

Since minerals, with possible rare exceptions, expand in melting, the effect of increased pressure is to raise the melting-point. It is here assumed that the pressure is the same for the solid and the liquid phase. If it be possible for uniform pressure to be applied to the solid phase only, the result, as shown by Poynting,[2] will be very different, the appropriate equation in this case being:

$$\frac{dT}{dp} = -v_1\frac{T}{L}.$$

[1] For a useful presentation of this subject see Johnston and Niggli, ' The General Principles Underlying Metamorphic Processes ', *Journ. Geol.*, vol. xxi (1913), pp. 481–516, 588–624.

[2] *Phil. Mag.* (5), vol. xii (1881), pp. 32–48.

Increased pressure will then *lower* the melting-point, and that to an extent much greater than the rise in the former case. By the same principle, the solubility of a crystal in contact with a solvent liquid will be much enhanced by uniform pressure affecting the crystal only, and this will be true whether solution involves increase or decrease of volume.

It is in this connexion that Johnston and Niggli use the term ' unequal pressure ', i.e. unequal as between solid and liquid phases,[1] and they show that the principle may have important applications to metamorphism. Consider a crystalline rock, supposed first for simplicity to consist of a single mineral, containing a certain amount of interstitial liquid solvent, and let it be subjected to external pressure. Even if the pressure so applied be uniform, the system of pressures set up internally will be of a complex kind, since the brunt will be carried by the crystals at those places where they bear against one another. Here, we must suppose, solution will take place. The interstitial liquid, already saturated, becomes thereby supersaturated, and there will be simultaneous crystallization, the material being deposited at places where the pressure is least. New crystals may perhaps be formed, but in general the net result is to transfer material from one part of a crystal to another, at the same time relieving internal stress.

It must be observed, however, that the mechanical conditions here are more complex than in the ideal case discussed by Poynting. When different parts of a crystal are under different external pressures, the crystal must be in a state of shearing stress, and a new factor is introduced, viz. that which is known as Riecke's principle. Of the problem in its generality we have as yet no adequate solution. Riecke [2] considered the case of a prism of ice in equilibrium with water, and supposed it subjected to a definite linear pressure, which, as we have seen, is equivalent to a uniform pressure in all directions together with shearing stresses (p. 142). He concluded that the melting-point will be lowered by an amount proportional to the square of the pressure. This apparently is meant to hold good only for a small pressure, and it can scarcely be supposed that such a formula is of general application. Some lowering of melting-point at least may be assumed, and by parity of reasoning we must infer that the solubility of a crystal is very sensibly increased by a state of shearing stress. A stressed

---

[1] This usage is not a convenient one, for the same term has sometimes been employed to signify pressure (in a solid) unequal in different directions. Such a condition, implying shearing stress, has no part in Poynting's principle.

[2] *Centr. Min.*, 1912, pp. 97–104.

crystal and an unstressed, in contact with a saturated solution, are to be regarded as two distinct solid phases, and therefore cannot be in equilibrium. Material will be dissolved from the one and added to the other until the former has disappeared. So, too, if a single crystal be unequally stressed in its different parts, differential solubility may cause a transference of material from one part to another, gradually changing the shape of the crystal. A sufficient degree of solubility in the mineral is of course presumed.

Temporary local liquefaction in the manner here pictured renders possible a relative displacement of the several crystals in a crystalline rock, always in the sense of relieving internal stress ; and this effect is cumulative as solution takes place now at one spot and again at another. The action has thus an important part in the gradual bodily deformation of rock-masses and the setting up of directional structures. In a limestone it is effective even at low temperatures, but the silicate-minerals in general require the increased solubility which comes with a notably elevated temperature. Under these conditions the behaviour of crystalline rocks is comparable with the ' flow ' of metals under great pressure and intense shearing stress, as illustrated by the well-known experiments of Tresca. Adams and Johnston have interpreted Tresca's results by the principle of ' unequal pressure '. They suppose that at the contact of particular crystals the lowering of melting-point was sufficient to cause actual fusion, and such local and temporary fusion, continually repeated at many different places, would make possible differential movement within the mass. Here the influence of shearing stress *per se* is not discriminated. In the case of a crystalline rock we have to do with solution instead of dry fusion, but this does not import any essential difference in behaviour.

INFLUENCE OF STRESS ON CHEMICAL REACTIONS

It may be assumed that, when solution and redeposition take place in a rock subjected to stress, what is deposited will be a phase stable under the actual conditions. So, in the case of a dimorphous body, any metastable form will be replaced by the stable one. The conversion of aragonite to calcite in a limestone comes under this head. Since aragonite is the denser of the two, mere hydrostatic pressure might be expected to resist this change. In all such transformations, however, the presence of a solvent must be supposed essential. The principle of Poynting can have no application to the case of two solid phases.

We have still to consider shearing stress as one of the factors controlling chemical reactions and determining the stability of different minerals and mineral-associations.   It is here that the inadequacy of our data, both on the theoretical and on the experimental side, is most severely felt.   It is safe to assert that, for a crystal in a state of stress, chemical activity, no less than solubility, is very sensibly augmented. Shearing stress in a rock will then increase the *rate* of chemical reactions, just as a rise of temperature would do, and will sometimes make effective a reaction which would otherwise be in abeyance.   But it is clear that wider possibilities are indicated.   We have already remarked that, from the point of view of the Phase Rule, a stressed and an unstressed crystal of the same mineral constitute two distinct solid phases ; and it is easy to believe that a state of shearing stress, by altering the atomic configuration of a crystal, changes also its chemical properties.   This possibility, however, has been generally disregarded in discussing the chemistry of regional metamorphism, and finds no expression in the equations which are written down to represent supposed reactions.   It would seem indeed that we have to do here with a *new chemistry*, differing—we know not how much—from that of the laboratory, and offering a new field for investigation.

Although research touching the part played by shearing stress in crystallization and chemical changes has not yet furnished any large body of information, some results have been established which are at least suggestive.   The experiments of Spring [1] have often been cited, and they are worthy of notice, both for the results themselves and as enforcing the need of caution in interpreting such results. Spring worked with pressures of some thousands of atmospheres.   He found it possible, not only to effect crystallization of amorphous substances, which had been demonstrated long before,[2] but also to bring about definite chemical reactions, evidently controlled by the dynamical conditions.   For example, from a mixture of sulphur and copper filings, under a pressure of 5,000 atmospheres, he obtained crystallized copper sulphide.   This work gave rise to some divergence of opinion ; the more so because other physicists failed to obtain like results.   The explanation of this, as subsequently pointed out, is instructive.   Spring's crucial experiments were conducted in a steel bomb with an ill-fitting piston, such that part of the material was squeezed out at its circumference.   The charge therefore was subjected to intense shearing stress, and to this, rather than to simple pressure and compression, the effects must be attributed.   Some

[1] *Bull. Acad. Roy. Sci. Belg.* (2), vol. xlix (1880), p. 323.
[2] Brewster, ' On the Production of Crystalline Structure in Powders by Compression and Traction ', *Proc. Roy. Soc. Edin.*, iii (1853), p. 178.   It is interesting that Brewster recognized the importance of ' traction ', i.e. shearing stress.

simple experiments by Carey Lea [1] emphasized the relative inefficiency of simple pressure and the potency of shearing stress as influencing certain chemical reactions. Changes such as the reduction of mercuric to mercurous chloride, which could not be initiated by a hydrostatic pressure of 70,000 atmospheres, were brought about by grinding in a mortar for fifteen minutes. It is well known to chemists that such trituration causes a certain loss of the combined water from a hydrated substance or of carbon dioxide from a carbonate in the ordinary procedure of chemical analysis.

These instances, it should be remarked, fall under a special case. It has already been observed that a reaction is strongly promoted when one of the products is a liquid or gaseous phase, and the conditions are such that this can escape out of the system (p. 9). Whether or no this latter proviso is satisfied in the metamorphism of a rock must depend upon the situation and circumstances. It may be said, however, that, while hydrostatic pressure always resists a reaction which yields a gaseous product, shearing stress has no such deterrent action. Further, apart from any consideration of stress, continued deformation in the manner of shearing must aid in the elimination of a liquid or gaseous product, and so promote the reaction by which this is produced. Even when no liquid or gaseous phase is involved (other than the pervading solvent), prolonged shearing, as distinct from shearing stress, will at least accelerate the rate of any possible reaction, since the internal rearrangement implied continually brings fresh surfaces of the reacting bodies into close contact.

More recently Bridgman [2] has conducted an investigation of the effects of high shearing stress combined with high hydrostatic pressure, but only at low temperatures. It was found that many substances normally stable become unstable, and conversely substances normally inert to one another may be made to combine. Attempts to synthesize certain minerals (mica and tremolite), however, gave, under these conditions, only negative results. The same lack of success attended the endeavours of Larsen and Bridgman [3] to bring about various low-grade mineral transformations by the agency of powerful shearing stress with high pressure. This indeed might be anticipated, since the experiments were conducted at room-temperature. The authors point out also that in nature the time element is probably an important factor.

[1] *Amer. Journ. Sci.* (3), vol. xlvi (1893), pp. 241–4, 413–20.

[2] *Phys. Review*, vol. xlviii (1935), pp. 825–47.

[3] *Amer. Journ. Sci.* (5), vol. xxxvi (1938), pp. 81–94.

Reverting now to general considerations : if we may assume that a stressed crystal differs in its chemical properties from an unstressed crystal of the same kind, consequences important for our purpose will follow. Shearing stress must be supposed to affect, adversely or favourably, the stability of different minerals. It is even possible that a particular compound, or a particular form of a dimorphous compound, may become definitely unstable under a sufficient measure of shearing stress. It is possible, on the other hand, that a particular compound or form has a stable existence *only* under more or less intense shearing stress, though it may persist in a metastable state when the stress has been removed. Since neither chemical theory nor experiment affords any specific information concerning this question, we must fall back on the *a posteriori* line of inquiry, viz. the study and collation of such petrological and geological evidence as bears upon the problem. Stress conditions during metamorphism leave an unmistakable impress upon the *structures* of the rocks so modified. The mineralogical constitution of these rocks, as we can examine them, has been acquired under the same stress-conditions which imparted the distinctive structures, subject only to some after-changes for which we can make allowance. It follows that whatever *mineralogical* characteristics, either positive or negative, are found to be peculiar to such rocks, or distinctive of them as compared with metamorphosed rocks which lack those peculiarities of structure, may be reasonably attributed to the influence of shearing stress.

We are led accordingly to institute a comparison, on mineralogical lines, between the crystalline schists and allied rock-types on the one hand and the products of purely thermal metamorphism on the other hand. The important minerals characteristic of thermal metamorphism in different kinds of rocks have already been noted. For the products of regional metamorphism a large body of observed facts is on record, and has been conveniently brought together in Grubenmann's treatise. We do not, however, find there the desired comparison, since simple thermal metamorphism is excluded from consideration.

Even a very general and cursory view shows us that numerous minerals are, in greater or less degree, characteristic of the crystalline schists as contrasted with thermally metamorphosed rocks. Some of these minerals are peculiar to the crystalline schists. Of others it may be said that they are of much more frequent and widespread occurrence there than as products of simple thermal metamorphism, or that they are promoted, in the sense of coming in at an earlier stage of progressive metamorphism. All these may conveniently be designated *stress-minerals*. If, on the other hand, we consider the minerals of thermal metamorphism, we find that some of them are peculiar to this mode of origin. Others, which are common in this association, are less prominent as constituents of crystalline schists, or appear there only in a more advanced grade of metamorphism. All these may be termed for the sake of convenience *anti-stress minerals*.

The distinction here based on petrographical data receives a certain measure of confirmation from the experimental side. Many attempts have been made to reproduce artificially silicate and other compounds analogous to minerals by what may be termed hydrothermal methods. With varying details, the practice has been on the same general lines ; viz. to operate at more or less elevated temperatures upon various materials enclosed in a bomb or other closed vessel, either with water alone or with other reagents in addition. High pressure is, of course, implied, though measurements of pressure are lacking. The conditions therefore are broadly comparable with those which rule in thermal metamorphism. When the results are brought together,[1] some rough general conclusions emerge, which are instructive if we recognize that negative, as well as positive, results have their significance. Numerous minerals have been obtained under the conditions described which are familiar in nature as products of metamorphism, and the temperatures used, 300°–500° C., are such as we may suppose realized in metamorphism of a medium grade. Quartz, felspars, certain micas, olivines, pyroxenes, wollastonite, spinel, corundum, ilmenite, haematite, and magnetite are among the products furnished by various experiments. It is instructive to find, for example, that orthoclase and albite can be formed thus at temperatures 350°–500°, not only from their component oxides, but at the expense of muscovite. The complete list includes most of the characteristic minerals of thermal metamorphism. Some of the commonest—anorthite, forsterite, diopside, wollastonite, magnetite, etc.—form very readily, and indeed can be obtained from simple fusion without the aid of mineralizers. In contrast with this, those minerals which are characteristic of the crystalline schists, and are thus marked as stress-minerals, never figure among the products of these experiments. For instance, while andalusite is reported to have been synthesized from silica and alumina at temperatures 350°–450°, artificial cyanite is unknown. Enstatite and the augite minerals form readily, but not the tremolite-actinolite series or the anthophyllites.

---

[1] Morey and Niggli, *Journ. Amer. Chem. Soc.*, vol. xxxv (1913), pp. 1086–1130.

So too glaucophane has never been reproduced, though the soda-pyroxenes are easily obtained. Of the lime-minerals, zoisite and epidote have defied all attempts at their artificial reproduction. The same may be said of other distinctively stress-minerals, such as chloritoid, ottrelite, and staurolite. This negative testimony is all that can be looked for, until chemists shall attempt the production of artificial minerals under conditions which include both elevated temperature and powerful shearing stress.

It is perhaps possible to frame a more definite conception of ' stress ' and ' anti-stress ' minerals as they are here to be understood. Taking first only temperature and pressure as the two controlling conditions, we must suppose that for any given mineral there is a certain field of stability, to be laid down on the p-t-diagram. If now the introduction of shearing stress as an additional condition causes an extension of the field, we have to do with a stress-mineral, if contraction an anti-stress-mineral. If a sufficiently intense measure of shearing stress causes the field to contract to a point and disappear, we have an anti-stress mineral in a very special sense. On the other hand, some form may make its entry only with the coming in of the stress factor, having otherwise no field of stability : this will be a stress-mineral in a very special sense. To picture the relations more clearly we must imagine a three-dimensional diagram with temperature, pressure, and shearing-stress as co-ordinates.

So regarded, every mineral would seem to fall theoretically into one or other of the two categories distinguished. Practically, however, we have to recognize a neutral class, including minerals which, within the range of conditions concerned, seem to be not notably affected, as regards their stable or metastable existence, by shearing stress. To consider minerals singly is indeed an inadequate treatment of the matter : paragenesis must also be taken into account. When two or more minerals are constantly found in company with one another, this is doubtless in many instances because they are joint-products of some reaction, and one could not be produced without the other in any rock of ordinary composition. Representing the probable reaction by an equation, we should compare the *mineral-associations* on the two sides, and recognize that increasing stress, just like rising temperature or increasing pressure, will drive a balanced reaction in one direction or the other.

One case there is which would seem to be simple ; viz. that of two dimorphous forms, one a stress- and the other an anti-stress mineral. It must be remembered, however, that not a few of the common minerals of metamorphism are merely metastable forms, in the sense

of having no theoretical field of stability, at least under stress-free
conditions, and that such metastable forms are found especially where
dimorphism or polymorphism enters. Thus the anti-stress mineral
andalusite and the stress-mineral cyanite are probably both metastable forms, monotropic towards the presumably stable sillimanite.
We must suppose, either that under shearing stress cyanite has a field
of true stability, or alternatively that the incidence of shearing stress
has the effect of reversing the *relative* stability of the two metastable
forms. It may perhaps be suggested that such rocks as cyanite-schists
are formed under great hydrostatic pressure, and that we are in danger
of attributing to shearing-stress effects which are really due to pressure.
Since the specific gravity of cyanite (at ordinary temperature) is
3·56 and that of andalusite only 3·16, it is clear that high pressure will
favour the formation of the denser cyanite wherever this form is a
possible one. If this factor, however, were of foremost importance,
it would still be difficult to account for the complete absence of cyanite
from aureoles of thermal metamorphism even under thoroughly deep-
seated conditions. Be this as it may, there are other cases in which
pressure and shearing stress tell in opposite directions, the stress
mineral being less dense than the anti-stress. The pyroxenes and
amphiboles afford the most familiar example. Thus, taking the
simple magnesian metasilicates, we find for enstatite the specific
gravity 3·175 and for anthophyllite 2·857, from which it is evident
that the formation of the amphibole rather than the pyroxene in
crystalline schists cannot be attributed to mere hydrostatic pressure.

If we try to assess the importance of hydrostatic pressure as a
factor in determining the mineralogical constitution of regionally
metamorphosed rocks, we shall find that—always excepting reactions
which involve a gaseous phase—it is only very great pressures that
are of prime concern. These enter in nature especially in conjunction
with very high temperatures and with a corresponding decline in the
magnitude and efficiency of shearing stress. The minerals which are
marked as high-pressure minerals in any significant sense are therefore
not many. The garnets are the most important. We may add
cyanite, as already remarked, the pyroxene group, more particularly
as compared with the amphiboles, the spinels, rutile, zoisite, and a few
others. As already intimated, a division of high- and low-pressure
minerals would traverse the division which has been made with reference to shearing stress, since both categories would include some
stress- and some anti-stress minerals. The influence of pressure on
solid solution—a favouring factor when diminution of volume is the
result—must have its importance in some cases.

CHAPTER XI

PURE DYNAMIC METAMORPHISM

Cleavage Structure in Rocks of Yielding Nature—Modifications of the Cleavage Structure—False Cleavage Structures—Mineralogical Transformations in Cleaved Rocks—Strain Effects in Crystalline Rocks—Cataclastic Effects in Crystalline Rocks—Solution Effects in Crystalline Rocks under Stress—Mineralogical Changes which Accompany Cataclastic Structures.

CLEAVAGE STRUCTURE IN ROCKS OF YIELDING NATURE

IN this chapter we shall examine the processes and the visible effects of what may be styled *pure dynamic metamorphism* ; that is, the class of changes produced in rocks which have been subjected to powerful mechanical forces at low temperatures. Here we include, in the first place, bodily deformation of rock-masses and the setting up of new structures by internal rearrangement of a mechanical kind ; but also, in most cases, some measure of solution and recrystallization, controlled by the stress-conditions, and mineralogical transformations which, if not caused, are at least promoted or precipitated by shearing stress.

Rocks differ greatly in respect of the resistance which they can offer to deformation, and in their manner of yielding when that resistance is overcome. Some, especially fine-textured sediments, yield freely, and are susceptible of unlimited deformation of a kind which is, to the eye, continuous. Others, including especially crystalline rocks, are of more stubborn nature, and cannot suffer any sensible deformation without visible fracture and internal displacement of a discontinuous kind. It will be convenient therefore to make a two-fold division on these lines, and to deal separately with the two categories.

We begin accordingly with *yielding rocks*, which, when deformed, behave in *quasi-plastic* fashion. Plasticity, in the strict sense, is a property which seems to depend upon the presence of a certain amount of colloid material, and this must always be lost at a very early stage of metamorphism. The rocks, however, still yield, under a sufficient differential stress, without fracture, and they acquire in the process, sometimes in a high degree of perfection, the fissile character of slates. The manner in which this distinctive property of *slaty cleavage* is conferred makes an interesting subject of inquiry.

The rocks which best illustrate this type of dynamic metamorphism are ordinary argillaceous sediments. Good examples are furnished by the roofing-slates of North Wales, of Cambrian and Ordovician age, and some of the Devonian slates of Cornwall. Calcareous slates and even some sediments purely calcareous may show a cleavage of a less perfect kind, as in parts of the Devonian of Devonshire. In the English Lake District good slates have been made from fine volcanic tuffs, and others, with a ruder cleavage-structure, from relatively coarse volcanic agglomerates.

Enclosed fossils, if not of too massive build, have shared in the deformation of the rock, and by their distortion afford some measure of it. A better indication is given by the ovoid spots and concretions of various kinds seen in many slates, such as green spots in a purple slate, due to reduction of ferric iron-oxide about certain centres. These, having been originally spherical, present the actual form of the strain-ellipsoid. In a Welsh slate Sorby found for the ratios of the semi-axes, $a : b : c$, the values—

$$1\cdot6 : 1 : 0\cdot27.$$

On the assumption that the mean semi-axis gives the radius of the original sphere, the rock occupies now only $0\cdot43$ of its former volume (p. 141). This great condensation of volume is normal in argillaceous rocks, and is explained by the squeezing out of the interstitial water of the original clay. Rough measurements on the ' bird's-eye ' slate of Kentmere in Westmorland, which is a fine tuff with flattened concretions, give for the axial ratios—

$$1\cdot5 : 1 : 0\cdot5 ;$$

which corresponds with a reduction of volume in the ratio $0\cdot75$.

The relation of the cleavage-direction to the strain ellipsoid is of a simple kind. The greatest axis a lies along the direction of cleavage-dip, the mean axis b along the cleavage-strike, and the least axis c is perpendicular to the cleavage-planes. In other words, as was first observed by Sharpe,[1] there has been a great compression of the rock in the direction perpendicular to the cleavage-planes with a certain elongation along the cleavage-planes in the direction of their dip. The proximate cause of the fissile property itself is evident upon an examination of the micro-structure, which reveals that, in a rock with

[1] *Quart. Journ. Geol. Soc.*, vol. iii (1847), p. 87.

good cleavage, a large part of the bulk consists of flaky or flattened elements set in approximately parallel position. From these facts it is possible to devise a theory of the origin of the cleavage-structure based on mechanical considerations alone.[1] Sharpe,[2] without the advantage of microscopical study, supposed that the individual elements of a fragmental rock share the deformation of the rock as a whole, and so became *flattened* with a regular disposition in common (Fig. 65). He instanced in particular the cleaved volcanic agglomerates of the Lake District, and for such rocks his hypothesis

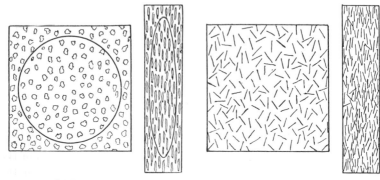

FIGS. 65 and 66.—DIAGRAMMATIC SCHEME OF THE MECHANICAL THEORY OF SLATY CLEAVAGE, firstly according to Sharpe and secondly according to Sorby.

doubtless affords in part the true explanation. Sorby[3] found from thin slices that an ordinary argillaceous sediment is composed very largely of minute scales of mica and other minerals of like habit, and that in a well-cleaved slate these are arranged in, or nearly in, the cleavage-planes. He pointed out that the several flakes, if originally without any regular arrangement, must, by the deformation proved, be *rotated* into an approximate parallelism ; and he showed that the degree of regularity so produced will be sufficient to give a direction or ready fissility (Fig. 66).

This is evidently true, and for some rocks, especially fragmental limestones, it affords a sufficient explanation of the cleavage-structure. As applied to those argillaceous sediments which furnish the best roofing-slates, Sorby's view embodies only part of the truth, and the actual structure is not to be explained as the result of any purely mechanical process. The rotation of the flaky constituents of the rock,

[1] For a fuller exposition of the mechanical theory see Harker, *Rep. Brit. Assoc.* for 1885 (1886), pp. 813–52.

[2] *Quart. Journ. Geol. Soc.*, vol. v (1849), pp. 111–15.

[3] *Edin. New Phil. Jl.*, vol. lv (1853), pp. 137–47, *Phil. Mag.* (4), vol. xii (1856) pp. 127–9.

as here pictured, presupposes that they were in existence prior to the deformation. A minute study of such slates as those quarried in North Wales [1] shows, however, that only a small part of the rock consists of the original clastic minerals and pseudomorphs after these. The main bulk is of new minerals, largely generated, or at least regenerated, in the rock under the conditions of dynamic metamorphism. Even in an uncleaved shale the main part is often made by a very finely divided aggregate, largely micaceous, which has been formed in place by chemical reactions ; but in the process by which such a shale becomes a slate, this material is recrystallized in more distinct flakes. The important minerals which enter in this way— mica, chlorites, haematite and ilmenite—are all of pronounced scaly habit, and, in accordance with a general principle to be discussed later, they set themselves, *as they grow*, in planes perpendicular to the maximum pressure. The parallelism so produced is of a more perfect kind than that brought about by rotation of pre-existing flakes, and the finest slates owe their highly fissile property to this mode of origin. Slates derived from fine-textured tuffs also consist largely of flaky minerals crystallized in place, with a parallel orientation due to the same cause.

MODIFICATIONS OF THE CLEAVAGE STRUCTURE

The actual disposition of the cleavage-planes, as seen in the field, is not always the same, and, for the sake of clearness, two typical cases may be distinguished. Where the deformation of the rocks is related to important overfolding and overthrusting, as in some parts of the Alps,[2] the cleavage-planes make usually *low angles* with the horizontal. We are to picture the rocks as being continually dragged over in a given direction with prolonged shearing along planes of low inclination, as in Case i of p. 141. The resulting deformation may therefore be very great. In this type the elongation or ' *stretching* ' becomes a prominent feature. When it is carried to the extreme, the strain-ellipsoid assumes a shape in which the semi-axis a is many times longer than b or c, and the rock tends to break into rods rather than parallel plates. Such ' *linear cleavage* ', however, is usually of a rude kind, and is of local occurrence.

Where, on the other hand, cleavage is uniformly well-developed over a considerable tract, as in North Wales, the cleavage-planes

[1] See especially a series of papers by Hutchings, *Geol. Mag.*, 1890, pp. 264–73, 316–22 ; 1891, pp. 164–9, 304–6 ; 1894, pp. 36–44 ; 1896, pp. 309–17, 343–50.
[2] See especially Heim, *Untersuchungen über den Mechanismus der Gebirgsbildung* (1878), part II and plates XIV, XV.

commonly make *high angles* with the horizontal, and are often nearly vertical. The external forces responsible for this arrangement may be pictured simply as a linear pressure acting horizontally in a definite direction. Compression has taken place in this direction and a partially compensating elongation in the upward direction, as in Case ii of p. 141.

We have been assuming that the whole mass of a rock which suffers continuous deformation is of the same yielding nature. If, however, some more stubborn body is enclosed, such as a crystal of some hard

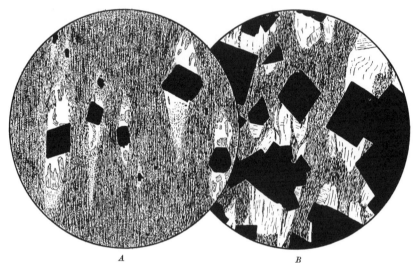

A   *B*

FIG. 67.—EYED STRUCTURES IN SLATES, ABOUT HARD ENCLOSED CRYSTALS ; × 25

A. Magnetite-Phyllite (*phyllade aimantifère*), Monthermé, Ardenne. The magnetite crystals before the cleavage epoch, were invested by a coating of chlorite, which has adhered to the matrix and been torn away during the deformation. The spaces are filled with clear quartz.

B. Pyritous Slate, Gallt-y-llan, near Snowdon. Here also is some chlorite. The quartz has crystallized with a parallel fibrous arrangement.

mineral, more peculiar phenomena are seen. The yielding matrix has flowed past the obstruction. Its consistency, however, did not allow it to follow closely the surface of the crystal ; so that a lenticular or spindle-shaped space is left about the crystal as centre. This space, where the protection of the resistant crystal caused a relief of pressure, has been filled by quartz, deposited from solution. In calcareous rocks the space, due to a massive fossil such as a belemnite, is filled by calcite. Renard,[1] in his account of the phyllites or glossy slates of the Ardenne,

---

[1] Bull., *Mus. Roy. Hist. Nat. Belg.*, vol. i (1882), pp. 215–49 ; ii. (1883), pp. 127–49 ; iii (1884), pp. 231–68. See also Mügge, *Neu. Jahrb. Min.*, Beil. Bd. lxi (1930), pp. 469–91.

remarked effects of the kind described about crystals of magnetite (Fig. 67, *A*) ; and good examples are seen in some Welsh slates about large crystals of pyrites.[1] In some instances the quartz shows a parallel fibrous structure, owing to its having crystallized during the progress of differential movement (Fig. 67, *B*).

A crystal of pyrites, though hard, is brittle, and, under a sufficiently intense stress, it may be shattered, the fragments then becoming parted and drawn out in the direction of elongation of the deformed matrix. From Alpine calcareous rocks exhibiting a linear cleavage, Heim [2] has figured broken belemnites with the fragments widely separated, affording a demonstration of the ' stretching ' and a means of estimating its amount. Here the spaces between the displaced fragments are occupied by crystalline calcite.

FALSE CLEAVAGE-STRUCTURES

In argillaceous rocks which have already become well compacted and hardened, the property of quasi-plasticity which has been discussed, permitting a continuous rearrangement of the intimate structure of the mass, is realized only when the rocks are subjected to great hydrostatic pressure. With a like stress-difference, but under a less total pressure, yielding does indeed take place, but it is effected in a different manner. The rock-mass as a whole does suffer a deformation which has roughly the effect of shearing, but the internal rearrangement now takes the form of folding or reversed faulting on a small scale, often microscopic. Where these minute structures are developed on a regular plan, they often impart to the rock a more or less pronounced fissility in a definite direction. We may recognize accordingly a class of structures which may be styled *false cleavages*. Here belong the ' close-joints-cleavage ' of Sorby, the ' Ausweichungsclivage ' of Heim, the ' strain-slip-cleavage ' of Bonney, etc. The fissility is less perfect than in a true cleaved slate, and there is an essential difference in kind. The property is here one of a series of parallel planes at a certain small distance apart ; but true cleavage, depending upon intimate structure, is a property of *any* plane parallel to a given direction. A false cleavage, however, may be a first step towards a true one.[3]

False cleavage-structures are very clearly exhibited in a rock which has been a well-laminated shale, the component scales of mica,

[1] Harker, *Geol. Mag.*, 1889, pp. 396–7.

[2] *Loc. cit.* and *Viert. Nat. Ges. Zürich*, vol. lxi (1916), pp. 516–28.

[3] Sorby, *Quart. Journ. Geol. Soc.*, vol. xxxvi (1880), Proc., pp. 72–3.

etc., lying mainly in the planes of deposition. This arrangement naturally arises, in the manner illustrated in Fig. 66, during the vertical settlement of the sediment.[1] Suppose such a rock to yield to lateral pressure by the formation of a system of minute regular folds. The folding is usually of an unsymmetrical type, and it often happens that there are parallel planes in the plicated rock such that the mica-flakes in the immediate vicinity of a plane lie very nearly parallel to it (Fig. 68). The rock therefore splits readily along these planes. Unsymmetrical folding often passes into a system of little parallel faults, and these too are naturally planes of weakness. Structures of

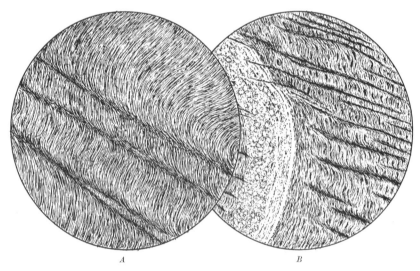

A   B

FIG. 68.—FALSE CLEAVAGES ;   × 23.

A.   Drws-y-Coed, near Snowdon.
B.   Rosevanion, Cornwall. The structure is interrupted by a gritty band.

this class show indeed great variety, and reproduce on a microscopical scale all the features of Alpine tectonics. For a figure illustrating the passage of an overfold into a reversed fault, Heim gives the scale as ' 200 to 1/10,000 of natural size '. A common case is for minute unsymmetrical folds to appear first, with shearing movement, at an angle of about 45°, to the general direction of schistosity. With continued movement the folds become compressed and pushed over

[1] If, for example, water makes four-fifths of the volume of a mud as first deposited, and only one-third of the resulting shale, the shale will have three-tenths of the vertical thickness of the original mud.

until, at about 30° they break into minute reversed faults.[1] These may number as many as two or three hundred in an inch.[2]

In a shale, which can offer no great resistance, yielding naturally takes the form of miniature folding, and internal faulting arises only as a further development of this. In a somewhat harder rock relief is found from the first in a series of minute parallel reversed faults. This is seen especially in rocks which are already in the state of well-cleaved slates, and are subjected anew to mechanical disturbance at a later epoch. The new structure may then cross the true cleavage at any angle. This is the nature of the so-called ' second cleavage ' described in the Ardenne, the Isle of Man, and other districts. Manifestly there cannot be two true cleavages in the same rock.

MINERALOGICAL TRANSFORMATIONS IN CLEAVED ROCKS

It remains to notice the *mineralogical changes* which accompany the setting up of the cleavage-structure, and with which indeed, in all typical slates, the cleavage-structure itself is closely bound up. The minute study of argillaceous sediments in general is a matter of no small difficulty, and it will be sufficient here to present the results in a summary fashion. We have remarked that Sorby's mechanical theory of cleavage is inadequate as applied to the case of normal argillaceous slates, for the reason that almost the whole substance of these rocks is of new formation. This latter observation is true in some measure of uncleaved shales and even of clays like those of the British Coal-measures. These, in addition to recognizable clastic constituents—quartz-granules, muscovite, and decaying biotite—contain more or less of an exceedingly finely divided micaceous material in the form of an interstitial ' paste '. This may constitute the greater part of the deposit, and has certainly been formed in place. It commonly contains a great number of minute needles of rutile, derived doubtless from the decomposition of biotite. What is described loosely as a micaceous substance does not, however, show the characters of any determinate species of mica. It seems to be less highly birefringent than muscovite, and has a smaller axial angle. It shows too a faint yellowish or greenish tint. As regards composition, it may be safely asserted that it carries more combined water than true micas, and that, in addition to potash, it contains magnesia and some iron. When heated to redness, it becomes coloured brown or red.

---

[1] Harker, *Geol. Mag.*, 1892, p. 343 (Start Point).

[2] Harker, *Quart. Journ. Geol. Soc.*, vol. xlvii (1891), p. 513 (Skiddaw Slates). On false cleavages generally, see further Harker, *Rep. Brit. Assoc.* for 1885 (1886), pp. 836–41.

Sediments of this kind, even without any visible chlorite, still yield on analysis a noteworthy percentage of magnesia, and Hutchings concluded that the substance in question is in some sense intermediate in composition between a potash-mica and a chlorite, being perhaps of the nature of a solid solution.

According to the same observer, the most important mineralogical change in the process which converts uncleaved sediments of this kind into typical slates is the conversion of the indeterminate micaceous substance into a mixture of sericite and chlorite, the latter usually

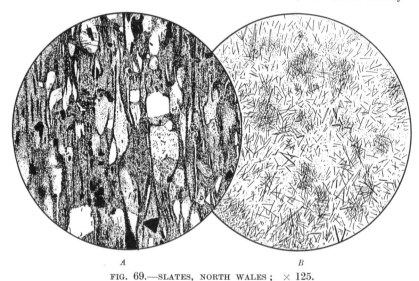

FIG. 69.—SLATES, NORTH WALES ; × 125.

A. Moel Tryfaen, Caernarvonshire : cut transversely to the cleavage-planes and showing a parallel arrangement of the elements which is sufficient to impart a perfect cleavage. The constituents are sericitic mica, chlorite, and haematite, with abundant clastic granules of quartz.

B. Morben, near Machynlleth, Montgomeryshire ; cut parallel to the cleavage to show the swarm of minute needles of rutile. The tendency to clustering indicates vanished shreds of biotite.

in subordinate amount. This goes on progressively, and concurrently with a general recrystallization of the micaceous constituents into larger and more distinct flakes. In this process clastic muscovite disappears, perhaps lost at first in the abundant new aggregate, but ultimately undergoing recrystallization. Among other chemical transformations, the iron-oxide in the rock, excepting such part as is contained in the chlorite, goes into haematite, which appears as minute scales, thin enough to be transparent. Often too ilmenite is produced, with a like form. By the changes noted the rock comes to consist finally almost wholly of minerals with flaky habit, which have been generated in place. Growing in a solid rock which is in a state of unequal stress, the flakes orient themselves—in accordance with a principle to be discussed later—in planes perpendicular to the direction of maximum pressure. In this way arises a more regular parallelism, and therefore a more perfect cleavage, than could result from a mere rotation of pre-existing flakes by the bodily deformation of the rock (Fig. 69, *A*).

The changes described are progressive, and of course have not always advanced to the final stage. Some slates which have acquired a pronounced cleavage-structure still contain some amount of the impure micaceous substance which has not yet split into true mica and chlorite, and muscovite of clastic origin may still be present. The ultimate stage of dynamic metamorphism in normal argillaceous sediments is well illustrated by the best Welsh roofing-slates.[1] Newly formed mica with the filmy 'sericitic' habit is the main constituent, and with it is a variable proportion of chlorite. Haematite is often present in minor amount, and becomes conspicuous in purple slates like those of Llanberis. Other new minerals which may occur are epidote, calcite, and sometimes interstitial secondary quartz from the recrystallization of amorphous silica. The original clastic granules of quartz seem, however, to be always intact. They are abundant in all ordinary slates, and with them may sometimes be seen granules of fresh felspar, probably albite. These are the only prominent original constituents remaining, but a careful examination reveals scattered minute crystals of such minerals as zircon and tourmaline, which figured among the minor derived constituents of the sediment. A high magnification shows also in many argillaceous slates a profusion of exceedingly minute needles of rutile in the micaceous matrix (Fig. 69, *B*). These, as we have seen, are formed as a secondary product at an early stage in the history of the sediment, and they remain apparently unchanged at the stage now reached.

The mineralogical changes illustrated by such rocks as these have been brought about under the influence of intense shearing stress, as well as very considerable pressure, but, so far as the evidence goes, without any noteworthy elevation of temperature. To some other roofing-slates, such as those in the Devonian of North Cornwall, and to many of the glossy slates styled phyllites, this latter qualifying condition is not applicable, and we have no longer to do with purely dynamic metamorphism. In the more complex type of metamorphism in which some definite elevation of temperature is a factor, to be

[1] Hutchings, *loc. cit.* ; also *Proc. Liverp. Geol. Soc.*, vol. viii (1900), pp. 464–73 ; vol. ix (1901), pp. 112–14.

discussed later, other mineralogical changes find a place. In the Devonian slates or phyllites, for example, the clastic quartz-granules have disappeared, and minerals, such as chloritoid, may enter which never figure in metamorphism of the purely dynamic kind. It should be remarked, however, that, not only these rocks, but any highly cleaved slates, such as those of North Wales, composed essentially of new minerals with a general parallel arrangement, may be regarded as crystalline schists of very fine texture. Between the structures which we designate ' cleavage ' and ' schistosity ', there is no essential difference ; and, as Darwin [1] long ago recognized, no real distinction can be drawn between the two as studied in the field.

STRAIN EFFECTS IN CRYSTALLINE ROCKS

We go on to examine the effects, of a mechanical kind, produced in *crystalline rocks* when they undergo dynamic metamorphism at a low temperature. The visibly heterogeneous constitution of such rocks and their relatively resistant nature imply a behaviour quite different from that of a plastic clay.

The individual crystals are elastic, in the proper sense, though only within narrow limits. Experiment shows that crystalline rocks can be regarded as possessing, in the gross, a similar elasticity.[2] The limit, however, is soon reached, permanent deformation is produced, and under sufficiently powerful forces the most stubborn rocks will yield freely to change of shape. The manner in which the deformation is effected depends upon the physical conditions which obtain, as well as upon the nature of the rock. At high temperatures, with pressure co-operating, actual fracture and visibly discontinuous displacement are in great part avoided. This is owing to concurrent molecular and atomic rearrangements, viz. to solution and recrystal-lization and to mineralogical transformations. In this way rocks of many kinds acquire, in the more advanced grades of metamorphism, a certain effective plasticity, and their manner of deformation, as viewed in the gross, resembles shearing as strictly defined. At low temperatures, however, relief of the kind indicated is much less operative, and deformation is not effected without internal fracture and finite slipping of one part over another. According to the nature of the rock and the magnitude of the forces concerned, the resulting *cataclastic structures*, as they are collectively termed, show much variety in detail. In rocks which have suffered dynamic metamorphism

---

[1] *Geological Observations on South America* (1846), pp. 163, 166, etc.

[2] Adams and Coker, *An Investigation into the Elastic Constants of Rocks*, etc., Washington (1906).

only, the degree to which fracture, crushing, and deformation have been carried affords some measure of the maximum intensity of the forces which have operated, due allowance being made for the more or less resistant nature of the rocks themselves. But it is to be observed too that cataclastic phenomena are very prevalent in rocks which have passed through a regional metamorphism of a high grade. Here we must infer a persistence or renewal of the stress-conditions after the temperature had greatly declined.

Crystalline rocks which have undergone severe stress, but not to

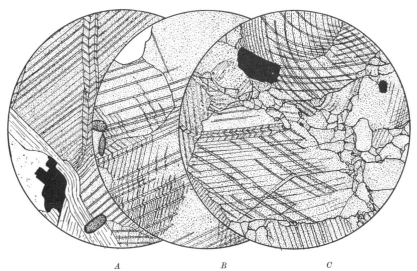

A   B   C

FIG. 70.—GLIDING-LAMELLAE IN CALCITE, Mont Gendres, Pyrenees ; × 25.

The rock is a coarse-grained marble containing some muscovite, sphene, pyrrhotite, and quartz.
A. Besides gliding-lamellae in the calcite, there is some bending of mica-flakes.
B. Shows the discontinuous character of the gliding-lamellae.
C. A rather more advanced stage. The calcite is in places bent, and is beginning to break up into smaller grains.

the point of yielding by actual fracture, have often received nevertheless a *permanent strain*, which shows itself in various ways. A crystal strained beyond the limits of elastic recovery, although outwardly intact, has its optical properties modified. Quartz especially is liable to be affected in this way, and then shows between crossed nicols the irregular extinction which gives the well-known appearance of ' strain-shadows '. Closer examination reveals that quartz crystals so strained have become biaxial.[1]

[1] For a closer study of strain-phenomena in quartz see Holmquist, *Geol. För. Stock. Förh.*, vol. xlviii (1926), pp. 410–28. The subject is also dealt with at length in numerous works on ' petrofabric analysis '.

A more obvious strain-effect is a visible bending of the crystals, as frequently seen in micas and less conspicuously in felspars. Distinct from mere bending is the setting up of twin-lamellae along ' gliding-planes ' (Fig. 70). Here the deformation of the crystal is of the nature of a simple shear, effected *per saltum*, the gliding-planes being parallel to some particular crystallographic direction for a given mineral. In calcite the direction is that of the form (110) bevelling the edges of the cleavage rhombohedron ; so that there are three possible sets of gliding-planes (Fig. 71, *A*). Here gliding

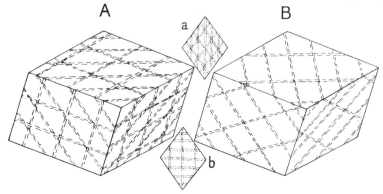

FIG. 71.—CLEAVAGE-RHOMBOHEDRA OF CALCITE AND DOLOMITE, SHOWING THE DIRECTIONS OF GLIDING-PLANES

A. Calcite : lamellae parallel to (110) (01$\bar{1}$2 of Bravais).
B. Dolomite : lamellae parallel to (11$\bar{1}$) (02$\bar{2}$1 of Bravais).
The smaller figures show the appearance in a section parallel to one cleavage and bounded by the other two.

is so readily effected that the structure is often produced in the process of grinding a thin slide of a crystalline limestone. In dolomite gliding-planes are much less commonly found, and they then follow a different crystallographic direction, viz. that of the form (11$\bar{1}$) which truncates the solid angles of the cleavage rhombohedron[1] (Fig. 71, *B*). In igneous rocks which have sustained considerable stress the plagioclase felspars are often affected by gliding, the chief direction being that of the albite-twinning. The secondary may be easily distinguished from the primary lamellae by their finer scale and inconstant occurrence. They often enter in evident relation to a slight bending of the crystal, and they are sometimes seen to end abruptly at a crack, which has relieved the strain. Gliding-planes are often seen in micas, augite (parallel to the basal plane), cyanite, and other minerals.

[1] Tilley, *Geol. Mag.*, vol. lvii (1920), p. 453 ; Rogers, *Amer. Min.*, vol. xiv (1929), pp. 245–50.

We have seen that solution, when it involves contraction of volume, is promoted by pressure. The permeation of a crystal by a liquid falls under this head. It may become effective under sufficiently great pressure, and in some cases secondary fluid-inclusions, especially in quartz, have arisen in this way. When the crystal is in a condition of unequal stress, the effect is controlled, not by the normal molecular structure of the quartz, but by the stress-system affecting it. Trains of minute fluid-pores are developed in parallel planes, which are perpendicular to the direction of maximum tension, and therefore to any ' stretching ' which the rock may have suffered. As seen in a section therefore the lines of inclusions have a common direction in all the grains of a grit or quartzite (see Fig. 114, *B*, below). Apart from unequal stress, solubility in a crystal is a vector property, and normally solution takes effect along certain crystallographic directions (' solution-planes '). Judd,[1] however, has shown that, when gliding-planes have been set up in a crystal, they supersede the normal solution-planes (if they do not coincide with these) as directions of easiest solubility. To solution and chemical reactions, induced by great pressure and guided by one or other of such structural planes, he attributed the various types of ' schiller ' structures which are seen in the crystals of many plutonic rocks, especially in pyroxenes and felspars.[2] It is not, however, to be supposed, nor did Judd contend, that schiller-structures have in all cases this secondary origin.

CATACLASTIC STRUCTURES IN CRYSTALLINE ROCKS

When differential stress in a crystalline rock passes the limits of strength, fracture must follow, and we see cataclastic phenomena, in the proper sense, which may take various forms. Since the individual crystals often have angular or irregular shapes, sometimes with partial interlocking, and the several minerals differ in strength, internal fracture commonly begins at the junctions of the crystals with one another, and especially by the breaking away of projecting parts. By losing their prominent angles, the crystals acquire shapes less vulnerable to attack, while the angular fragments flaked off are very liable to further comminution (Fig. 72, *A*). So we often see, at a certain stage of the breaking-down process, relatively large and rounded relics of the original crystals embedded in a matrix of much smaller elements (Fig. 72, *B*). This is known as the ' mortar-structure ' (*Mörtelstruktur*). In a more advanced stage these relics

[1] *Min. Mag.*, vol. vii (1886), pp. 81–92 ; vol. ix (1890), pp. 192–6 ; *J. Chem. Soc.*, vol. lvii (1890), pp. 404–25.

[2] See also Judd, *Quart. Journ. Geol. Soc.*, vol. xli (1885), pp. 374–89.

themselves are gradually ground away. This cataclastic process may be modified in detail in many ways, in accordance with the more or less heterogeneous constitution of the rock, the relative hardness and brittleness of the several component minerals, and the possession by some of them of a direction of easy cleavage.

In what is properly termed *brecciation*, the first stage in the breaking-down process is a fragmentation upon a rather larger scale, the individual elements of the breccia being not separate crystals but pieces of rock. The fragments are at first necessarily angular ;

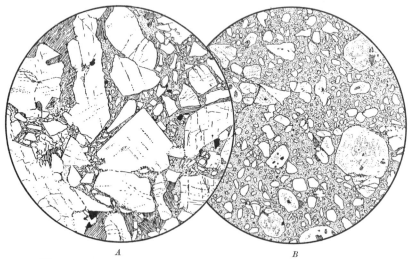

A   *B*

FIG. 72.—CATACLASTIC STRUCTURES, in Tertiary plutonic rocks, Skye ;   × 23.
   A.   Gabbro, Belig ;   partly crushed.
   B.   Granophyric Granite, Strollamus ;   showing a more advanced stage of crushing, with mortar-structure.

but, if the deformation of the rock is of the kind which involves rotation of the fragments, the angles are reduced by mutual attrition. Perfect rounding may be the result, while the finer material ground away furnishes a matrix. A rock in this condition is rather a crush-conglomerate than a crush-breccia. Brecciation of one kind or another is often seen locally in connexion with faulting, but it may also attain a considerable extension beneath a surface of overthrusting.

Where resistant rocks are intimately associated with others of yielding nature, the consequences of crushing are more peculiar. In the Isle of Man [1] crush-conglomerates have been extensively developed

[1] Lamplugh, *Geology of the Isle of Man* (*Mem. Geol. Sur. U.K.*, 1903), pp. 55–8. Van Hise has described identical structures in America : 16*th Ann. Rep. U.S. Geol. Sur.* (1896), part I. p. 679.

in a banded series of alternating grits and shales. The former have yielded by fracture, the latter mainly by flow. The fragments of grit have acquired by mutual attrition rounded or lenticular or especially spindle-like shapes, and the shale has been pressed in between the fragments, receiving in the process a new cleavage-structure.

The most striking phenomena of the cataclastic kind are found when the external forces have been of great magnitude, and have maintained a given direction, so as to produce a cumulative effect. The best illustrations are seen in a region of great over-thrusts, such as those which in Caledonian times affected the North-West Highlands of Scotland.[1] In some of these the displacement amounts to many miles. Both igneous and sedimentary rocks of various types are involved, and dynamic metamorphism can be studied in successive stages, reaching a maximum in the rocks immediately above or below the principal surfaces of discontinuity. The external forces having a constant direction, the system of stress set up in the rocks had a definite character, and was renewed on the same lines as often as it was relieved. Deformation of the rock-masses, as viewed in the gross, was analogous to shearing, but with the difference that internal movement was largely of the discontinuous kind.

Normally stress has been relieved in the first place by the formation of oblique gently curved surfaces of discontinuity, along which slipping took place. The mass is thus broken into lenticular pieces, of larger or smaller dimensions, with one general orientation, and may become subdivided in the same way. The lens form is indeed highly characteristic of this type of dynamic metamorphism, and recurs on every scale of magnitude. It is seen on the geological map ; it is seen in the field ; and it is seen under the microscope. The lenticle may be taken as the symbol of discontinuous, as the ellipsoid is of continuous, deformation (Fig. 73).

In a more advanced stage of breaking-down, when a more intimate fragmentation has followed, this has been guided on the same lines ; so that the uncrushed relics remain as lenticles, with a common orientation, set in a finer-grained matrix. The lenticular shape is retained as being that best adapted to resist further crushing under the given type of stress. Here too the shape may be seen on a small as well as on a larger scale, giving a characteristic 'eyed' structure (*Augenstruktur*), in which the 'eyes' may be made by

[1] See numerous descriptions in *The Geological Structure of the North-West Highlands* (*Mem. Geol. Sur. G.B.*, 1907) and in various memoirs of the Geological Survey of Scotland.

single crystals of felspar or other minerals. A somewhat different case is presented where hard crystals or aggregates are embedded in a matrix of notably less resistant nature. A crystalline limestone enclosing nests of silicates is the best example. Under stress not too intense, while the general matrix is crushed down, the more stubborn inclusions remain intact, each making the main part of a lenticle. The corners of the ' eyes ' consist of uncrushed relics of limestone protected by their position [1] (Fig. 75, *A*).

When the process is carried to the extreme, it may be pictured

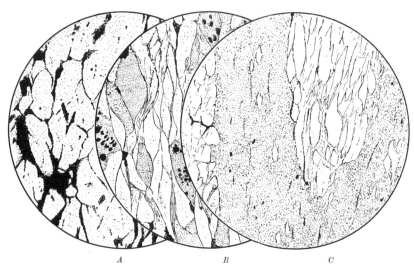

A   *B*   *C*

FIG. 73.—CATACLASTIC STRUCTURES, ILLUSTRATING THE LENTICULAR MODE OF FRACTURE ; × 23.

A. Lewisian Gneiss, near Kinlochewe, Ross-shire ; the surfaces of discontinuity picked out by an opaque infiltration.
B. Torridon Sandstone, near Kinlochewe. Here the fracture has often followed the boundaries of the individual grains of quartz and felspar.
C. Quartzite, Eiribol, Sutherland ; a more advanced stage of breaking-down. The figure shows parts of two residual lenticles, which are breaking into smaller lenticles. Smaller relics of the same shape are scattered through the crushed matric, which is in great part recrystallized.

as a forcible rolling out of the rock, which, under intense differential stress as well as great hydrostatic pressure, is constrained into a kind of flowing movement. The deformation is analogous to that which in a more easily yielding rock gives a highly developed cleavage at a low inclination (p. 155), but here it involves an intimate crushing down of the solid rock with internal slipping. A laminated structure results, closely resembling the flow-structure in a rhyolitic lava, emphasized by trains of magnetite-dust or thin streaks of colour

[1] Bonney, *Geol. Mag.*, 1889, p. 485.

representing the breaking down of some particular mineral of the original rock. Any lenticular relics ('augen') which may survive are oriented in the same direction, but these are gradually ground away and ultimately disappear. To rocks having these characters, presenting the appearance of having passed through a powerful mill, Lapworth gave the name *mylonites* (Fig. 74). In some parts of the North-West Highlands mylonites have been formed from various members of the Lewisian complex, from Torridon sandstone, and from Cambrian quartzite, and all in the extreme phase have the same

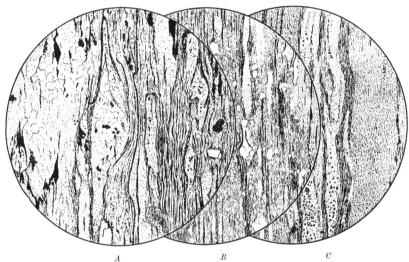

FIG. 74.—STAGES OF MYLONITIZATION, in Lewisian Gneiss, near the Moine Over-thrust, Eiribol, Sutherland ; × 23.

The pale lenticular parts in *A* are relics shattered but not yet ground down. They are much reduced in *B*, and have disappeared in *C*, where the whole is of exceedingly fine grain.

general aspect. They are exceedingly close-textured rocks with a flow-structure like that of a rhyolite, often with colour-banding, and having a more or less platy fracture.

Pressure, in the hydrostatic sense, has its importance as a controlling factor in the cataclastic process. Mylonitization, as contrasted with brecciation of a less drastic kind, requires a notable pressure, implying a considerable depth of cover. At such a depth the temperature too must doubtless be above the ordinary, but it does not appear that the elevation of temperature is such as to remove the process from the category of purely dynamic metamorphism as here understood. Its most important effect will be to increase solubility, and so to promote changes which are still to be discussed.

Although, for convenience of description, we have separated merely mechanical effects from those which involve molecular and atomic rearrangements, there are in fact no dynamically metamorphosed rocks in which changes of the latter kind are wholly negligible. In particular, we must presume in all cases a certain amount of *re-crystallization* of the more soluble minerals present, proceeding con-

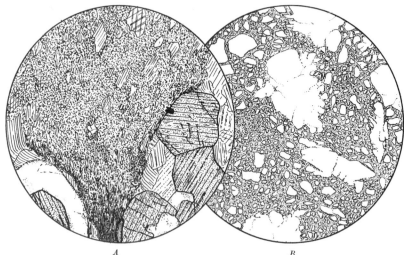

A   B

FIG. 75.—CATACLASTIC STRUCTURES, MODIFIED BY EFFECTS OF SOLUTION ; × 25.

  A. Crystalline Limestone enclosing aggregates of augite, felspar, etc., Tiree. The limestone, excepting scattered relics, has been broken down but recrystallized on a finer scale. The hard silicates have resisted crushing, and have protected small patches of the original coarse limestone. On the other hand the crushed matrix where squeezed between two silicate aggregates has become schistose.
  B. Crushed Quartzite, near Inchnadamph, Sutherland. Relics are shown in different stages of breaking down, controlled by some amount of solution, and the crushed matrix is recrystallized.

currently with the cataclastic process proper. It is evident that without such continual recementation a thoroughly crushed rock would be reduced to a powder. We must inquire accordingly how the mechanical disintegration is modified by this factor. Since we have to do with the two correlative actions of solution and recrystal-lization, it is not to uniform but to differential pressure that we must look for explanation of what is seen. The general principle is that which has already been laid down, solution taking effect at places of greater pressure with concurrent redeposition at places of less pressure.

  In crystalline rocks other than calcareous the most freely soluble mineral is usually quartz. When such a rock as a granite or gneiss is subjected to great shearing stress, the quartz elements, by solution at some points and concurrent addition at others, are capable of gradually changing shape, always in the sense of yielding to and relieving the internal stress. This readjustment, co-operating with cataclastic effects in the other minerals, is a factor of importance in the quasi-plastic deformation of the rocks. In the crushed Lewisian gneisses overlying the Moine overthrust we often see the felspars shattered, while the quartz, yielding without apparent fracture,

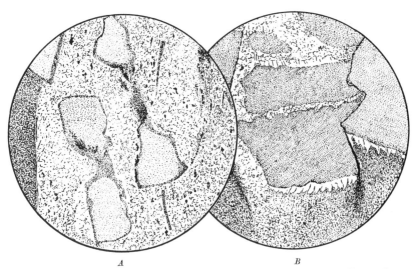

A   B

FIG. 76.—HIGHLY SHEARED CRINOIDAL LIMESTONE, Devonian, at Ilfracombe, North Devon ; × 25.

The direction of maximum pressure was right and left, and deformation has been effected largely by solution and recrystallization.

A. The effect of crushing in the general matrix has been almost obliterated by concurrent recrystallization, but there are still marks of some internal faulting. Ossicles of crinoids lying at right angles to the maximum pressure have been attenuated by solution.

B. Here distinct ossicles are seen interpenetrating, as a consequence of solution at the places where they were pressed together. The dissolved calcite has been deposited in places of relative relief from stress.

assumes curving shapes as if flowing round and between the fragments of felspar.

Again, crushed quartzites often present an appearance which shows that solution has played an important part in the breaking down. Relics of the original rock have outlines indicative of corrosion, and the crushed matrix has been more or less completely recrystallized (Fig. 75, *B*).

It is, however, as might be expected, the calcareous rocks that exhibit most clearly the effects of differential solubility as determined by differential stress. In particular, fragments of calcite organisms often show by their outline that material has been dissolved from those parts which bore the brunt of the pressure, to be redeposited on parts where there was relative relief [1] (Fig. 76). In a simple crystalline limestone, in which the constituent grains are all of the same character, this effect is much less in evidence, though we can often verify a tendency of the grains to have their longest axes in one plane. When limestones possess a good cleavage-structure, it has arisen mainly in the mechanical fashion described by Sorby,[2] viz. by fragments of flat or linear shape being rotated into positions more nearly at right angles to the direction of chief compression. This goes with a bodily deformation of the rock, and the effect is enhanced by the mechanical deformation of the individual grains by bending and 'gliding'. On the other hand, the free solubility of calcite (aided by some excess of carbonic acid) enables it to be transferred to considerable distances within a rock-mass under unequally distributed stress. When a bed of limestone is involved in a system of closely appressed folds, it may become entirely dissolved away in the middle limbs of the folds, the material being transferred to the crests and troughs, which were places of relative relief.[3]

MINERALOGICAL CHANGES WHICH ACCOMPANY CATACLASTIC STRUCTURES

The mechanical effects in dynamic metamorphism are often complicated, not only by recrystallization of some of the original constituent minerals, but also by chemical reactions, giving rise to *new minerals*. These new minerals, moreover, often have an important share in determining the characteristic structure of the resulting metamorphosed rocks. Being generated in a rock which is in a condition of unequal stress of more or less clearly defined type, they tend to take on a definite orientation ; and, since some of the commonest are minerals of flaky habit, they are chiefly responsible for the schistosity which often characterizes the extreme products of dynamic metamorphism. In some cases mineralogical changes provoked by stress make themselves felt in rocks which have not yet yielded sensibly to crushing (Fig. 77). More generally they proceed *pari passu* with cataclastic processes and recrystallization. The new minerals themselves do not often show the effects of crushing, because they possess considerable power of regeneration.

Since the mineralogical changes here are the same that we shall have to observe in the lowest grades of regional metamorphism, a

[1] Sorby, *Quart. Journ. Geol. Soc.*, vol. xxxv (1879), Proc., pp. 87–89.
[2] *Phil. Mag.* (4), vol. xi (1856), pp. 20–37.
[3] Sorby, *loc. cit.*

brief notice will suffice in this place. The crystalline rocks which are most susceptible to change are those of igneous origin and those which represent the results of high-grade thermal metamorphism. Many of their component minerals represent combinations which are unstable or only metastable at low temperatures.

A highly characteristic change is the sericitization of the potash-felspars. For example, the reduction of a crystal of orthoclase to a lenticular shape is often as much a chemical as a mechanical effect : the proof is seen in streaks of sericitic mica at the corners of the ' eye '.

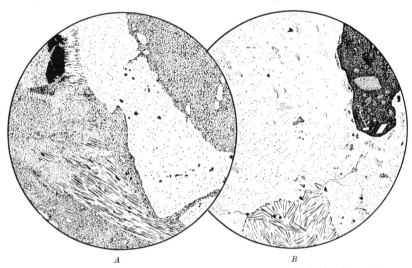

A   B

FIG. 77.—EARLY STAGES OF DYNAMIC METAMORPHISM, Sutherland ; × 25.

Here cataclastic effects are little in evidence, but mineralogical changes have already begun.
   A. Lewisian Gneiss, near Inchnadamph : showing sericitization of felspar in progress.
   B. Borolanite, near Loch Borolan. The orthoclase is intact, but the nepheline is completely replaced by an aggregate of white mica. Abundant granules of sphene have been developed in the melanite.

Where a felspathic rock is intersected by surfaces of discontinuity, along which slipping has taken place, these surfaces are marked out by thin micaceous films. In a more advanced stage of crushing and rolling out the felspar may be wholly destroyed, and the final product is a sericitic schist (Fig. 79, A). Nepheline also, in dynamic metamorphism, is replaced by white mica, instead of the usual zeolites. We have no knowledge of the nature of the mica in this case, and the chemistry of the change remains obscure. Analyses show that few nephelines contain, in themselves, enough potash to make muscovite, and it is noteworthy that the sericitization of nepheline may be complete while orthoclase in the same rock is still quite fresh (Fig.

77, *B*). It is not impossible that this, like some other common changes of degradation in pyrogenetic minerals, is referable to some stage in the original cooling down of the igneous rock, and in that case lies outside the province of simple dynamic metamorphism, though it may still be promoted by stress.

An important case is the saussuritization of the plagioclase felspars, which has often been attributed to the influence of stress, and it is necessary therefore to make some remarks on the question. In saussuritization the sodic and calcic components of an intermediate

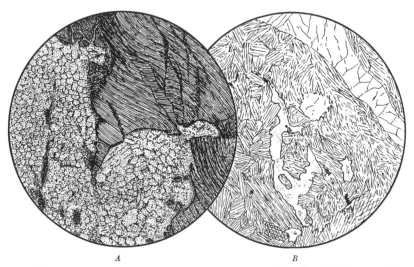

A   *B*

FIG. 78.—SAUSSURITIZATION IN BASIC PLUTONIC ROCKS, New South Wales; × 25.

A.   Saussurite-Eucrite, Moonbi.   The felspar here was anorthite, and has been simply replaced by zoisite in a granular aggregate.   The diallage is bent and fissured, but not otherwise altered.
B.   Saussurite-Actinolite-Gabbro, Bingara.   This rock has suffered total reconstruction, as well as shearing.   It is now composed of blade-like actinolite and clear albite together with prehnite, which is seen making a conspicuous vein.
These rocks have been described by Benson, *Proc. Linn. Soc. N.S.W.*, vol. xxxviii (1914), pp. 684 and 687.

plagioclase became divorced.   The former separates out as clear granules of albite, while the latter gives rise to some new aluminosilicate of lime.   Most usual are zoisite and epidote, but prehnite is also found and more rarely grossularite.   Saussuritization of the felspar is commonly accompanied by uralitization of the augite of the rock, and actinolite may enter as a constituent of the saussuritic aggregate itself (Fig. 78).   The epidote minerals may be formed at ordinary temperatures, but this can scarcely be true of prehnite and actinolite, still less of grossularite.   The observations of numerous petrologists indeed lead to the conclusion that saussuritization and uralitization alike belong to a late stage in the cooling down of an igneous rock. Shearing stress, if it operates, is doubtless to be reckoned as a factor favouring these changes; but it is noteworthy that in such an area as the Lizard district of Cornwall,[1] where the gabbros, etc., often show saussuritization and often cataclastic effects, no relation is clearly discernible between the two kinds of phenomena.

The characteristic transformation of the pyroxenes in simple dynamic metamorphism is not uralitization but chloritization. Chlorite is also the most conspicuous product of the breaking down

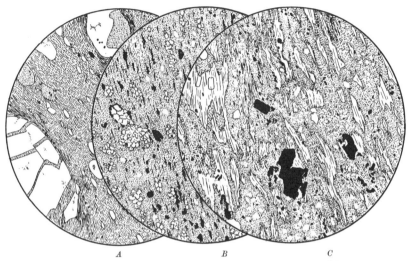

A B C

FIG. 79.—ADVANCED DYNAMIC METAMORPHISM; × 25.

A, Schistose Quartz-porphyry with sericitized ground-mass, Llanberis. The porphyritic quartz-grains have been broken, and the parts slightly separated by ' stretching '.

B. Crushed and Sheared Dolerite, Isle of Man; with abundant secondary carbonates.

C. Crushed and Schistose Dolerite, Garth, near Portmadoc. Here abundant chlorite has been formed at the expense of the augite: other new products are quartz and leucoxene.

of hornblende and (together with sericite) of biotite. The pyrogenetic minerals named yield, of course, other products in addition—quartz, carbonates (calcite, dolomite, chalybite), epidote, iron-oxides (usually magnetite), etc. (Fig. 79, B and C)—but the final result of dynamic metamorphism in an ordinary basic igneous rock may be styled a chloritic schist. The olivine of a crushed and sheared rock gives rise to serpentine, magnetite, and carbonates. It is not necessary to discuss the behaviour in dynamic metamorphism of all the less important constituents of igneous rocks and crystalline schists. The common red garnet yields chlorite and magnetite; the melanite of alkaline

[1] Flett, *Geology of the Lizard* (*Mem. Geol. Sur.*, 1912), pp. 88–90.

rocks sphene and magnetite. Sphene (in the form of ' leucoxene ') is formed also at the expense of titaniferous iron-ores.

Regarding these various mineralogical transformations collectively we see that they are in the main the same kind of changes that high-temperature minerals undergo spontaneously at low temperatures in contact with the atmosphere, i.e. without the intervention of stress. Here, however, they are effected more rapidly and completely. Just as a state of stress increases the solubility of a body, so it increases also its chemical activity.

Again, the changes noted involve a large element of metasomatism. Many of the new products cannot be formed without taking up a considerable amount of water, or oxygen, or carbon dioxide. In other cases there is a loss of material, as in the partial elimination of potash in the sericitization of orthoclase and the frequent removal of lime, doubtless as carbonate, from basic igneous rocks. The addition of material necessary to form such minerals as sericite, chlorite, and calcite in such abundance can come only from an atmospheric source. Reactions of the kind indicated are possible, therefore, only in comparatively shallow levels of the earth's crust, which are not beyond at least some remote and indirect communication with the surface. In so far as this is true of particular examples, these lie outside the strict limits of metamorphism as we have defined it ; but the same reactions proceed at greater depths in so far as the normally limited supply of water permits. In the most typical mylonites there has been little or nothing of sericitization and analogous changes, and these rocks have accordingly no sensible schistosity.

CHAPTER XII

REGIONAL METAMORPHISM : GENERAL CONSIDERATIONS

Distribution of Regional Metamorphism—Grades of Regional Metamorphism —The Scottish Highlands as a Typical Area.

DISTRIBUTION OF REGIONAL METAMORPHISM

THE epithet *regional*, as applied to metamorphism of the most general kind, betokens primarily the wide extension which characterizes effects of this class, but it carries also a further implication. While very different grades of metamorphism may be exhibited within the limits of one natural region, the variation is of a gradual and orderly kind, and is related to the region as a whole, rather than to such incidents as igneous intrusion. In the ideal case there is a steady advance in the grade of metamorphism from the borders of the region to the central tract. Often this is not to be verified, because only a part, usually the interior part, of the whole region is exposed to view. Again, the original regularity of disposition may be obscured by the effects of subsequent great dislocations. None the less it may be laid down that, not only the extent of area involved, but the *distribution of metamorphism* within that area is on a regional scale, pointing clearly to causes of a large order.

A discussion of ultimate causes is not within the scope of the present inquiry. In general, mountain-building and regional metamorphism come about as the logical consequences of prolonged sedimentation in a slowly subsiding geosynclinal basin. As the accumulation of sediments proceeds, the temperature-distribution within the globe gradually adjusts itself to the increased thickness of that part of the earth's crust. The isothermal surfaces rise through the rocks, or, what is the same thing in effect, the rocks become depressed into a domain of higher temperature. The rise of temperature affects all that region of the crust down to an indefinite depth, and there is in consequence a very great expansion of volume. In the shallower levels of the crust expansion can take effect only in the upward direction ; and this is true in a measure at greater depths, until the effective rigidity of the rocks is destroyed by very high temperatures, or, in other words, down to what American geologists name the 'zone of flowage'. In the upper crust, thus subjected to powerful lateral pressure, great shearing stresses are necessarily developed, and these play, together with the elevated temperature, an essential part in the metamorphism of the region. Folding of an extreme type, affording some measure of relief, is a natural accompaniment of regional metamorphism. This bodily yielding normally precedes internal rearrangements of a more intimate kind, as is proved by the relation of cleavage and foliation to the main folding.

In so far as this may seem to imply that the mechanical forces concerned are generated within the region itself, it is only a partial presentation of the matter. Crustal stress taking the form of lateral pressure may be propagated through long distances ; and to consider a circumscribed area merely by itself is to disregard the part which it sustains in relation to the larger events of earth-history. In many regions of metamorphism a marked tendency to uniformity in the direction of strike of the structures, large and small, points to a stress-distribution of a large order, often transcending the limits of the region itself. While it is impossible for the rocks throughout an extensive tract to become greatly heated without bringing in also the operation of shearing stress, the converse is by no means true. To pursue the subject further would carry us beyond the limits laid down in the scheme of this work.

There is another consideration which it is proper to notice in this place, though a discussion of it is deferred to a future chapter. The conjunction of high temperature with intense shearing stress, which is the essential condition of regional metamorphism, may also in certain circumstances be realized as a *local* incident, the requisite heat being not conducted from below but generated in place. This gives rise to metamorphism which, being controlled by the same physical conditions, is of the regional type, although it may be very narrowly localized.

In simple thermal metamorphism, as illustrated in an ordinary 'aureole', heat is carried up from the deeper levels of the globe by an intruded igneous magma. In regional metamorphism, as here conceived, we are to picture rather a *direct* invasion of the earth's internal store of heat. The former factor, however, must be credited also with a share in the result. The regional uprise of the isogeotherms is in general accompanied by intrusion of molten magmas, and such intrusion is likely to be localized in the hottest places. Plutonic rocks therefore are very frequently found associated with crystalline schists, and especially with those in a high grade of metamorphism. Where a tract of regional metamorphism is of limited extent, it sometimes borders an important intrusion, or train of intrusions, in such fashion that it may be described as an aureole. The Stavanger district, as described by Goldschmidt,[1] is a good example. In the Scottish Highlands Barrow considered the high temperatures to have been due to the invasion of the igneous magmas, though the intrusions actually seen are by no means confined to the most highly metamorphosed country-rocks. In an extensive region such as this the grade of metamorphism does not at least show any close relation to the distribution of plutonic rocks, and it seems more natural to regard intrusion as a frequent incident of regional metamorphism rather than as its prime cause.[2]

A more immediate connexion between intrusion and metamorphism is, however, indicated in certain districts, where intrusive granites assume a ramifying habit, and have permeated the country-rocks in a peculiarly intimate manner. Such is notably the case in parts of Central Sutherland,[3] where the rocks so injected show a rapid rise in their metamorphism (from the garnet grade to the sillimanite grade). According to Read, the more highly metamorphosed rocks (with their permeating granite) may actually overlie the less metamorphosed. Even here, however, the peculiarly intimate type of intrusion proves the country-rocks to have been already raised to a high temperature before the invasion of the granite magma.

The intricate fashion in which the intrusions in such a district have penetrated the solid rocks implies a degree of fluidity realized only in magmas rich in the volatile fluxes. It bears witness also to intrusion effected under very great stress. Since igneous rocks so intruded have not only consolidated under quite special mechanical conditions, but have also shared in so much of the metamorphism as post-dates their intrusion, they must not be omitted from our study of regional metamorphism.

One important consideration has not yet been noticed, viz. the energy-change involved in regional metamorphism. The reactions in thermal metamorphism are essentially endothermic, those in simple dynamic metamorphism exothermic, the quantity of heat absorbed in the one case and liberated in the other being very considerable. In the most general kind of metamorphism with which we are now con-

---

[1] *Vidensk. Skr.*, 1920, No. 10 (1921).

[2] Compare Horne and Greenly, *Quart. Journ. Geol. Soc.*, vol. lii (1896), p. 645 ; Clough, *Geol. of Cowal* (*Mem. Geol. Sur. Scot.*, 1897), p. 91 ; Elles and Tilley, *Trans. Roy. Soc. Edin.*, vol. lvi (1930), p. 622.

[3] Read, *Geology of Central Sutherland* (*Mem. Geol. Sur. Scot.*, 1931), p. 146.

cerned, the thermal and dynamic factors enter together and, as we must suppose, with opposite effects as regards energy-change. The relations are not of the kind which would allow us to evaluate the two effects separately and subtract the one quantity from the other, but it is clear that the case of regional metamorphism is in a general sense intermediate between the two special cases of thermal and dynamic metamorphism. It will be shown in what follows that the dynamic factor is the dominant one at low temperatures and the thermal at high temperatures. We may infer with confidence that the reactions in regional metamorphism are in the lower grades exothermic in decreasing measure and in the higher grades endothermic in increasing measure. In the medium grades the energy-change, positive or negative, will be relatively small. Regional metamorphism, therefore, if of no very high grade, may attain a wide extension without making any great demand upon the earth's internal store of energy. We find also vast tracts of country in which the metamorphism is throughout of a very high grade ; but in this case the main bulk of the rocks is commonly of igneous origin, representing molten magmas which brought with them a large store of heat.

GRADES OF REGIONAL METAMORPHISM

We have seen that the physical conditions controlling metamorphism in the most general case are determined by three factors : viz. temperature, pressure, and shearing stress. These three are *prima facie* independent of one another, and a function of three independent variables presents triply infinite possibilities. It is clear that if the assumption here made were true without qualification— i.e. if any temperature, high or low, could occur in conjunction with any measure of shearing stress, great or small, and again with any magnitude of hydrostatic pressure—the complexity would be such as to render any systematic study of metamorphism impossible. When, however, we turn from this ideal presentation to the actual conditions as normally realized, we shall see that the problem is greatly simplified. The temperature at any particular place is not an isolated or arbitrary thing, but is part of a temperature-distribution affecting a certain area and thickness of the earth's crust. So too there is a distribution of pressure and of shearing stress. In so far as the three have resulted from common causes of a large order, we must expect a certain general correspondence among them ; and in the ordinary case of regional metamorphism on an extensive scale such correspondence is the rule. It is a rule subject to exceptions, but, allowing for these, it is found possible in practice to arrange most regionally metamorphosed rocks, of a given initial type of composition, in a single linear series, from less to more highly metamorphosed. We will consider more particularly how this comes to hold good, at least as a first approximation to the truth, and we may conveniently begin by examining the scheme devised by Grubenmann.[1]

In Grubenmann's classification of the crystalline schists the conditions which control metamorphism are conceived as functions of a single variable, which is, ostensibly at least, *depth within the earth's crust*. After division into twelve groups on the basis of bulk chemical composition, the rocks in each group are ranged in three zones of depth, distinguished, in descending order, by the prefixes *epi-*, *meso-*, and *cata-*. The dividing lines are conventional, for the transition is gradual from one zone to the next ; and it is recognized also that the divisions do not necessarily connote the same depths in the different groups. The characteristics assigned to the several zones are shown in the table. The system elaborated on this basis has been largely

	Temperature	Pressure	Stress
Epi-zone . . .	moderate	low	strong
Meso-zone . . .	higher	higher	very strong
Cata-zone . . .	very high	very high	weaker

justified in practice, in the sense that most types of crystalline schists readily find their places in the scheme. Certain anomalies are revealed, for the simplication implied in the adoption of a single variable factor can be gained only at the expense of some discrepancies. The theoretical basis itself, however, is open to criticism, and it is evident that a given rock must often be assigned to its classificatory position upon petrographical evidence rather than from any actual knowledge of the depth at which it has been metamorphosed.

Pressure necessarily increases downward, and the law of increase is sufficiently simple ; for, beyond a moderate depth, the pressure cannot permanently differ much from that due to the depth of ' cover '. The case of temperature is less simple. Our limited knowledge suggests that, even beneath an undisturbed region, the downward rate of increase may vary considerably, and metamorphism implies a departure from that relatively ordered status. Further complication is introduced by the intrusion of molten magma, bringing heat from lower levels. It may perhaps be reasonably assumed that such local variations of temperature-gradient are lost in the general rise of the

[1] *Die Kristallinen Schiefer* (1906) ; 2nd ed. (1910).

M.—7

isogeotherms which goes with regional metamorphism upon an extensive scale. It follows that, under ordinary conditions at least, temperature increases with depth, and—what is more to the purpose—that temperature and pressure increase together.

The intensity of shearing stress in a rock undergoing metamorphism has not primarily any relation to depth or to temperature and pressure, but depends upon the nature and magnitude of the external forces to which the rock is subjected. The material of the rock, however, is not capable of sustaining shearing stress beyond a certain limit, but will eventually yield in one way or another, so affording relief. The stress may be continually renewed, so as to rise to the limit so fixed, but cannot surpass it. For a given type of rock, under given conditions of temperature and pressure, there is then a limiting value of possible shearing stress set by the nature of the rock itself and the given conditions. Resistance to shearing is a complex property, only in part mechanical. We have seen that yielding may be effected, not only by fracture, gliding, etc., but also by solution and chemical changes ; and that these latter factors become increasingly important in the higher grades, molecular and atomic rearrangements, being powerfully promoted by rise of temperature. In this way the resistance of the rock to shearing is continually weakened, and may ultimately become of little account. We see then that—not the shearing stress—but the *maximum possible shearing stress* is a function of temperature, diminishing with rise of temperature. This is true for any particular kind of rock : the maximum shearing stress at a given temperature will be different in rocks of different composition.

All this is in accordance with the petrographical evidence. It is in the lower grades of metamorphism that shearing stress, as a factor controlling the mineralogical constitution of the rocks, is most potent. In the medium grades it still exerts a powerful influence, as is seen on comparing the rocks with corresponding terms among thermally metamorphosed products, but this influence is gradually waning. At the highest temperatures of metamorphism the distinctive stress-minerals are no longer found. The rocks now differ little, if at all, in mineralogical constitution from those in the innermost ring of some thermally metamorphic aureoles. Such differences as are observed are referable, not to shearing stress, but to very great hydrostatic pressure.

At low temperatures the possible shearing stress has a wide range, from zero to the maximum value, but with rising temperature the range is continually contracted. It is for this reason that a classificatory scheme based on a single variable factor, and therefore giving in each lithological group of rocks a single linear arrangement, encounters difficulties especially in the lower grades of metamorphism. For theoretical discussion at least, we will propose instead a scheme involving two independent variables. They are firstly temperature with pressure, these being assumed to vary together, and secondly shearing stress. The relations can then be represented by the ideal diagram, Fig. 80, which illustrates the various points already noted. The curve shows the maximum shearing stress, in rocks of a given nature, as a function of temperature. The area within the curve

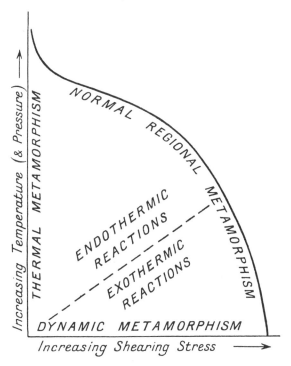

FIG. 80.—IDEAL DIAGRAM OF THE CONDITIONS CONTROLLING METAMORPHISM.

The curve shows maximum shearing stress as a function of temperature. The area enclosed by it shows the range of variation of conditions in the general case. Three special cases are indicated :
(i) Thermal Metamorphism (no shearing stress).
(ii) Dynamic Metamorphism (no rise of temperature).
(iii) Normal Regional Metamorphism (shearing stress maintained continually at its maximum value).

corresponds with metamorphism of the most general kind, while purely thermal and purely dynamic metamorphism appear as particular cases. They are the limiting cases, in which either shearing stress or rise of temperature is negligible.

The diagram, however, clearly indicates another limiting case, viz. that in which *shearing stress is constantly maintained at its maximum value* with the changing temperature. The actual stress, being the same as the maximum possible stress, is then a function of the temperature ; and, if we assume pressure also to vary with temperature, the conditions controlling metamorphism come to depend upon a single variable. This variable is temperature, though doubtless it generally corresponds with depth of cover also. Now it can be safely asserted that the condition here premised is very widely satisfied in regional metamorphism upon any extensive scale ; the external forces being such that, during the time of metamorphism, shearing stress is constantly maintained at or about its maximum value at each temperature. So commonly is this realized, that we are warranted in regarding it as the *normal case of regional metamorphism*, and in setting apart as exceptional those cases in which shearing stress has been notably less than that which the rocks were capable of sustaining. The proof of this is to be found in a systematic study of the rocks themselves, and is indeed implicit in such a classificatory scheme as that of Grubenmann. That the great majority of crystalline schists, within any one group defined by total composition, are found to fall into a single linear arrangement, admits of no other explanation. It also appears that, in so far as regional metamorphism can be viewed as dependent upon a single factor, that factor is temperature, since, in addition to its direct influence, it normally determines the other principal factor, shearing stress.[1]

While the normal relation of shearing stress to temperature is of the inverse kind, the one diminishing as the other increases, pressure, on the other hand, may be presumed to vary with temperature directly. This at least will be approximately true in the case, as conceived by Grubenmann, in which temperature and pressure are alike direct functions of depth. It is otherwise in simple thermal metamorphism, in which heat is introduced into the shallower levels of the crust by a molten magma. There, in the inner part of an aureole, a very high temperature may be reached at no great depth below the surface, and therefore under no great pressure. It follows that, comparing like temperature-grades of thermal and of regional metamorphism, the pressure must generally be very considerably greater in the latter than in the former. We must therefore be prepared to recognize more clearly the influence of pressure, according to the ' Volume Law ', on the mineralogical constitution of the rocks, and especially so in the higher grades of metamorphism.

[1] See Harker, ' Normal Regional Metamorphism ', *Fennia*, 50, No. 36 (1928).

THE SCOTTISH HIGHLANDS AS A TYPICAL AREA

It will be instructive to discuss the matter in terms somewhat more precise, and we shall take for illustration the Highlands of Scotland, which will also furnish much of the material for succeeding chapters. Here erosion has exposed rocks in every grade of metamorphism. The south-eastern section of the country, bounded by the important dislocations of the Highland Border and the Great Glen, constitutes a natural unit, in which the relations are clearly displayed. Starting from low-grade slates and more or less schistose grits at the border, it is possible to follow the advance of metamorphism step by step to a central area of very highly metamorphosed sediments. Among these especially intrusions of granite, of the age of the metamorphism, are of frequent occurrence. Their most usual habit is that of sheets of no great dimensions, running out into innumerable tongues and veins, and in certain ' permeation-areas ' pervading the country-rocks in an intricate plexus of veinlets. It is likely that the granite injections exposed are offshoots from much larger masses concealed below, and Barrow ascribed the thermal element of the metamorphism to these subterranean intrusions. A somewhat different view has been adopted here, but the difference is not of a kind to affect materially the question at issue.

In this model area the dome-like uprise of the isothermal surfaces is capable of exact demonstration. Barrow [1] has shown that, having regard to argillaceous sediments of ordinary type, advancing metamorphism is marked by the appearance in turn of particular index-minerals. By means of these he was able to distinguish a number of well-marked successive *zones of metamorphism,* and to lay down their boundaries on the map. Tilley,[2] carrying on the mapping of Barrow's zones, has found it possible to fix the boundary-lines with remarkable precision. We may reasonably suppose that advancing metamorphism corresponds here with increasing depth of cover prior to the general erosion ; but it is clear that the zones recognized are primarily temperature zones, and that the bounding lines, marking the first appearance of particular minerals, are, in a definite sense, *isothermal lines.* They are comparable with those that might be laid down in a thermally metamorphosed aureole round a large igneous intrusion (p. 24), but the temperature distribution is now of a regional kind.

The principal index-minerals are stress-minerals, and, where not

[1] *Proc. Geol. Assoc.,* vol. xxiii (1912), pp. 274–90.

[2] *Quart. Journ. Geol. Soc.,* vol. lxxxi (1925), pp. 100–10. See also Elles and Tilley, *Trans. Roy. Soc. Edin.,* vol. lvi (1930), pp. 621–46.

complicated by later purely thermal events, the metamorphism is everywhere of the kind involving the stress-factor. We have seen that the development of shearing stress is, in part at least, a direct consequence of the regional rise of temperature : the distribution of more and less intense stress through the region makes a distinct question. The petrographical evidence argues a definite stress-distribution, such that the influence of stress has been least felt in the most highly metamorphosed rocks. We may imagine then a series of isodynamic lines laid down on the map to indicate different measures of shearing stress. If these lines crossed the isothermals, we should find in the constitution of the metamorphosed rocks that complexity or confusion which must result from a dependence on two independent variables. Where well-characterized and sharply defined zones of metamorphism can be made out, we have proof that the stress-distribution is conformable with, and has been controlled by, the temperature-distribution. The *isodynamics are also isothermals*.[1]

The orderly arrangement described prevails over the greater part of the Highlands, and is very typically exhibited in the tract lying between the Highland Border and the latitude of the Dee. It illustrates ' normal regional metamorphism ' in the sense already defined. Beyond this tract lies an area, including a large part of Aberdeenshire and the north-eastern portion of Banffshire, which cannot be brought under so simple a scheme. Here is exemplified the exceptional case, in which shearing stress was not, during the metamorphism, constantly maintained at its maximum possible intensity at the actual temperatures. The precise delimitation of successive zones of metamorphism becomes more problematical, though lower and higher grades are still to be recognized in a general way. The rocks often show, from the point of view here adopted, decided anomalies in their mineralogical constitution, distinctive stress-minerals being found side by side with others which are normally associated with simple thermal metamorphism.[2] The deficiency of the stress-factor is made manifest. It is not difficult in this instance to suggest a possible explanation, having regard to the unilateral character of the crust-movements and forces concerned. The area in question, situated on the north side of the dome-like uprise, which corresponded doubtless with an extensive elevated tract, was presumably protected in some degree against the thrust from the south or south-east.

---

[1] Isograds of Tilley, *Geol. Mag.*, vol. lxi (1924), p. 169.

[2] It is, of course, necessary to discriminate between this case and that in which normal crystalline schists have become modified by thermal metamorphism of much later date. This will be considered in a subsequent chapter.

The rocks selected for tracing the successive zones of metamorphism are ordinary argillaceous sediments, practically free from calcareous admixture. Such is the nature of numerous distinct members of the Dalradian and Moine series, and the type recurs as frequent intercalations in other members which are mainly arenaceous or calcareous. For this reason, and also as a result of much repetition by folding, the observer seldom has far to seek for rocks carrying the significant index-minerals. There is another reason why these rocks are especially suitable for the study of progressive metamorphism. In ordinary argillaceous sediments metamorphism starts from a state of chemical equilibrium, and shows therefore from the beginning a steady gradual progress (p. 48). For some other classes of rocks, e.g. semi-calcareous sediments, a scheme of zones of metamorphism, marked by appropriate index-minerals, is doubtless practicable, though it has not yet been so completely established. A further step should be a correlation between such scheme and the standard set by Barrow ; and the correspondence may be imperfect as regards the lowest grades of metamorphism. In this connexion it should be observed that the maximum possible shearing stress at a given temperature may be much greater in one type of rock than in another, and that this difference is not necessarily in accordance with the relative rigidity of the rocks as observed at ordinary temperature. In limestones especially, at temperatures at all elevated, local solution and recrystallization will enable the rock to yield with comparative freedom to slow deformation, all the shearing stress actually sustained must be considerably less intense than that experienced by an argillaceous rock at like temperature. The same is true of lime-silicate rocks when they occur as nodular masses embedded in limestone or as intercalated bands of no great thickness.

It is much to be desired that the zonal method of mapping, initiated by Barrow, should be confirmed and further elaborated by its application in other countries. Its value has already been proved in the Caledonian chain of Norway, especially by the investigations of Goldschmidt [1] and of T. Vogt,[2] who seem to be unacquainted with Barrow's previous researches ; and it has been applied with success in New Zealand [3] and other areas. Alpine geologists, while recognizing of course different grades of regional metamorphism, have hitherto

[1] ' Geologisch-petrographische Studien im Hochgebirge des Südlichen Norwegens, III (Trondhjem), *Vidensk. Skr.* (1915), and V (Stavanger), *ibid.* (1921).

[2] ' Sulitelmafeltets Geologi og Petrografi,' *Norg. Geol. Unders.*, No. 121 (1927).

[3] Turner, *Geol. Mag.*, vol. lxxv (1938), pp. 160–74, summarizing the results of earlier papers.

shown little disposition to adopt the conception of definable zones marked by the incoming of specified new minerals. In a region greatly complicated in structure by successive crust-movements such zones may not be easily established. The sequence of mineralogical changes with advancing metamorphism must, like a stratigraphical succession, be worked out in the first instance in areas where the original relations have not been seriously disturbed. Faulting or overthrusting of later date may bring into juxtaposition rocks in different grades of metamorphism, and this has happened in some parts of the Highlands. This being so, a knowledge of the normal metamorphic sequence may afford useful assistance in unravelling the tectonics of the country.

For the most part, metamorphism in the Highlands can be studied quite independently of the difficult and disputable questions of stratigraphy and tectonics. The Dalradian series includes examples of all the common types of sediments, and each is to be seen in very different grades of metamorphism. The Moine series, which occupies so large a tract beyond the Great Glen, is less varied in lithological constitution. Most of the ground has been covered by the *Geological Survey of Scotland*, and much valuable information is contained in the Memoirs explanatory of the several published sheets of the Geological Map. Of contributions from other quarters a few, which possess general interest, are mentioned below [1] : others, of more special application, will be cited in their places.

The rocks of the Lewisian plutonic complex, with associated metamorphosed sediments, have been most closely studied in the North-Western section of the Highlands, beyond the Moine overthrust. [2] According to the view most commonly entertained, these rocks are of greater antiquity than the sediments in the interior of the Highlands, and belong to an earlier period of metamorphism. The question is not revelant to our purpose. It is sufficient to observe that

[1] Barrow, *Quart. Journ. Geol. Soc.*, vol. xlix (1893), pp. 330–56 (Glenesk); Hill, *ibid.*, vol. lv (1899), pp. 470–92 (Loch Awe); Cunningham-Craig, *ibid.*. vol. lx (1904), pp. 10–28 (Loch Lomond); Barrow, *ibid.*, pp. 400–46 (Moine Gneisses); Bailey, *Geol. Mag.*, vol. lx (1923), pp. 317–31 (South-West Highlands); Read, *Trans. Roy. Soc. Edin.*, vol. lv (1927), pp. 317–53 (Deeside); Horne and Greenly, *Quart. Journ. Geol. Soc.*, vol. lii (1896), pp. 638–48 (foliated) granites); Flett, *Summ. Progr. Geol. Sur.* for 1905 (1906), pp. 155–67 (inliers of Lewisian); Phillips, *Quart. Journ. Geol. Soc.*, vol. lxxxiii (1927), pp. 622–51 (Shetland Isles); Hutchison, *Trans. Roy. Soc. Edin.*, vol. lvii (1933), pp. 557–92 (Deeside Limestone); Wiseman, *Quart. Journ. Geol. Soc.*, vol. xc (1934), pp. 354–416 (epidiorite sills); Read, *ibid.*, pp. 637–88 (Unst, Shetland Isles).

[2] *The Geological Structure of the North-West Highlands of Scotland* (*Mem. Geol. Sur. Gt. Brit.*, 1907).

erosion has there exposed deeper levels of the crust, where plutonic rocks largely predominate. Lewisian rocks constitute also almost the whole of the Outer Hebrides [1] and the smaller islands Iona,[2] Coll,[3] and Tiree. Of more isolated areas, Anglesey [4] is the one which has received most attention.

All these are rocks of high geological antiquity. The same is probably true of certain detached areas of metamorphic rocks, of which the relations are less clearly exhibited, the Manx Slates,[5] the Start district of South Devon,[6] and the Lizard in Cornwall.[7] We have, however, to recognize in the Highlands metamorphic effects related to the Caledonian crust-movements and in Cornwall and Devon a rather widespread regional metamorphism, though of no high grade, which must be referred to a Hercynian time. The belief that crystalline schists belong necessarily to the earliest chapter of geological history is a relic of the Wernerian geology, which finds no support in the results of later investigations. On the west coast of Norway the crystalline schists include some which apparently represent Palaeozoic sediments, and part of the metamorphism has been assigned to a Caledonian date. Saxon geologists recognize an important metamorphism of Hercynian age in the Erzgebirge and Granulitgebirge. In the Alps crystalline schists of high grade have been made from Mesozoic sediments. These and other European areas of regional metamorphism have a copious literature, which space does not allow us to notice.

[1] Jehu and Craig, *Trans. Roy. Soc. Edin.*, vol. liii (1923), pp. 419–41 ; (1925) pp. 615–41 ; vol. liv (1926), pp. 467–89 ; vol. lv (1927), pp. 457–88 ; vol. lvii (1934), pp. 839–74.

[2] Jehu, *ibid.*, vol. liii (1922), pp. 165–87 ; *Geol. of Staffa, Iona, etc.* (*Mem. Geol. Sur. Scot.*, 1925).

[3] *Geol. of Ardnamurchan, etc.* (*Mem. Geol. Sur. Scot.* (1930)), pp. 4–27.

[4] Greenly, *The Geology of Anglesey*, vol. i (*Mem. Geol. Sur. Eng. and Wales*, 1919) ; also *Quart. Journ. Geol. Soc.*, vol. lxxix (1923), pp. 334–50.

[5] Lamplugh, *The Geology of the Isle of Man* (*Mem. Geol. Sur. U.K.*, 1903).

[6] Tilley, *Quart. Journ. Geol. Soc.*, vol. lxxix (1923), pp. 172–203.

[7] *Geol. of the Lizard* (*Mem. Geol. Sur.*, 1912).

CHAPTER XIII

STRUCTURES OF CRYSTALLINE SCHISTS

Ruling Physical Conditions in Regional Metamorphism—Obliteration of Earlier Structures—Crystal-Growth in a Solid under Stress—Influence of Specific Properties of Minerals—Characteristic Types of Structure—Nature and Origin of Foliation.

RULING PHYSICAL CONDITIONS IN REGIONAL METAMORPHISM

THE crystalline schists and allied products of regional metamorphism have acquired their distinctive characters as the result of reconstitution of their substance, which took place in response to continued rise of temperature, but was further determined by definite mechanical conditions. In specifying these conditions two distinct factors are to be noted, which have not always been clearly discriminated. Firstly, the reconstitution was effected by means of solution and recrystallization, of a very limited and local kind, at many isolated points in succession, in such a way that the rock as a whole *remained a solid body* throughout the process. Secondly, the solid rock was subjected during the metamorphism, not only to hydrostatic pressure, but to more or less *powerful shearing stress*. The main features of crystal-growth in the solid—without the added condition of shearing stress—have been sufficiently discussed in an earlier chapter. The *crystalloblastic structure*, as there characterized, and the various special types which have received such names as porphyroblastic, poeciloblastic, etc., are all reproduced in the rocks now under discussion, but are modified in their development, in greater or less degree, by the influence of the stress-factor (Fig. 81). The properties of a rock which is in a state of shearing stress are, from a mechanical point of view, different in different directions. It is in a medium thus *mechanically anisotropic* that the growing crystals must exert their innate force of crystallization, and the direction in which they push their growth is more or less influenced by this condition. There results a certain *directional* element in the crystalloblastic structures of the rocks, which is normally lacking in simple thermal metamorphism. It thus becomes the special mark of regionally metamorphosed rocks as a class, though in particular types it may be little apparent.

190

OBLITERATION OF EARLIER STRUCTURES

Before examining this matter more closely, it will be convenient to make here some remarks concerning *residual structures*. With continued rise of temperature the original constituent minerals of a rock, one after another according to their nature, yield to recrystallization or are drawn into the sphere of chemical reactions. Very few survive to even a medium grade of metamorphism, and they are merely accessory minerals, such as zircon, of no importance in this connexion. With the remaking of the main substance of the rock

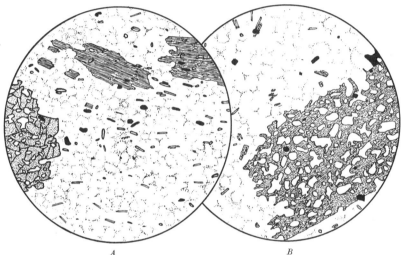

FIG. 81.—PORPHYROBLASTIC AND POECILOBLASTIC STRUCTURES, in quartzite, Glen Roy, Inverness-shire ;   × 18.

A. Porphyroblasts of biotite and garnet, the latter enclosing quartz in poeciloblastic fashion.
B. A large porphyroblast of green hornblende showing typical poeciloblastic or sieve-structure.

must disappear also all original structures other than those of a large order. Even these, such as pebbles or pronounced alternations of bedding, lose something of their sharpness. The dynamic element in regional metamorphism has its part here. Internal differential movement, often in the lower grades of a discontinuous kind, promotes the breaking down of structures ; and in the setting up of a new directional structure in the rock all pre-existing relations on a small scale are superseded.

There is, however, another kind of residual structures to be noticed, viz. those which, originating in a lower grade of metamorphism, still persist, or are not wholly obliterated, in a more advanced grade. Of special interest are *pseudomorphs* of a higher-grade mineral after one proper to a lower grade, but these are not often to be observed. The value of these to the petrologist is that they afford direct evidence of the derivation of one mineral at the expense of another. It is reasonable to suppose that when, with advancing metamorphism, one mineral is formed mainly from the substance of a pre-existing one, the new mineral is likely to take at first the shape of the old one (Fig. 82). So too, when a certain mineral breaks up, yielding two new products,

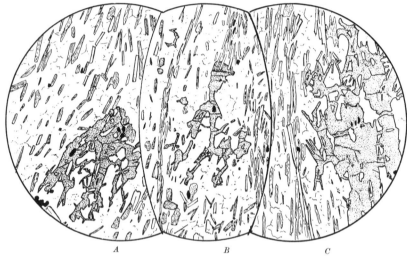

FIG. 82.—GARNET PSEUDOMORPHOUS AFTER CHLORITE, Balquhidder, Perthshire ; × 23.

The minerals shown are chlorite, muscovite, garnet, and quartz. The garnet preserves in many places the shapes of the flakes of chlorite from which it has been largely formed.

we might look for a composite pseudomorph, in which the two would be intimately associated in poeciblastic or diablastic fashion. The fact is that such structures, when produced, are very quickly obliterated by continued reconstruction of the rock with further rise of temperature. Most of the high-grade minerals are strong in their force of crystallization, and hasten to take on idioblastic shape (Fig. 83). A pseudomorph of this kind therefore is preserved only when the transformation which it records belongs to the extreme stage of metamorphism reached in the given rock : thus pseudomorphs of garnet after chlorite will be found only on the outer edge of the garnet zone. The kind of pseudomorph which is most frequently found, both in crystalline schists and in thermally metamorphosed rocks, has a different origin and significance. It is related to the decline of temperature which follows upon metamorphism proper, marked often by mineralogical changes of the nature of ' retrograde metamorphism '. Here accordingly we find a higher mineral replaced by a lower ; and, since there is no longer that continual reconstitution of the rock-substance which went with constantly rising temperature, more or less perfect pseudomorphs are now the rule.

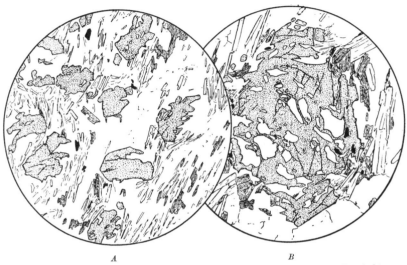

A   B

FIG. 83.—DEVELOPMENT OF GARNET in Mica-schists, near Pitlochry, Perthshire ;
× 23.

A. Ben Lui Schists. Some of the garnet crystals still retain something of the shapes of chlorite which they have replaced.
B. Killiecrankie Schists. The large garnet seems to represent a plexus of chlorite flakes with interspaces of quartz, which are gradually losing their angular shapes.

CRYSTAL-GROWTH IN A ROCK UNDER STRESS

We have next to consider how crystal-growth, whether of re-crystallized minerals or of new products of metamorphism, is affected by a state of shearing stress in the rock. It has been seen that, even at ordinary temperatures, a regular arrangement of newly crystallized elements is an important factor, with others purely mechanical, in producing the perfect cleavage-structure of the more highly developed slates. We did not then inquire strictly how this regular arrangement is brought about. It was tacitly assumed, for the sake of simplification, that the rocks themselves play a merely passive part. In so far as new crystal-growth entered, this was an incomplete presentation of the actual conditions. In the more general case of regional metamorphism, now to be considered, it is no longer possible, even as a first approximation to the truth, to ignore a factor which acquires enhanced importance with higher temperature. We have seen how in purely thermal metamorphism the growth of crystals in the heart of a solid rock gives rise potentially to shearing stresses of a high order, but that in effect these are greatly minimized by being set off one against another. The strong tendency to structures of the decussate type gives indeed an actual picture of how in that case the compensation is effected. In regional metamorphism the principle of minimizing internal stress still holds good ; but its application is to the complete system of stresses, including those set up in reaction against the external forces as well as those internally generated by crystal-growth. The former have a common direction, imposed from without, while the latter are susceptible of adjustment. The desired result will therefore be attained by setting off the latter against the former. In this way the total stress will be minimized, though by no means annulled.

In order that the internally generated stresses may be mobilized to this end, crystal-growth must now be directed, so far as is possible, in conformity with a definite plan. In the common case of regional metamorphism related to orogenic thrust, the external force-system, and therefore the internal reaction which it provokes, may be conceived for the sake of clearness as a single linear pressure in a given direction. The shearing stresses so set up will have directions at 45° to the given line. If then the growing crystals assume such a posture that they exert their forces mainly in planes perpendicular to the given direction, the shearing stresses due to this cause (at 45° to those planes) will be in position to cancel in part the externally provoked stresses, and the total stress will be reduced to a minimum. Accordingly a rock metamorphosed under the conditions supposed shows in its component new crystals a strong tendency to *parallelism along planes perpendicular to the direction of orogenic pressure*. In proportion as it is more or less perfectly developed, this arrangement imparts to the rock a schistose or fissile property, and to this tendency to a common parallel orientation of the crystal-elements the name *schistosity* is appropriately given (Fig. 84, *A*). It is to be distinguished from the allied structure, foliation, to be discussed below. With reference to its function, it should be remarked that the parallel orientation of crystals in a crystalline schist and the deliberate non-orientation (decussate structure) in a hornfels have the same ultimate significance, both being devices, adapted to the different conditions, for minimizing internal stress.

The controlling influence of the dynamic factor in crystallization is well illustrated by an experiment devised by Wright.[1] Cubes of glass having the composition of different minerals (wollastonite, anorthite, diopside) were subjected to unequal stress, and were then raised to a temperature high enough to induce devitrification without seriously impairing the effective rigidity of the glass.   In every case the crystals so generated, in the form of slender fibres, tended to set themselves perpendicular to the direction of maximum pressure.

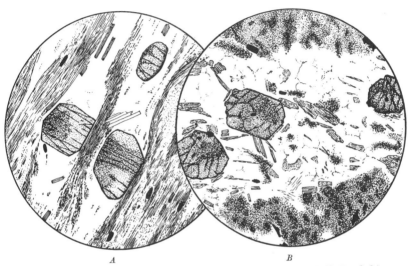

A   B

FIG. 84.—GARNETIFEROUS AND GRAPHITIC MICA-SCHIST, Blair Atholl, Perthshire ; × 18.

To illustrate typical schistosity.   This is shown (in *A*) by the parallel orientation of the flakes of sericitic white mica and biotite and crystals of iron-ore.   Cracks in the garnet are at right angles to the direction of 'stretching' or relative tension.   This section is taken perpendicular to the plane of schistosity and *B* for comparison parallel to that plane.   The latter shows the clotted patches of carbonaceous matter, not so clearly seen in a cross-section.

INFLUENCE OF SPECIFIC PROPERTIES OF MINERALS

The chief part in determining the schistose structure belongs to those numerous minerals which tend naturally either to tabular or to columnar forms.   We have seen that such habit becomes more pronounced when the crystals grow in a solid environment.   It is still further exaggerated when crystal-growth takes effect mainly in a certain plane under the influence of an external pressure perpendicular to that plane.   According to measurements by Trueman,[2] the

[1] *Amer. J. Sci.* (4), vol. xxii (1906), pp. 224–30.
[2] *Journ. Geol.*, vol. xx (1912), pp. 235–43.

long diameter of boitite flakes averages in igneous rocks about one and a half times the thickness, but in crystalline schists six times or more. For hornblende crystals Leith [1] found that in igneous rocks the length is on the average two and a half times the breadth, but in crystalline schists four or five times. The fissility of such a rock as a mica-schist is enhanced by the fact that the cleavage of the mica itself, parallel to the broad faces, coincides in direction with the general schistosity of the rock. When the principal constituents of a rock have not this flaky habit, a parallel elongation of the indi-

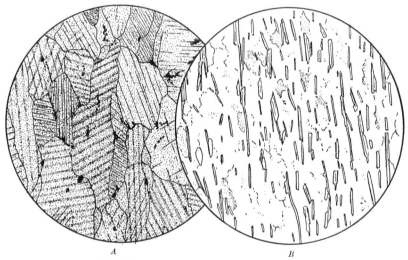

FIG. 85.—IMPERFECT SCHISTOSITY ;   × 23.

A. Crystalline Limestone, Blair Atholl, Perthshire. There is a well-marked parallel elongation of the crystals of calcite, but the mineral has no master-cleavage in the given direction, and the rock is only rudely fissile.

B. Micaceous Quartzite, Glen Lochy, Argyllshire. The flakes of white mica have a very regular parallel arrangement, but the quartz and subordinate felspar, which make the chief bulk, show only a certain tendency to elongation in a common direction.

vidual crystals gives no more than a rude fissility (Fig. 85, *A*). So too flakes of mica scattered sparingly through a quartzose rock, with parallel orientation, do not suffice to impart any very perfect schistosity (Fig. 85, *B*). If, on the other hand, as in many mica-schists, the slender flakes are numerous and closely set, the weaker quartz, hemmed in between parallel flakes, is constrained into tabular and lenticular shapes, so that all the principal elements of the rock come to share in the general schistose structure.

A distinction may be drawn between minerals of tabular and those of columnar habit. To make this clear, it should be remembered

[1] *Bull.* 239, *U.S. Geol. Sur.* (1905), pp. 30-1.

that there is a direct relation between stress and strain. The direction of maximum pressure is also that of maximum compression, and corresponds with the shortest axis c of the strain-ellipsoid, so that the plane of schistosity is that containing a and b. In this plane, when the structure is regularly developed, lie both flakes of mica and needles of actinolite ; but the needles tend further to lie towards the direction of a rather than that of b. If this tendency is strong, in a rock composed largely of columnar or acicular crystals, it gives, instead of a plane-schistosity, a *linear or fibrous structure*, analogous

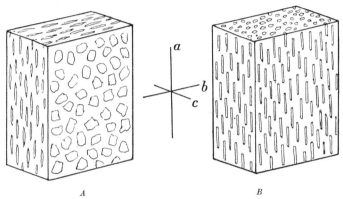

FIG. 86.—IDEAL ARRANGEMENT OF THE COMPONENT CRYSTALS IN CRYSTALLINE
SCHISTS

A. Plane schistosity.   *B*.   Linear schistosity.

to the linear cleavage already noticed (Fig. 86). Since, however, the difference between the axes a and b (and between the corresponding stresses) is often inconsiderable as compared with the difference between either of these and c, a linear arrangement may be but little developed. A section taken parallel to the plane of schistosity will then show no perceptible parallel disposition of columnar crystals (Fig. 87).

A mechanical factor contributes to the setting up of a schistose structure in regional as in purely dynamic metamorphism ; viz. the rotation of already existing flat or linear crystals into approximate parallelism as a result of plastic deformation of a rock-mass (p. 154). Such deformation normally accompanies the mineralogical changes, and is indeed made possible by those changes. In this way crystals generated at earlier stages of progressive metamorphism, in so far as they survive, are drawn more closely into conformity with the general scheme of parallelism. This action, however, ceases to be effective in proportion as continual reconstruction of the rock is quickened by rising temperature ; the earlier formed crystals giving rise to new minerals, or at least becoming recrystallized, usually in larger individual elements.

The development of a schistose structure is indeed by no means a steadily cumulative process. There are, on the contrary, various causes which, when a very high grade of metamorphism is reached, work together to replace the schistosity characteristic of lower and medium grades by other types of structure. One cause is the weakened resistance of the rocks at very high temperatures, which, while facilitating shearing movements, sets a constantly narrowing limit to

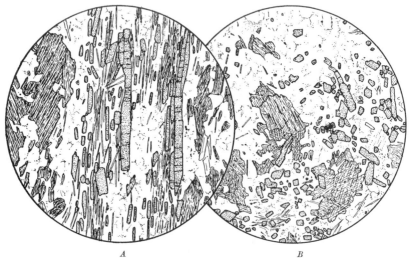

A   B

FIG. 87.—ZOISITE-AMPHIBOLITE, Arne, near Bergen ;   × 23.

The section A, taken perpendicular to the plane of schistosity, shows a regular parallel arrangement of the slender flakes of mica and prisms of zoisite.   This is less clearly seen in the hornblende, with its rudely columnar habit, while the felspar and quartz have at most a certain tendency to elongation in the common direction.   A section B, taken parallel to the plane of schistosity, shows no perceptible orientation of crystals.

possible shearing stress. Increasing coarseness of grain of the metamorphosed rocks, due to the freer diffusion which goes with higher temperature, is another factor which tells in the same direction. So too is the altered mineralogical constitution of the rocks. The minerals of flaky or acicular habit—chlorites, micas, zoisite, amphiboles, and the like—belong for the most part to the lower and middle grades of metamorphism, and are ultimately replaced largely by other minerals having no such pronounced habit. At the same time, the characteristic minerals of the highest grades—garnet, cyanite, diopside, forsterite, etc.—have for the most part a strong crystallizing force, and are little amenable to direction. For all these reasons the most highly metamorphosed rocks, of almost any composition, although their natural affinities cause them to be classed with the crystalline schists, commonly show little or nothing of the schistose character. Their structure is often described as *gneissose*. The term, in this usage, may be taken to imply a relatively coarse grain, with at most a very imperfect orientation of the component crystals, but often showing none the less a parallel structure of a different order (foliation).

It will be seen that both the development of schistosity and its ultimate decline are closely bound up with that innate force in virtue of which crystals are enabled to push their growth in a solid rock ; a force which is itself a function of temperature, and is further subject to being controlled in direction by unequal stress. That it is also a specific property, differing widely in different minerals, has been made clear in our study of the micro-structure of thermally metamorphosed rocks, and an attempt was made in that place to range the more important species in order with reference to that fundamental property. Since the minerals of regional metamorphism are only in part the same as those included in the former list, we set down here a rough indication of the *crystalloblastic series* as adapted to the present case. It must be understood, however, that, as the result of various disturbing causes, the apparent mutual relations of the several minerals are not invariably the same. It is possible that the amount of solvent present, bringing into play differential solubility, is a factor in the result.

In Argillaceous and Arenaceous Rocks

Sphene, rutile ;
Haematite, ilmenite, magnetite ;
Garnet, tourmaline ;
Staurolite, cyanite, sillimanite, chloritoid ;
Albite, muscovite, biotite, chlorite ;
Quartz, cordierite, orthoclase and microcline.

In Calcareous and Igneous Rocks

Sphene, rutile, spinel ;
Pyrites, garnet, tourmaline, apatite ;
Epidote and zoisite, forsterite, augites ;
Hornblende, dolomite, glaucophane, albite ;
Muscovite, biotite, tremolite, chlorite, talc ;
Calcite, quartz, orthoclase and microcline.

Subject to various exceptions, the more or less idioblastic or zenoblastic development of any mineral depends upon its place in the crystalloblastic series relatively to other minerals with which it is brought in contact. The stronger mineral, viz. that which stands higher in the list, tends to assert itself at the expense of its weaker neighbours. But, as already intimated, specific force of crystallization has its importance also in relation to directional structures. Those constituents of crystalline schists which chiefly determine the schistose character are for the most part of not more than moderate crystallizing strength. When notably strong minerals enter, they may pay scant regard to the prescribed orientation, and their crystals

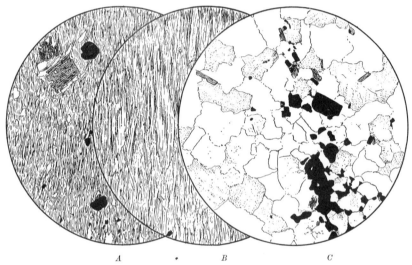

FIG. 88.—STRUCTURES OF METAMORPHOSED ROCKS ; × 18.

A. Lepidoblastic. Sericite-phyllite with porphyroblastic biotite and magnetite, Appin Phyllites, Onich, Loch Leven.
B. Nematoblastic. Tremolite-schist, a sheared peridotite dyke, Lochinver, Sutherland.
C. Granoblastic. Magnetite-Biotite-Granulite, Moine Series, Glenelg, Invernessshire.

are often seen to lie at all angles to the plane of schistosity. Thus, in a mica-schist, where micas and quartz show a regular parallelism, large crystals of staurolite or cyanite may be scattered through the rock with no approach to any common orientation.

Such considerations, it may be remarked, are not without significance for the field-geologist, and to ignore them is to incur the risk of serious error. When, for example, a mica-schist encloses large crystals of chloritoid, which lie at all angles and forcibly thrust aside the laminae of schistosity, we see only the normal behaviour of a relatively strong mineral, and must not take it as evidence of a second metamorphism. Again, an eclogite, a rock composed essentially of strong minerals, may be found intercalated among thoroughly schistose rocks, while itself showing no trace of such structure. An observer not familiar with this branch of petrology might erroneously infer that the eclogite has been intruded subsequently to the metamorphism of the associated rocks.

CHARACTERISTIC TYPES OF STRUCTURE

Different types of schistosity, determined mainly by the mineralogical composition of the rocks, have received appropriate names.

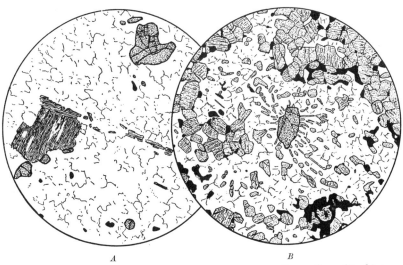

FIG. 89.—GRANOBLASTIC STRUCTURES IN IGNEOUS ROCKS, Saxon Granulitgebirge ; × 23.

A. Garnet-Granulite, Rohrsdorf : an acid type, with garnet and biotite.
B. Pyroxene-Granulite, Hartmannsdorf : a basic type, mainly of hypersthene, plagioclase, and magnetite. A special structure is a radiate arrangement of pyroxene about an occasional crystal of garnet.

Of chief importance are the *lepidoblastic* or flaky, due to an abundance of minerals like micas and chlorites with a general parallel arrangement, and the *nematoblastic* or fibrous, seen in rocks composed largely of such minerals as glaucophane and actinolite (Fig. 88, *A* and *B*). In crystalline schists as a class the degree in which the ideal parallelism of elements is actually developed varies much. Even when thin slices show very little suggestion of a regular arrangement, the average orientation of the component crystals often suffices to impart a decided fissility to the rock in mass (Fig. 98). Regionally metamorphosed rocks composed essentially of equidimensional elements, and therefore incapable of any schistose structure, are styled *granoblastic*. This is a common type of structure in highly metamorphosed quartzo-felspathic sediments (Fig. 88, *C*), and occurs also simply or with modifications, in rocks of igneous origin (Fig. 89). The suffix *-blastic* denotes always a new structure induced in the solid rock. For residual structures, more or less modified by metamorphism but still recognizable, the prefix *blasto-* is employed, as in blastoporphyritic, blastophitic etc.

The general fabric of the rock may be modified by the coming in of more special new structures, and especially by the development of

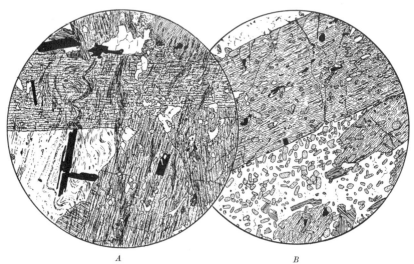

A   *B*

FIG. 90.—HETEROBLASTIC (PORPHYROBLASTIC) STRUCTURE in Hornblende-schist ('Hornblendegarbenschiefer') ; × 23.

A. Hohsandhorn, Binnental, Valais. The original lines of sedimentation, much corrugated, pass unbroken through the porphyroblasts of hornblende. Other minerals shown are biotite and ilmenite with a quartzitic ground-mass.
B. Val Piora, Ticino. In addition to the large hornblendes, there are smaller crystals of biotite and abundant epidote in a ground-mass of quartz and felspar.

some particular mineral in relatively large crystals. To rocks in which the essential constituents are of two distinct orders of magnitude Becke gave the name ' heteroblastic ', in contradistinction to ' homoeo-blastic '. The simplest type is the porphyroblastic, already mentioned, in which large crystals of some one mineral are embedded in a matrix made up of smaller elements. Exceptionally the ' porphyroblasts ' may build the main part of the rock, the finer-grained portion merely occupying the interspaces : the rocks styled ' Hornblendegarben-schiefer ' afford good examples (Fig. 90). Often the large crystals enclose numerous small crystals of another mineral, giving the poeciloblastic or ' sieve structure ' (Fig. 81, above).

NATURE AND ORIGIN OF FOLIATION

To be carefully distinguished from schistosity as defined in the foregoing pages is *foliation*. The two structures [1] are very commonly found in association and, taken together, characterize the whole class of regionally metamorphosed rocks with few exceptions. Darwin,[2] who first used the term foliation, clearly distinguished this type of structure from schistosity (cleavage), while recognizing a close relation between the two. Owing, however, to the concordance shown by these two directional structures when associated, and to the fact that foliation conduces, together with schistosity, to the characteristic fissile property, a certain laxity of usage is often met with. Foliation consists in a more or less pronounced aggregation of particular constituent minerals of the metamorphosed rock into lenticles or streaks or inconstant bands, often very rich in some one mineral and contrasting with contiguous lenticles or streaks rich in other minerals. All these show, at any one place, a common parallel orientation, which agrees with the direction of schistosity, if any, and is manifestly related to the same system of stress and strain in the rock. It is to be observed, however, that whereas schistosity declines as the highest grade of metamorphism is approached, foliation on the other hand often becomes more salient, and is developed on a larger scale. From the point of view of progressive metamorphism, the one type of structure is then superposed upon the other, and comes to supersede it as the dominant structural characteristic of the most highly metamorphosed rocks. Foliation is thus marked as related to continued elevation of temperature, with all that is thereby implied.

Regarded merely as a physical phenomenon, foliation appears as a *selective* effect, in which like is drawn to like, and to that extent separated from unlike, by a process which is described rather than explained by the name *segregation*. The process is doubtless, in the main, one of continued local solution, diffusion, and recrystallization, following that order at any one spot. The diffusion, by which the transference of material is brought about, is most effective in a certain plane, viz. that perpendicular to the maximum pressure, being presumably influenced by the unequal distribution of stress and doubtless guided too by such schistose structure as the rock at that stage may

---

[1] In Grubenmann's terminology they are not structures but textures. He defines *structure* as depending upon the shape and size of the constituent minerals of the rock, and *texture* upon their arrangement in space ; but this distinction has not been generally adopted.

[2] *Geological Observations in South America* (1846), p. 141.

possess. Elevation of temperature, while enlarging the amplitude
of diffusion, also relaxes continually the resistance of the rock ; and the
directional effect of the physical process described is thus further
accentuated by bodily deformation, viz. compression in the direction
perpendicular to the plane of foliation and extension in that plane.

Segregation may be taken to imply a nucleus about which accretion
has gathered ; and, if so, the heterogeneous constitution of a foliated
rock points to some original heterogeneity, which has been exaggerated
and extended by the segregatory process. Where, for instance, we
see a lenticle very rich in biotite, we may infer that it originated in a
part of the rock already relatively rich in that mineral. It is clearly
suggested therefore that foliation is based ultimately upon pre-
existing structures in the rocks. In a remote sense it is of the nature
of a residual structure, though we should strain the meaning of that
term by applying it where nothing remains of original outlines. In
the case of rocks of sedimentary origin the prime cause of heterogeneity
must be sought in stratification and in the various accidents, such as
folding and faulting on a small scale, by which the regular stratification
is liable to be disturbed. An examination of the structures due to
this cause as seen in the less advanced grades of metamorphism
should therefore throw some light upon the origin of the more problematical structures in which these are ultimately merged.

Many argillaceous and calcareo-argillaceous sediments, as first
deposited, are composed of successive or alternating layers, which
differ more or less in composition. In metamorphism such differences
translate themselves into different mineral-associations, or at least
different relative proportions of the minerals, as between one layer
and another. Banding of this nature is a familiar feature of many
thermally metamorphosed sediments (Figs. 13, *A* ; 33, etc., above).
Only in a very high grade does it sometimes become indistinct or
disappear as a result of the enlarged scope of diffusion at very high
temperatures. With increased diffusion, however, a new action
comes into force, viz. a stronger segregation of particular minerals
into lenticles and streaks, which, in simple thermal metamorphism,
follow the direction of original stratification (pp. 61, 75). In regional
metamorphism all this is modified by the intervention of the dynamical factor, viz. by a mechanical breaking down of the regularity of
stratification and by the directional element imported into the process
of recrystallization. Very common in rocks which have been fine
textured sediments is a close puckering or plication, sometimes passing
into a system of miniature reversed faults. Such structures, which in
the lowest grade often figure as false cleavages, are often recognizable

too in a higher grade.[1] When they are most clearly displayed, we
see the highly sinuous lines of bedding, as indicated by mineralogical
differences, cut at varying angles by the planes of schistosity (Figs.
91, *A* ; 96, *B*). Much more commonly the bedding, as seen in any
one thin section, is *approximately* parallel to the schistosity (Fig.
91, *B* ; see also Fig. 92, etc.). In such a region as the Scottish
Highlands the rocks of more yielding nature are largely affected by
acute plication on a small scale, either of the nature of isoclinal
folding or of the unsymmetrical type to which Barrow [2] gave the

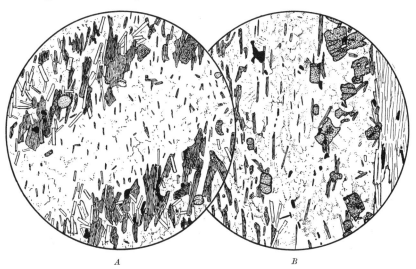

A   *B*

FIG. 91.—BANDED STRUCTURE IN MICA-SCHISTS : × 23.

A. South of Muchalls, near Stonehaven. The original stratification is indicated by
evident banding, and this is obliquely cut by the schistosity as marked by the orientation
of the mica-flakes.
B. Near Aberfeldy, Perthshire (Ben Lui Schists). The stratification is given by the
distribution of the various new minerals : quartz and felspar, biotite, garnet, and muscovite. The schistosity has sensibly the same direction.

name ' buckling '. In another place the same writer describes a
quartzite as ' shut up on itself like the bellows of a concertina (concertina structure) '. It results from this that, while the formations
as mapped may lie at low angles to the horizontal, the laminae of
sedimentation are in general highly inclined, and tend to be parallel,
or *nearly* parallel, with the planes of schistosity.

The peculiar features of foliation, as seen in crystalline schists of

[1] It must be remarked, however, that these structures may also be impressed
upon the rocks at a much later stage, when the temperature has declined, but the
region is still under the influence of lateral pressure.
[2] *Rep. Brit. Assoc.* for 1912 (1913), pp. 473–4.

an advanced grade, seem to be due in the main to a process of segregation affecting rocks which were already in this highly disturbed condition, and had acquired a strong schistosity. The segregation was directed primarily along the constant direction of the schistosity, while the initial mineralogical differences, upon which it operated, were distributed in narrower or broader bands making variable, and often small, angles with the schistosity. The bedding was often broken also by earlier displacements of the discontinuous kind. Where bedding and schistosity locally coincided, segregation served

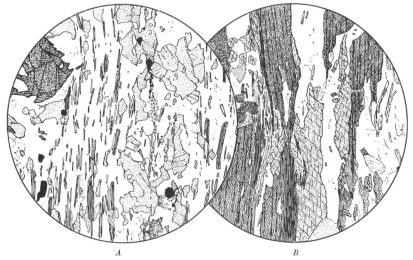

A   B

FIG. 92.—CALCITE-BEARING CRYSTALLINE SCHISTS ;   × 23.

A. Calc-Mica-schist, with garnet (Ardrishaig Phyllites), Glen Lochy, near Dalmally, Argyllshire. The relatively soluble calcite and quartz tend to segregate into bands parallel with the general schistosity.
B. Calc-Hornblende-schist (Ben Lawers Schists), Ben Vrackie, Perthshire. The minerals are green hornblende, biotite, muscovite, quartz, and calcite. Here the segregation of the calcite and quartz is more complete, giving the rock a well-marked foliation.

merely to emphasize the banded disposition ; elsewhere this became blurred and gave place increasingly to trains of lenticles and folia. Often it is possible to discern in the elements of the foliation a certain arrangement *en echelon*, such as we may picture by supposing Fig. 91, *A*, to be much enlarged and the separate flakes of biotite replaced by lenticles rich in that mineral.

Since solution is an essential part of the mechanism of segregation, the chief part in setting up a foliated structure devolves upon the more soluble of the minerals present. The solubility of the ordinary rock-forming silicates, except at very high temperatures, seems in general to be very limited, and in most crystalline schists the important mineral in this connexion is quartz. To this must be added calcite, more soluble but less widely distributed. The two minerals have often segregated together into bands and lenticles, and it is especially in the calc-mica-schists that foliation is often a prominent feature even in a moderate grade of metamorphism (Fig. 92).

Rocks which in their original state offered a strong resistance to deformation present a somewhat different case. In a low grade of metamorphism they yielded by way of fracture and differential displacement, and there was a strong tendency to lenticular shapes, on a large or small scale, with an approximate common orientation (p. 167). With this mechanical effect there went, however, in most cases mineralogical changes, especially along the surfaces of sliding, where films and streaks of such minerals as sericite and chlorite were often abundantly developed. Recrystallization in higher grades of metamorphism, which effaced actual fractures, did not obliterate the mineralogical heterogeneity thus set up, and this remains as an important element in the foliation of rocks of this kind. Many igneous, as well as sedimentary, rocks fall under this head, but only such as have passed through the lower grades of metamorphism before attaining the higher. Igneous rocks intruded during the time of metamorphism stand upon a different footing. They too show frequently a foliated structure, but the significance of this will be more conveniently considered later.

CHAPTER XIV

REGIONAL METAMORPHISM OF NON-CALCAREOUS SEDIMENTS

Successive Zones of Metamorphism—The Chlorite-Zone—The Biotite-Zone—The Almandine-Zone—The Staurolite and Cyanite Zones—The Sillimanite-Zone.

SUCCESSIVE ZONES OF METAMORPHISM

IN discussing regional metamorphism in rocks of various lithological types, we shall keep in view primarily the conception of something *progressive*, viz. of a succession of changes which follow in response to a continued rise of temperature. It was thus that we reviewed the phenomena of purely thermal metamorphism, but we have now to take account of new conditions. The additional factors are two : shearing stress, normally pushed to the limit of endurance of the rocks, and a hydrostatic pressure notably greater than that operative in the former case. Both have an important influence upon the mineralogical changes induced, and the former profoundly affects the structural and textural development of the metamorphosed rocks. The natural manner of approach then is to inquire in what way the course of metamorphism is modified by the introduction of these new conditions, and we shall accordingly take as our guide a *comparison between thermal and regional metamorphism.* This point of view is different from that of petrologists of the descriptive school, and leads to a different manner of treatment.

It is in rocks having the composition of ordinary argillaceous sediments that the successive stages of advancing metamorphism are most clearly indicated by corresponding mineralogical changes. As already explained, distinct zones can be recognized on the ground, such that the outer limit of each zone is marked by the first appearance of some particular index-mineral. We will set down here for reference the sequence of zones, as worked out by Barrow in the South-Eastern Highlands of Scotland. It must be understood that an index-mineral is not confined to the zone to which its name is attached, but usually

208

persists into higher zones. Within its own zone it is found in all rocks of appropriate bulk-composition.[1]

(0) *Zone of clastic micas.* Here we have purely dynamic metamorphism, the effect of rise of temperature being inappreciable. This zone is faulted out along the greater part of the Highland Border.

(1) *Zone of digested clastic mica.* Since recrystallization of white mica doubtless proceeds in dynamic metamorphism at ordinary temperature, the division between this and the preceding zone is of doubtful significance. A prominent mineral in most of the rocks is chlorite, and Tilley uses the name ' chlorite zone '.

(2) *Biotite-Zone.*

(3) *Garnet-Zone.* The distinctive mineral is one with a composition near that of almandine, and garnets of a different kind may come in at an earlier stage : hence Tilley prefers the name ' almandine-zone '. The preceding zones are relatively narrow, but garnet continues to be the significant mineral over a wide range of country.

(4) *Staurolite-Zone.* The limit of this is laid down only approximately, and the zone is not everywhere recognized. The reason is that staurolite is not formed unless the rocks have a certain peculiarity of total composition.

(5) *Cyanite-Zone.*

(6) *Sillimanite-Zone,* exposed over a large tract in the interior of the mountainous country.

A zone of higher-grade metamorphism normally underlies one of lower grade, but this relative position may become inverted by subsequent disturbance. It results from the dome-like disposition of the isotherms that the most highly metamorphosed rocks, laid bare by deep erosion, make a wide spread, while the low-grade zones usually outcrop in narrow bands.

THE CHLORITE-ZONE

Pure dynamic metamorphism may be regarded as registering the zero of regional metamorphism, and our starting-point is therefore given by the argillaceous type of slate already noticed. The chief constituent minerals of such a rock are white mica, chlorite, and quartz, in varying proportions. The mica has the sericitic habit, and a high magnification often discloses an immense number of minute rutile needles scattered through it. The chlorite likewise is in very small scales, almost impossible to isolate from the rock, and our know-

---

[1] Barrow, *Proc. Geol. Assoc.*, vol. xxiii (1912), pp. 274–90 ; Tilley, *Quart. Journ. Geol. Soc.*, vol. lxxxi (1925), pp. 100–10. See also sheets 65, 66, 67, of the Geological Survey of Scotland.

ledge concerning its actual nature is very deficient. In the Stavanger district, according to Goldschmidt,[1] the chlorite is of a highly aluminous variety, and has a ratio FeO : MgO somewhat higher than unity. Mica and chlorite are authigenetic, but the quartz is still mostly in the form of detrital granules. With it can sometimes be detected granules of felspar, which are of albite. Free iron-oxide is at this stage normally in the state of haematite, and ilmenite is another common constituent. Many black slates contain a noteworthy amount of finely divided graphite, though its actual quantity is less than might be judged from

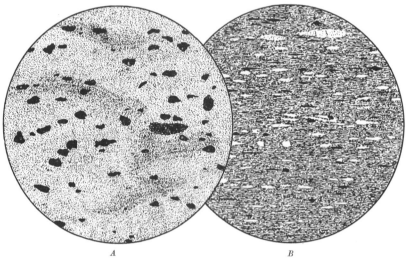

A   *B*

FIG. 93.—CHLORITE-SERICITE-SCHIST (Devonian), Delabole, North Cornwall ; × 60.

Showing sections parallel and perpendicular to the schistosity or cleavage. The main constituents are sericitic mica, chlorite, ilmenite, magnetite, and quartz.

its conspicuous appearance. In non-calcareous slates these are the only constituents to be observed, if we except rare and minute crystals of such minerals as zircon and tourmaline, of detrital origin, and sometimes pyrites, formed in place.

We have already remarked, however, that many rocks which are equally styled slates in common usage belong to a somewhat higher grade of metamorphism, and in these we see the first effects of a certain elevation of temperature. The principal change is a recrystallization of the detrital quartz. At the same time, or soon after, rutile and tourmaline likewise recrystallize, and the last trace of the clastic origin of the rock is obliterated. At this stage too haematite is reduced to magnetite. The important difference, in comparison with

[1] *Vidensk. Skr.* 1920, No. 10 (1921), pp. 59, 64. The composition is calculated from bulk-analyses of the rocks.

the lowest grade of slates, is that the quartz now appears in little lenticles or streaks conforming with the general parallel orientation. Schistosity is strongly pronounced (Fig. 93, *B*), though often affected by corrugation on a small scale (Fig. 94, *B*). The authigenetic constituents, mica and chlorite, have undergone regeneration, and figure usually in rather larger elements. The chlorite especially, in virtue of its freer solubility, tends to gather into patches and knots, which are sometimes seen to be related to the corrugation of the rock. Not uncommonly we see little lenticles of chlorite consisting of one or more

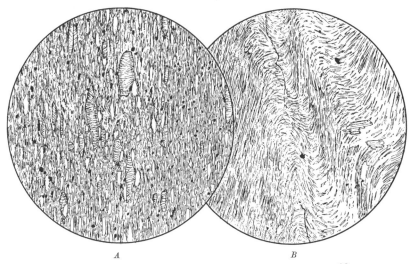

A   *B*

FIG. 94.—CHLORITE-SERICITE-SCHISTS OR PHYLLITES ;   × 23.

A. Dunoon Phyllite, Port Bannatyne, Bute. The flakes of chlorite show the transverse arrangement often seen at this early stage.
B. Ardrishaig Phyllite, Loch Awe : rich in sericitic mica, but with a smaller proportion of chlorite. Besides quartz there are also little granules of albite.

flakes set transversely to the schistosity, while the lenticles themselves have the regular orientation (Fig. 94, *A*).

These rocks, which may be named *chlorite-sericite-schists*, show considerable variation in respect of the relative proportions of the chief constituent minerals, and often vary much in alternating beds. When the rock, or part of a banded rock, is composed almost entirely of mica, the flakes take the form of thin films, often with a wavy arrangement (Fig. 94, *B*). Such highly micaceous rocks, with a very perfect schistosity and a glossy sheen on the surfaces of splitting, are distinguished as *sericite-phyllites*.

As compared with simple thermal metamorphism, the recrystallization of the stress-mineral muscovite has been *promoted*, in the sense of taking effect at an earlier stage. It is to be observed, too, that, despite the rising temperature, muscovite and chlorite continue during the earlier stages of regional metamorphism to recrystallize side by side, without any mutual reaction. We have seen that in simple thermal metamorphism the first important mineralogical change in argillaceous sediments of ordinary type was a reaction between white mica and chlorite (with iron-oxide) producing biotite and other new minerals. It appears therefore that the influence of strong shearing stress *hinders or defers* such reaction, in the sense that a greater elevation of temperature is now needed to bring it about. The aluminous silicates andalusite and cordierite, which were often joint products with biotite of that reaction, are not merely delayed in their appearance, but are completely inhibited, so long as shearing stress continues to exert its full influence, and they are clearly marked as typical anti-stress minerals.

An important constituent of many of these rocks remains to be noticed. There is abundant evidence that albite, in contrast with the potash- and lime-felspars, is stable under stress-conditions at low as well as at higher temperatures, and must be ranked with the associated muscovite and chlorite among the most typical stress-minerals. It is a prominent constituent of different members of the Dalradian series in different grades of metamorphism. So abundant is albite in some parts of the Cowal district of Argyllshire, that Clough [1] contemplated the possibility of 'a large impregnation with soda', but there are no good grounds for such suggestion. The albite-chlorite- and albite-quartz veins, which are not uncommon in the phyllites of Cowal, doubtless derive their material from the rocks which they traverse. The abundance of soda-felspar is merely one indication among others that alkaline igneous rocks were rather widely exposed on the old land-surface from which these sediments drew their material. Pebbles of such rocks are conspicuous in the Boulder Bed of Islay and Schichallion, and grains of albite are abundant in many of the Dalradian grit-formations. Detrital albite is a common constituent likewise in the argillaceous deposits, but occurs there usually in a finely divided state which makes its identification difficult. A close examination, however, will often detect, even in rocks of the lowest grade, minute grains of pellucid felspar. With the advance of metamorphism they become larger, and are at last very evident to the naked eye. Indeed albite often attains a conspicuous porphyroblastic development even before that stage at which biotite makes its appearance, and this

[1] *Geol. of Cowal* (*Mem. Geol. Sur. Scot.*, 1897), p. 39.

continues, in rocks of appropriate composition, in the succeeding biotite- and garnet-zones.

We have then to recognize a series of albitic rocks, which are represented in this low grade by *albite-chlorite-sericite-schists*.[1] These rocks consist of albite, chlorite, and sericite in varying relative amounts, with quartz, often in rather subordinate quantity, magnetite, and such accessory constituents as recrystallized rutile and tourmaline. The relatively large crystals of albite, idioblastic but usually rather rounded, give a characteristic appearance to the rock (Fig. 95, *A*). They are

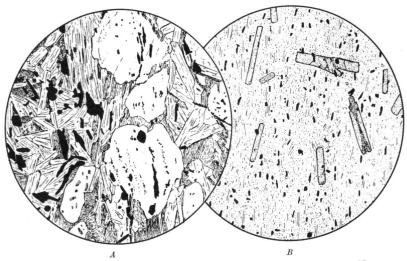

A   B

FIG. 95.—SPECIAL TYPES OF SCHISTS IN THE CHLORITE-ZONE; × 23.

A. Albite-Chlorite-Sericite-schist, Ben Ledi Group, Loch Lomond. Composed essentially of albite, chlorite, muscovite, magnetite, and quartz. The albite is in large porphyroblastic crystals with inclusions of magnetite and chlorite, partly in trains reminiscent of the original bedding.

B. Chloritoid-Sericite-schist, Devonian, Tintagel, North Cornwall. Shows crystals of chloritoid and ilmenite set in a sericitic matrix, without chlorite.

quite clear, but carry inclusions of magnetite, chlorite, etc. Twinning is not common, and the close twin-lamellation so general in igneous rocks is not found.

Another well-characterized type, but one having a more restricted occurrence, is a *chloritoid-sericite-schist*. Chloritoid is produced only in rocks which, with an abundance of alumina and a sufficiency of iron-oxide, are relatively poor in magnesia, lime, and potash. In a

---

[1] Sericite-albite-gneiss of Grubenmann and others. In modern usage the term ' gneiss ' is commonly employed with reference to structural and textural characters, and it is therefore confusing to apply the name to all felspathic crystalline schists.

M.—8

sediment of this kind the mineral forms very readily, and persists through a wide range of temperature. In many occurrences it has been designated ottrelite, but the two minerals, with similar manner of occurrence, appear to be quite distinct.[1] The formation of chloritoid is probably dependent, in this early stage of regional metamorphism, upon the presence of kaolin. The reactions which would give rise to andalusite or cordierite being inhibited, kaolin reacts with magnetite to give the stress-mineral chloritoid :

$$3H_4Al_2Si_2O_9 + Fe_3O_4 = 3H_2FeAl_2SiO_7 + 3SiO_2 + 3H_2O + O.$$

That this mineral may be produced in a very low grade of metamorphism is shown by its occasional presence in the Skiddaw Slates.[2] It has a wide distribution in the Devonian slates of North Cornwall.[3] The rocks here are highly sericitic, with or often without chlorite, and contain a considerable amount of ilmenite. The chloritoid is in little flakes, intergrown with sericite, or often in conspicuous crystals which give a porphyroblastic effect, and lie at various angles with the plane of schistosity (Fig. 95, B). The chief inclusions are minute crystals of rutile. In the Scottish Highlands chloritoid is not a very prominent mineral, either in this or in higher grades. It is found rather plentifully in some chlorite-sericite-schists of the district north and west of Stonehaven. The best-known low-grade chloritoid-schist of the Alpine region are in the St. Gotthard district, and have been fully described by Niggli.[4]

THE BIOTITE-ZONE

The coming in of biotite marks, as has been shown, a definite stage in the advance of metamorphism. That its first appearance registers a determinate temperature cannot be asserted, in view especially of the doubtless variable composition of the mineral. It is formed largely at the expense of the stress-minerals chlorite and muscovite, and we may infer that its formation is retarded here in regional as compared with simple thermal metamorphism. As before, it is typically of a variety rich in ferrous iron, the ratio FeO : MgO being in general greater than unity. Since the material for its formation is drawn from muscovite, chlorite, magnetite, quartz, and rutile,

---

[1] Analyses of the original ottrelite of the Ardenne give the formula $H_2RAl_2Si_2O_9$. It is characteristically a manganiferous variety.

[2] Rastall, *Quart. Journ. Geol. Soc.*, vol. lxvi (1910), p. 126.

[3] Hutchings, *Geol. Mag.*, 1889, pp. 214–20 ; Tilley, *ibid.*, vol. lxii (1925), pp. 314–17 ; Phillips, *ibid.*, vol. lxv (1928), pp. 546–55.

[4] *Beitr. geol. Karte Schw.* (N.F.), Lief. xxxvi (1912).

the biotite cannot be expected to present itself as pseudomorphs after any particular parent-mineral, but in fact it is always the chlorite that provides the starting-point. If the chlorite was finely disseminated, biotite figures at first as very numerous minute scales, while a patchy distribution of chlorite leads to an early porphyro-blastic development of biotite (Fig. 96, *A*). The large flakes are sometimes set athwart the schistosity, and this too seems to be determined by the arrangement of knots of chlorite. In most mica-schists the biotite-flakes have the typical habit, lying more or less closely parallel

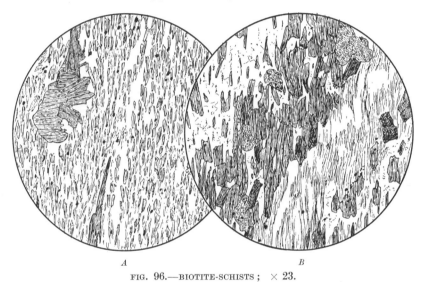

A   *B*

FIG. 96.—BIOTITE-SCHISTS ; × 23.

A. Biotite-Chlorite-schist, Ardrishaig Phyllites, Dalmally, Argyllshire : showing the early formation of biotite (with porphyroblastic habit) in a chlorite-schist.
B. Biotite-Sericite-schist, Ballachulish Slates, Loch Leven. Alternating bands, richer in dark or light mica or quartz, run obliquely to the direction of schistosity.

to the plane of schistosity (Figs. 96, *B* ; 97, *A*). The other chief constituents are white mica, chlorite, and quartz in very variable proportions, and we may recognize accordingly *biotite-schist, biotite-sericite-schist*, and *biotite-chlorite-schist*. In addition there are *albite-biotite-schists* with conspicuous porphyroblasts of pellucid albite.

In this and in most higher grades biotite is a characteristic product of metamorphism in argillaceous sediments of ordinary type, but its formation is necessarily dependent on the presence of the requisite materials. So we find, both in the biotite-zone and in the succeeding garnet-zone, *sericite-phyllites* carrying no biotite (Fig. 103). These represent highly micaceous rocks which orginally contained little or no chlorite. On the other hand, there are certain richly chloritic sediments originally devoid of potash-mica, though possibly containing kaolin. Rocks of this composition still figure as *chlorite-schists* even in an advanced grade of metamorphism. They show a relatively coarse-textured aggregate of chlorite-flakes with neither light nor dark mica, but often with some other aluminous mineral such as chloritoid or garnet (Fig. 98). The chlorite here has a pronounced colour and pleochroism, with fairly high birefringence. It is a variety rich in alumina, and these chlorite-schists of sedimentary origin are thus sharply distinguished chemically from others which represent ultra-

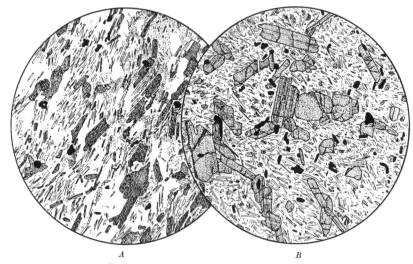

A   B

FIG. 97.—BIOTITE- AND CHLORITOID-SCHISTS ; × 23.
A. Biotite-Sericite-schist, Glenesk, Forfarshire.
B. Chloritoid-Sericite-schist, Drumtochty, Kincardineshire.

basic igneous rocks. The absence or scarcity of magnetite is characteristic in the same sense.

In still another type of sediment chloritoid is found, to the exclusion of biotite. Barrow [1] has described a *chloritoid-schist* of this kind from the biotite-zone in Kincardineshire (Fig. 97, *B*), and good examples are found on the coasts north of Stonehaven but the type is not a common one in the Highland region.

The gradual advance of metamorphism within the biotite-zone in rocks of more ordinary composition is marked by an increasing proportion of biotite, often in larger flakes, and by a corresponding diminution of muscovite, chlorite, and iron-ore, though these minerals are by no means exhausted. Probably too the composition of the

[1] *Quart. Journ. Geol. Soc.*, vol. liv (1898), pp. 149–55.

biotite itself undergoes a change, in the sense of becoming more ferriferous. This is indicated in many cases by a deeper colour and higher refractive index.[1]

In the Highlands, as already noted, garnet makes its appearance only at a more advanced stage of metamorphism than that indicated by the production of biotite. This is undoubtedly the usual sequence. In the Stavanger district of Norway, however, Goldschmidt [2] found that garnet appears before biotite. It is a *spessartine-almandine*, having the manganese and iron components in the proportion of about

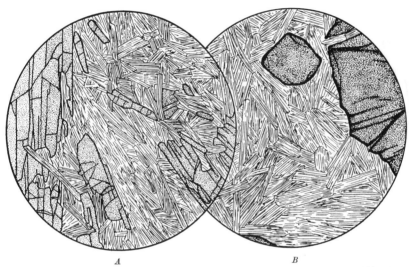

A   B

FIG. 98.—CHLORITE-SCHISTS ;   × 18.

The interlaced arrangement of the chlorite-flakes shows here little approach to an ideal parallelism. Magnetite is conspicuously absent.
A. Chloritoid-Chlorite-schist, Val d'Aosta, Piedmont.
B. Garnet-Chlorite-schist, Tirol.
These rocks belong to an advanced grade of metamorphism.

2 : 3. Eight specimens of these garnetiferous rocks, without biotite, gave an average of only 0·41 per cent. of manganous oxide : so it appears that a very moderate content of manganese suffices to determine the observed result. Goldschmidt found further that, when biotite in turn makes its appearance, the garnet which accompanies it is less manganiferous, and with advancing metamorphism gives place to an ordinary almandine. We have to recognize accordingly

[1] Kunitz has shown that the refractive index increases with the ratio FeO : MgO ; *Neu. Jahrb.*, Beil Bd. 1 (1924), p. 365.

[2] *Vidensk. Skr.* 1920, *No.* 10 (1921), p. 48. The same order seems to hold in the Mögster district, farther north : N. H. Kolderup, *Bergens Museums Aarbok,* 1923–24.

a special type of *garnet-chlorite-muscovite-schist*, without biotite, the garnet being of a variety rich in manganese (Fig. 99). Perhaps we should see here rather a postponement of biotite than a promotion of garnet; for we have already remarked that manganese-bearing garnet forms readily, under suitable conditions, in an ordinary thermal aureole, and there comes after biotite (p. 54). The sequence recorded by Goldschmidt has not been observed in the Scottish Highlands, but is exemplified in the Start Point district of South Devon [1] and in the Devonian slates of Cornwall.[2] In both areas, although the meta-

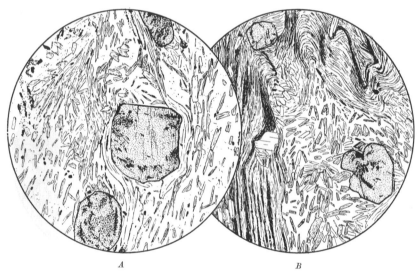

A   *B*

FIG. 99.—GARNET-CHLORITE-MUSCOVITE-SCHIST, Aamö, near Stavanger; × 23.

A. The principal constituent minerals are spessartine-almandine, chlorite, muscovite, quartz, and albite.
B. A more phyllitic variety, rich in muscovite and containing graphite.

morphism has not risen to the biotite stage, garnet is locally found, and in both it is of a variety containing 12 to 13 per cent. of the spessartine molecule.

THE ALMANDINE ZONE

The red garnet which is so widely distributed in the Dalradian and Moine mica-schists, and is taken as the index-mineral of a well-marked zone, is the 'common garnet', approaching *almandine* in composition, i.e. having a strong preponderance of iron in the

[1] Tilley, *Quart. Journ. Geol. Soc.*, vol. lxxix (1923), pp. 180, 189–90.
[2] Phillips, *Geol. Mag.*, vol. lxv (1928), p. 550.

dioxides.[1]   Unlike the manganiferous varieties, it is definitely a stress-mineral, as well as a high-pressure mineral.   It is formed mainly at the expense of the remaining chlorite in the rock (Fig. 82, above), but must also draw much of its iron from magnetite.   There is not at this stage any breaking down of the biotite, which would involve the simultaneous production of a potash-felspar.   A large crystal of garnet may be formed at the expense of a group of chlorite-flakes, and is then likely to contain numerous inclusions (Fig. 83, B).   Inclusions, chiefly of quartz, are very frequent, and doubtless arise in more than one way.   Sometimes they are irregularly distributed, giving the ordinary poeciloblastic or ' sieve ' structure.   Sometimes the inclusions have an orderly parallel arrangement, related to the structure, not of the crystal, but of the rock, being conformable with the general direction of schistosity (Fig. 100, B).

More rarely we may see a graphic intergrowth of thin plates of quartz with a regular arrangement in the garnet crystal [2] (Fig. 100, A). This is of considerable interest, for it brings out the fact that the crystal is, in such instances at least, polysynthetically twinned on the plan which is so general in the lime-garnets (p. 82).   In those it was made evident owing to the birefringence, and we have no means of telling how frequent the twin-structure may be in common isotropic garnets.   We have already made mention of the regular disposition sometimes observable in the minute opaque inclusions in garnet (p. 44). The planes along which these are distributed correspond with the planes which divide the twelve individuals of the complex twin-crystal, and the two kinds of orderly inclusions may be found together.

Garnet possesses a much greater force of crystallization than the other principal minerals present at this stage, and modifies the structure of the rock accordingly.   The crystals forcibly thrust apart the folia of mica, etc. ; and when they are of considerable size, as is often the case, a garnetiferous mica-schist shows an ' eyed ' structure like that of grained wood containing knots.   Quartz, as the most easily soluble mineral, often becomes collected into the corners of the eyes, under the protection of the garnet (Fig. 102, B).   A more or less perfect crystal outline is very general : it is that of the rhombic dodecahedron, sometimes modified on the larger crystals by narrow faces representing

---

[1] Heddle records manganous oxide amounting in one instance to $4\frac{1}{2}$ per cent., but it is well known that in these older analyses manganese is often greatly over-estimated : *Trans. Roy. Soc. Edin.*, vol. xxviii (1878), pp. 312–14.

[2] Heddle has figured what seems to be an example of this from granitic veins : *loc. cit.*, p. 299.   It is seen also in the well-known garnet of Bastogne in the Ardenne.

the trapezohedron. Even irregularly-shaped crystals seldom show any tendency to elongation in the plane of schistosity. It may be remarked, however, that the brittle garnet is very liable to be shattered in a later chapter of its history, the cracks often showing a regular arrangement, which has no relation to a crystallographic direction (Fig. 102, *B*). In extreme cases the fragments may be parted, and dragged out in a common direction. Effects of this kind are referable to the waning phase of metamorphism, when, with declining temperature, the crystals had lost their power of recuperation.

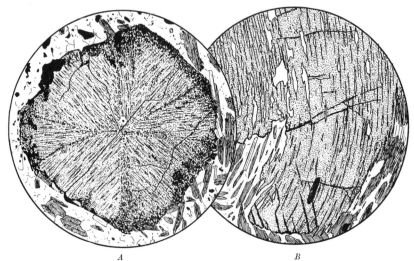

A   *B*

FIG. 100.—INCLUSIONS IN GARNET ; × 23.

A. Glen Lyon, Perthshire ; showing slender plates of quartz intergrown with garnet with parallel arrangement in different sectors. The arrangement makes evident the polysynthetic twinning of the garnet crystal. There are also opaque inclusions ? graphite), which tend to lie along the twin-planes and on the margin.

B. Piano di Segno, Lukmanier Pass. The form of the garnet and the arrangement of inclusions give a deceptive appearance of bending, being in fact an effect of gradual rotation during growth.

Porphyroblasts of garnet and other minerals, however, not infrequently exhibit irregularities and peculiarities which have a different significance, and belong to the period of growth. The growth of a large crystal under these conditions doubtless covers a certain duration of time, during which the enclosing matrix is undergoing progressive deformation. As a result of the differential movement the growing crystal is often sensibly rotated. This is most clearly seen when the crystal contains inclusions oriented (when originally enclosed) in the plane of schistosity. The inclusions in the border of the crystal conform with the actual plane of schistosity, but those in the interior have suffered rotation in varying degree, most in the core of the crystal.

This gives a deceptive appearance of the crystal having been forcibly bent (Fig. 100, *B*). In the case of a birefringent mineral like staurolite, it can be verified that the curved crystal gives uniform extinction between crossed nicols (Fig. 101, *A*). A crystal grown under such conditions may present a very irregular outline, always with a twisted appearance, **S**-shaped or suggestive of a vortex (Fig. 101, *B*), affording a visible picture and an actual measure of the rotatory effect.[1]

The chief minerals found in company with garnet in the stage of metamorphism now reached (almandine-zone) are, broadly, the same

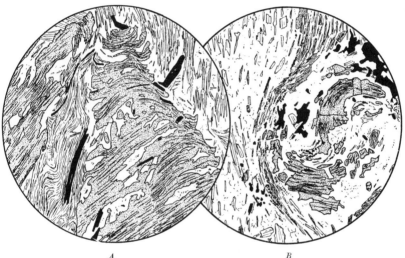

A   *B*

FIG. 101.—CURVED SHAPES OF CRYSTALS CAUSED BY ROTATION DURING GROWTH ; × 23.

A. Staurolite in phyllite, Somascona, near Lukmannier Pass. Despite their twisted appearance, each crystal extinguishes as a whole between crossed nicols.
B. Spiral Garnet in mica-schist, Airolo, St. Gotthard. The arrangement of inclusions indicates for the core of the crystal a rotation of more than 180 degrees.

that have been noticed in a lower grade, but tending in process of recrystallization to larger dimensions. It is clear, however, that some of the early minerals can survive only in diminishing quantity, inasmuch as new minerals have been continually forming at their expense. This change in the relative proportions of the minerals is, of course, not to be verified in random examples, since there was a wide variety of composition among the original sediments. The commonest type of *garnetiferous mica-schist* contains abundant light as well as dark

[1] See a discussion of this subject by Becke with abstracts of papers by Sander, Schmidt, and Backlund, *Fortschr. d. Min.*, vol. ix (1924), pp. 194–8. See also Mügge, *Neu. Jahrb. Min.*, Beil. Bd. lxi (1930), pp. 469–510 : Bailey, *Geol. Mag.*, vol. lx (1923), pp. 328–30.

mica (Figs. 91, 102).   Indeed, in some of the highly micaceous rocks which may be styled *garnetiferous phyllites*, biotite is scarce or absent (Fig. 103, *B*).   Some chlorite may remain, at least in the less advanced part of the garnet-zone ; but a certain ambiguity sometimes attaches to this mineral, for it is the most common product of alteration both of biotite and of garnet, belonging then to a later time of ' retrograde metamorphism '.   When chlorite is seen evidently replacing one of the higher silicate-minerals its late origin is manifest, but otherwise the manner of occurrence may fail to afford a decisive criterion.   Gold-

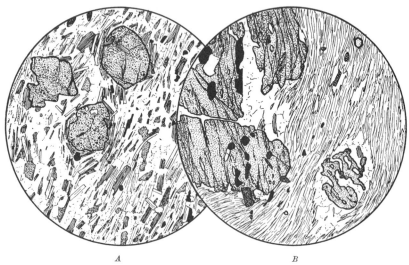

A   *B*

FIG. 102.—GARNETIFEROUS MICA-SCHISTS, Perthshire ;   × 18.

A.   Ben Lui Schists, Pitlochry.   The principal constituents are garnet (almandine), two micas, and quartz.
B.   Ben Bheula Schists, Aberfeldy : rich in muscovite.   Magnetite is enclosed in the garnet.   Small crystals of tourmaline represent detrital grains recrystallized.

schmidt [1] states that, in the rocks studied by him, original chlorite is always feebly negative in optical character, while derivative chlorite is as a rule either strongly negative or feebly positive.   Chlorite derived from biotite often encloses little needles of rutile, representing the titanium content of the destroyed mica, and this, when seen, may be regarded as a conclusive test.   Magnetite is still in noteworthy quantity in many rocks of this zone, and is often enclosed in the garnet. Here and in the succeeding staurolite-zone the biotite is often of a less ferriferous variety than that found in a lower grade.   This is due to the distinctive minerals, almandine and staurolite, taking up iron in preference to magnesia.   In the highest zones, characterized by

[1] *Vidensk. Skr.* 1920, No. 10 (1921), pp. 56–7.

cyanite and sillimanite, the biotite is often of a deeply coloured and presumably highly ferriferous variety.

A distinctive type, well represented in the Highlands, is a *garnetiferous albite-mica-schist*. The albite—an original constituent of the sediment, as already remarked—tends to come out in porphyroblasts, comparable in size with the garnet.

The *chloritoid-schists* of the garnet zone make a very distinctive type, the characteristic mineral being developed in crystals of conspicuous size. These are sometimes clear, sometimes poecilitic from

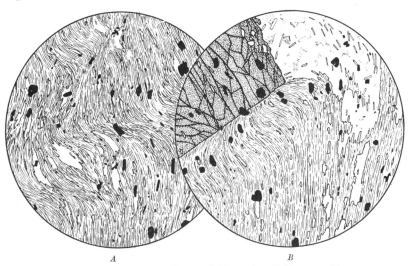

FIG. 103.—PHYLLITES, Leven Schists, Argyllshire ; × 18.
A. Ardmucknish : composed essentially of filmy white mica and quartz with some magnetite.
B. Glencoe : a higher grade of metamorphism, showing garnet in addition.

crowded inclusions of garnet, magnetite, graphite, etc. As usual, they lie at all inclinations to the plane of schistosity. We may mention here the well-known examples of the Ile de Groix, off the coast of Brittany,[1] though these probably belong in part to a slightly more advanced grade.[2] The chief component minerals are chloritoid, usually garnet, a white mica which is perhaps paragonite, and quartz, while the accessory constituents include magnetite, rutile, graphite, tourmaline, etc. Biotite is absent from the most typical rocks, but there are transitions to more ordinary mica-schists containing some chloritoid.

In this, as in lower and higher grades, there are rocks sufficiently

[1] Barrois, *Ann. Soc. Géol. Nord*, vol. xi (1884) pp. 22–45.

[2] Bonney mentions the occurrence of staurolite ; *Quart. Journ. Geol. Soc.*, vol. xliii (1887), p. 302.

rich in graphite to assume a black colour, but not otherwise differing from the common types (Fig. 84, above). We find also, in the same general grade of metamorphism, *graphitic mica-schists* without garnet; and it is possible, as maintained by Vogt,[1] Bailey,[2] and others, that the presence of carbonaceous matter in quantity tends to inhibit or delay the normal reactions.

In all these rocks the schistose structure, depending upon an approximate parallel orientation of the micas especially, is well marked, though often modified by the porphyroblastic development of particular minerals. At the same time foliation, in the proper sense, that is a partial segregation of the several minerals into parallel streaks, becomes increasingly noticeable. In this the mobility of the quartz, due to its relatively free solubility, plays an important part.

THE STAUROLITE AND CYANITE ZONES

Where, as in the south-eastern division of the Highlands, erosion has exposed a succession of definable zones in due order, the garnet-zone is a broad one. In other words, from the first appearance of garnet a considerable further rise of temperature must take place before the formation of any new mineral which will serve as index to another zone. The advance of metamorphism within the garnet-zone is marked at first by an increased production of the distinctive mineral at the expense of others, but afterwards chiefly by coarser texture and more evident foliation, as distinguished from schistosity. It is to be remembered that we are considering what is to be regarded as the normal case in regional metamorphism, viz. that in which shearing stress is maintained at or near its limiting value (p. 184). It is for this reason that the aluminous silicates andalusite and cordierite, so prominent in simple thermal metamorphism of like rocks, are here lacking. With more elevated temperature, and the weakening of shearing stress which goes with it, we come now to a stage at which the formation of notably aluminous silicates is no longer inhibited. The stable forms under these conditions are, however, not the same that present themselves in the absence of stress. Instead of andalusite we have *cyanite* (disthene), while the place of cordierite is to some extent taken by *staurolite*. Both minerals tend to a conspicuously porphyroblastic habit, and the coming in of one or other is easily

[1] *Norges Geol. Und.*, No. 121 (1927), pp. 198, 484.

[2] *Quart. Journ. Geol. Soc.*, vol. lxxviii (1922), p. 93 (Ballachulish Slates). The carbonaceous Blair Atholl Schists, however, contain abundant garnets. The retarding action is presumably of a mechanical kind, the particles of carbon forming a protective coating on the crystals. In the laboratory carbonaceous shales are only with difficulty dissolved by hydrofluoric acid.

marked in the field. The two are very often seen in association, but Barrow found that staurolite is the earlier one to appear, and he was able to lay down in the Highlands the limits of a definite staurolite-zone.[1]

Staurolite, though recorded from some thermal aureoles, is a typical stress-mineral. Being an iron-aluminosilicate very rich in alumina, it is to be looked for only in rocks which fall within certain limits of bulk-composition.[2] Richness in ferrous iron is the distinguishing character of staurolite, as it is of the lower type of silicate chloritoid, and accordingly staurolite forms rather than cyanite in that type of sediment which in a lower grade carries chloritoid :

$$5Al_2SiO_5 + 4H_2FeAl_2SiO_7$$
Cyanite   Chloritoid
$$= 2HFe_2Al_9Si_4O_{24} + SiO_2 + 3H_2O.$$
Staurolite

Pelikaan[3] has observed in some Alpine schists the actual process of conversion of chloritoid to staurolite. The reaction may be represented by the equation :

$$9H_2FeAl_2SiO_7 = 2HFe_2Al_9Si_4O_{24} + SiO_2 + 5FeO + 8H_2O,$$

the iron-oxide set free contributing to convert part of the white mica present to biotite. In the same rocks crystals of staurolite may sometimes be seen changing to chloritoid, proving that the reaction is a reversible one. In less aluminous rocks chloritoid, by taking up silica alone, may yield staurolite and garnet. At this stage accordingly chloritoid disappears, except only in rocks deficient in silica, where it may persist into the zones of cyanite and sillimanite.

Staurolite, appearing at first in small grains, soon develops into large idioblastic and often porphyroblastic crystals, which are commonly interpenetrating twins. They are often crowded with inclusions of quartz which has presumably become entangled in the building up of large staurolite crystals from numerous small grains. Other inclusions are found, but a regular arrangement of these within the crystal is rare (p. 44). The common associates of staurolite are garnet, biotite, muscovite, and quartz ; often also

[1] On the distribution of staurolite in the Alps see Weiss, *Zeits. Ferd. Tirol* (3), Heft 45 (1901), pp. 129-71.

[2] For a discussion of this matter see Suzuki, *Schw. Min. Pet. Mitt.*, vol. x (1930), pp. 121-26 : for analysis of a staurolite-schist from Forfarshire see Barrow, *Quart. Journ. Geol. Soc.*, vol. xlix (1893), p. 355.

[3] *Geologie der Gebirgsgruppe des Piz Scopi*, Inaug. Diss., Amsterdam (1913), pp. 23-4.

felspar of some sodic variety (Fig. 104, *A*). A schistose structure is sometimes well pronounced in the more micaceous types, though the staurolite crystals themselves show little or no regularity of orientation. In this and the succeeding zones, however, the influence of higher temperature and declining stress is increasingly evident in the textural and structural characters of the rocks, which eventually assume a relatively coarse grain and a gneissic structure (Fig. 105, *A*).

Judging by the very moderate width of the staurolite zone, where it can be distinctly laid down it appears that this mineral has not a

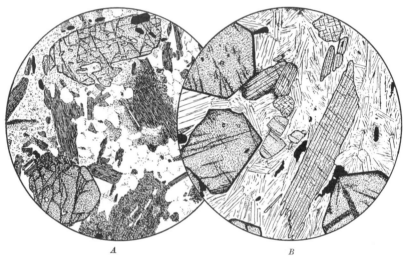

A   *B*

FIG. 104.—STAUROLITE- AND CYANITE-SCHISTS ; × 18.

A. Staurolite-Garnet-Mica-schist, Glen Clova, Forfarshire. The mica is mainly biotite. Besides quartz, there is a certain amount of felspar.
B. Cyanite-Garnet-Mica-schist, Glen Fernait, Perthshire : rich in muscovite. In the centre of the field some tourmaline is seen.

wide temperature-range of stability. It gives place, presumably, to cyanite and garnet :

6 Staurolite + 11 Quartz = 23 Cyanite + 4 Almandine + $3H_2O$.

Cyanite is, like staurolite, a mineral of strong crystallizing force, and develops idioblastic crystals, often of large size. Its common habit in the crystalline schists is a very simple one, though in segregation-veins with quartz it shows a considerable variety of forms. The associated minerals are the same as before, including at first staurolite, but the relative proportions of the minerals vary rather widely. There are silvery mica-schists, in which muscovite is, with cyanite, the main constituent (Fig. 104, *B*). From these phyllitic types cyanite-bearing rocks range to others coarsely gneissic (Fig. 105, *B*). The latter are often strongly foliated, the cyanite being gathered into aggregates instead of single crystals, while some seams are composed almost wholly of mica, and the quartz tends to collect in lenticles.

THE SILLIMANITE-ZONE

The increasing coarseness of texture in reconstructed argillaceous rocks in the highest grades of metamorphism completes the obliteration of any surviving traces of original structures.   In Norway, in the Alps, and elsewhere fossil remains of the larger and more robust organisms

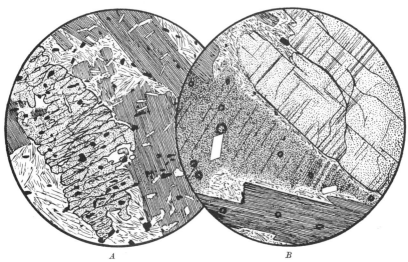

FIG. 105.—STAUROLITE- AND CYANITE-GNEISSES ;   × 18.

A.   Staurolite-Gneiss, Glenesk, Forfarshire.   A large staurolite crystal, with inclusions of quartz : the other chief constituents are biotite, muscovite and quartz.

B.   Cyanite-Gneiss, Loch Assapol, Ross of Mull.   The cyanite shows a cross-fracture connected with gliding-planes.   The biotite has dark haloes surrounding inclusions of zircon.

are exceptionally preserved up to the garnet stage.   For example, the Bündnerschiefer (Lias) in certain localities contain recognizable belemnites in rocks rich in garnet.   In higher grades such evidence is lost, owing to the enlarged amplitude of diffusion as well as to actual deformation of the rocks.

In the highest grade of all cyanite gives place to *sillimanite*.   The relation between these two dimorphous minerals is still obscure,[1] and there is, it would seem, no theoretical ground for believing that the

[1] Vernadsky believed that he had converted cyanite to sillimanite by heating to a high temperature, but this is shown to be an error.   Greig finds that under laboratory conditions, both cyanite and sillimanite when heated gradually change to mullite ($3Al_2O_3.2SiO_2$) with separation of silica.   *Journ. Amer. Ceram. Soc.*, vol. viii (1925), pp. 465–83.

conversion of the lower to the higher form takes place at a precise temperature (for given pressure). None the less it appears that the one mineral does in fact give place to the other with considerable regularity, so that it is possible to lay down a boundary-line on the map. It is to be remarked, however, that the sillimanite is derived in part from another source. These most highly metamorphosed rocks often contain a considerable amount of potash-felspar, a mineral not found (in rocks of argillaceous composition) in any lower grade of regional metamorphism. This indicates that the dissociation of

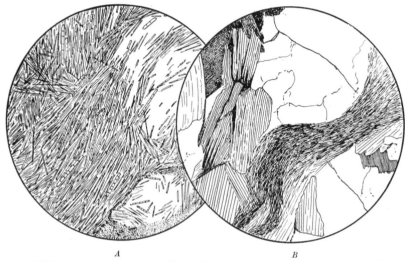

A   *B*

FIG. 106.—SILLIMANITE-GNEISS, Clova, Forfarshire ; two slices cut from the same specimen ; × 18.

A. A lenticle rich in sillimanite, which occurs as a dense crowd of little needles embedded in quartz.

B. A micaceous part, with muscovite and some biotite. Quartz is present and a compact felt of sillimanite needles enclosed in mica. Another part of the slice has abundant garnet.

muscovite is no longer inhibited by the declining stress, and a corresponding quantity of alumina is set free in the process. It is possible that biotite also sometimes suffers at this advanced stage a like dissociation, but yielding garnet instead of sillimanite.

Sillimanite, in its most usual habit, differs from the other high-temperature silicates already noticed, presenting itself commonly in the form of numerous slender prisms or needles with a distinctive felted or quasi-fluxional arrangement. This has no necessary relation to any direction of schistosity, which indeed is often little in evidence. Swarms of needles are sometimes embedded in quartz or muscovite, sometimes wind stream-like between the larger crystals (Fig. 106).

The more highly micaceous type, when it persists into this highest grade, may be styled *sillimanite-schist*, but for the most part schistosity has now given place to a gneissic structure with rude but strongly marked foliation. The *sillimanite-gneisses* of the Highlands are often of as coarse grain as the gneissic granites which are in many places intruded among them ; and, where the latter have been injected in ' lit-par-lit ' fashion, it is not always easy to discriminate the two rocks, except by the presence of the distinctive streaks of sillimanite. The other common minerals are quartz, garnet, light and dark micas, oligoclase or some near variety of felspar, and not infrequently ortho-clase. These, however, are not uniformly mingled, but show a strong tendency to segregate into lenticles and streaks with a common orientation. There are lenticles and knots of quartz crowded with sillimanite needles, others mainly of garnet with interstitial quartz, and inconstant bands very rich in mica, which also often contain sillimanite. On a weathered surface the quartz-sillimanite-knots show in relief with a characteristic dead white colour and silken lustre (' Faserkiesel ' or ' quartz sillimanitisé ').

It is manifest that the replacement in the most advanced grade of metamorphism of the denser cyanite by the less dense sillimanite cannot be attributed to great hydrostatic pressure. We have to recognize here the influence of elevated temperature, including especially its indirect effect in bringing about a declension of shearing stress. Garnet, a high-pressure as well as a stress-mineral, normally remains stable to the last. Not all argillaceous sediments are rich enough in alumina to yield the distinctive index-minerals staurolite, cyanite, and sillimanite ; and we find therefore some rocks in these advanced grades carrying no mineral higher in the scale than garnet. A *garnet-gneiss*, without sillimanite but showing the same coarse texture and the same tendency to a marked segregation-foliation, is represented in some Highland districts.

Another well-characterized type of the highest grade is *cordierite-gneiss*. The production of cordierite, often in abundance, in regional metamorphism must be interpreted, in the light of what we have seen, as indicating a very decided relaxation of shearing stress at the high temperatures reached. There is, however, another consideration which seems to be relevant. When we examine the actual distribution of cordierite-bearing rocks in the Scottish Highlands, outside thermal aureoles, it appears that such rocks are found in force especially in those districts where we have reason to believe that shearing stress did not at any stage attain its full measure. For this reason it will be convenient to defer notice of the cordierite-gneisses to the next chapter.

CHAPTER XV

REGIONAL METAMORPHISM OF NON-CALCAREOUS SEDIMENTS (*continued*)

Metamorphism under Deficient Shearing Stress—Regional Metamorphism of Aluminous and Ferruginous Deposits—Metamorphism of the Purer Arenaceous Sediments—Metamorphism of More Felspathic Types—Metamorphism of Impure Arenaceous Sediments—Pneumatolysis and Metasomatism in Regional Metamorphism.

METAMORPHISM UNDER DEFICIENT SHEARING STRESS

IN attempting in the preceding pages to follow the metamorphism of ordinary argillaceous sediments through successive grades, we have supposed the conditions to be such that the dynamic factor exerts its influence to the full. This rule appears to be so widely applicable, that we have been able to regard it as representing, in a sufficiently definite sense, the normal régime (p. 184). It remains now to notice, though only summarily, the less common case, in which shearing stress has been gravely in defect, as compared with its maximum possible value at the actual temperatures realized. In respect of the ruling physical conditions, this case is therefore intermediate between purely thermal and normal regional metamorphism, and this mixed character is reflected in the mineralogical constitution of the metamorphosed rocks. Illustrations are to be found at many places throughout an extensive tract of the north-eastern Highlands of Scotland, comprising the greater (northerly) part of Aberdeenshire and extending into north-eastern Banffshire (p. 186). Here the metamorphosed sediments have the characteristic structures of crystalline schists and (in the highest grades) gneisses, but, considered mineralogically, present some striking anomalies.

A very noteworthy feature in the lower grades of metamorphism is the relatively early appearance of biotite. In normal regional metamorphism (as compared with purely thermal) the effect of intense shearing stress was to defer the production of this mineral (p. 212). This influence is now much less operative, and biotite may figure in rocks which still retain much of their original clastic structure.

230

Abundant brown mica is found in company with angular or only
partially dissolved grains of quartz, and in more gritty sediments
with fragments of felspar, while any surplus of cericitic material that
may remain has not yet been built up into larger flakes of muscovite.
This is well seen even in the biotite-schists of the Stonehaven district,
which are thus contrasted with those of Perthshire and Argyllshire
(Fig. 107).   From the zonal point of view, the biotite-zone is extended
at the expense of the chlorite-zone.

A more striking peculiarity is the widespread and abundant

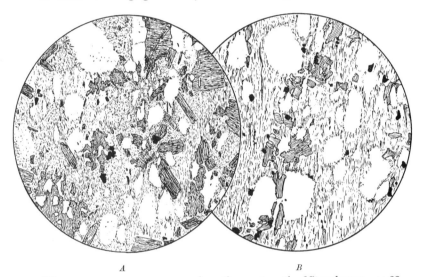

A   B

FIG. 107.—BIOTITE-SERICITE-SCHISTS, from the coast north of Stonehaven ;   × 23.

The detrital quartz-grains still retain their shape, or show only partial corrosion of
their borders, although biotite has been abundantly developed.

presence of andalusite and cordierite.   These minerals, so commonly
found in contact-aureoles and conspicuously absent from normal
crystalline schists, are clearly marked as anti-stress minerals ; and
their occurrence here can be attributed only to a decided relaxation
of shearing stress during the metamorphism which gave birth to
them.   Their wide distribution is quite distinct from their local
occurrence in aureoles about the Caledonian intrusions, which is also
illustrated in the same tract ; nor does it show any relation to visible
intrusions of the ' older granites ', belonging to the period of the
regional metamorphism itself.[1]   The critical minerals in question are
found in rocks belonging to every grade of metamorphism.   In the

[1] See Memoirs of the Geological Survey, especially Read, *Geology of the
Country round Banff, etc.* (1923).

biotite zone the Banff and Fyvie Series include ' knotted schists ', in which the knots are nascent crystals of andalusite and cordierite. In a more advanced grade andalusite often makes conspicuous porphyroblasts. It is found in association with biotite, with garnet, with staurolite, and with sillimanite (Fig. 108). It may be observed that sillimanite, in the form of scattered slender needles, is not always restricted to rocks having all the characteristics of the highest grade, but may have a rather wider distribution in a sporadic fashion. This accords with the generalization already advanced, that in an area

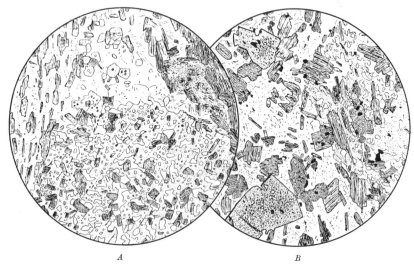

A   B

FIG. 108.—ANDALUSITE-BEARING MICA-SCHISTS ; × 23.

   A. Andalusite-Biotite-schist (' knotted schist '), Braes of Gight, near Fyvie, Aberdeenshire. The knots are large poeciloblastic crystals of andalusite, enclosing quartz and biotite. A few needles of sillimanite are present. The general mass of the rock has a schistose structure.
   B. Andalusite-Staurolite-schist, near Banff. The conspicuous minerals are staurolite, biotite, and andalusite : the rest is principally quartz and muscovite.

characterized by deficient shearing stress the distinction between successive zones, so well marked in normal regional metamorphism, is less clearly traceable. We may recognize, however, as distinctive types, *andalusite-biotite-schist, andalusite-garnet-schist,* and *andalusite-staurolite-schist,* in ascending order as named.

    A review of the literature of crystalline schists shows that the andalusite-staurolite-association is decidedly rare and that of andalusite and cyanite even more so. The former has been recorded by Streckeisen [1] in the Flüela district, near Davos, and by Suzuki [2] at

---

    [1] *Schw. Min. Pet. Mitt.,* vol. vii (1928), pp. 124–9.
    [2] *Ibid.,* vol. x (1930), pp. 117–32.

Piodina, near Lago Maggiore. At the Flüela locality cyanite is like-wise found. It is suggested that the formation of andalusite belongs to a later phase of metamorphism, when the normal physical condi-tions had undergone a change. The decline of temperature, where it has a regional distribution, cannot be other than a very gradual process, but this does not necessarily hold good of the stress-con-ditions. If we suppose, after the production of staurolite, a decided falling off of shearing stress, while the temperature still remained high, the conditions would seem suitable for the formation of andalusite.

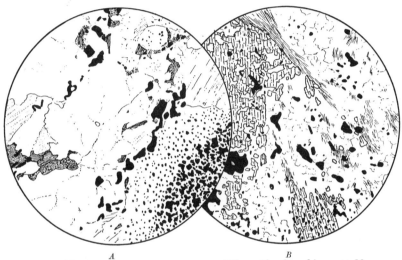

A   B
FIG. 109.—CORDIERITE-GNEISSES, near Ellon, Aberdeenshire ; × 18.

A. The cordierite contains abundant inclusions of pleonaste. The other minerals shown are biotite, magnetite, quartz, and felspar, while sillimanite and muscovite occur elsewhere in the slice.
B. Sillimanite-Cordierite-Gneiss, with grains of pleonaste.

It could come only at the expense of staurolite and other aluminous silicates, but the bulk of the staurolite, though theoretically unstable, might well survive in virtue of chemical inertia. Suzuki's explana-tion is not very different from this, and Streckeisen definitely regards the andalusite as a late product from staurolite, sillimanite, and cyanite. This, it should be observed, is quite different from a second metamorphism, as e.g. when a staurolite-schist is invaded by a later granite intrusion.

Many of the andalusite-bearing schists mentioned above carry cordierite in addition, but it is especially in certain rocks of the highest grade, with gneissic habit, that the latter mineral becomes a constituent of the first importance (compare p. 229). Such *cordierite-gneisses*, often of coarse grain and with strong foliation, are abundantly represented at Banchory in Deeside, Ellon in eastern Aberdeenshire, Cabrach on the border of Banffshire, and elsewhere. In some of these rocks, originally richly chloritic sediments, cordierite makes a large part of the total bulk. Very characteristically it is crowded with little grains of pleonaste (Fig. 109) or in other examples of magnetite (Fig. 110). Except where it is in contact with a potash-felspar, the cordierite is xenoblastic, but the large crystals sometimes assume a lenticular habit with trains of mica, etc., winding about them. In these coarse-grained rocks the place of andalusite is usually taken by

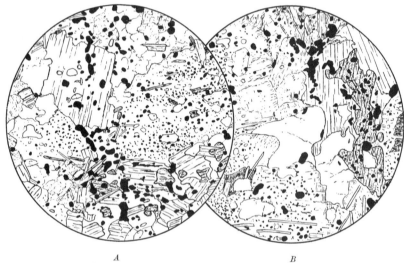

A   *B*

FIG. 110.—CORDIERITE-MUSCOVITE-GNEISS, Buck of Cabrach, Banffshire ; × 23.

A. The principal minerals are cordierite, muscovite, and quartz, with very numerous little grains of magnetite, especially as inclusions in cordierite. There are also present microcline (top and left side) and andalusite (in lower part, associated with muscovite), besides a little biotite and a few small crystals of tourmaline.
B. The same minerals are shown, including a large crystal of andalusite on the right. The muscovite occurs both in large crystals and in aggregates of small scales.

sillimanite. As a rule, the mica is biotite only, muscovite having given place to potash-felspar, which may become an important constituent—a feature doubtless related to the diminished influence of the stress-factor. The Cabrach rocks, however, which must have been exceptionally rich in original sericite, still contain abundant muscovite as well as microcline (Fig. 110). Since the conversion of a mica to a felspar takes up a considerable amount of silica, the proportion of quartz present is variable, being lowest in the most felspathic rocks.

In many of these rocks, e.g. in those of Cabrach, garnet is conspicuously wanting. Its absence is easily explained. The cordierite being apparently of a richly ferriferous variety, we can write down such equations as :

4 Garnet + 2 Musc. + 3 Qtz. = 3 Cordt. + 2 Biot.
10 Garnet + 5 Musc. = 6 Cordt. + 3 Pleon. + 5 Biot.

or again :

2 Garnet + 4 Musc. + 9 Qtz. = 3 Cordt. + 4 Ortho. + 4H$_2$O.
Garnet + 2 Musc. = 3 Pleon. + 2 Ortho. + 2H$_2$O.

Here the left side may be taken to represent the normal stress-association and the right that which replaces it when shearing stress is no longer effective. In other rocks of this kind, however, and notably in the Ellon district, garnet is present in varying amount. These *garnet-cordierite-gneisses* must therefore represent, with respect to the physical conditions governing their production, an intermediate case between that of simple cordierite-gneisses and that of ordinary garnetiferous gneisses. It appears that the balance represented by the above equations is one which is very sensitive to variation in the intensity of shearing stress. The garnet is often to be observed in process of destruction, being replaced by biotite, cordierite, and pleonaste, indicating a decline of the stress-factor while high temperature still continued.

The foregoing observations are enough to show that the tract extending northward from Deeside presents some striking peculiarities in comparison with other parts of the Highlands.[1] They are such as can be explained only by supposing a very decided relaxation of the normal stress conditions during regional metamorphism. It should be added that anomalies of this kind are not confined to the pelitic sediments. They have been noted in the same region by Hutchison [2] in semi-calcareous rocks and by Wiseman [3] in the group of basic sills known as epidiorites.

REGIONAL METAMORPHISM OF ALUMINOUS AND FERRUGINOUS DEPOSITS

In discussing the simple thermal metamorphism of argillaceous rocks, we saw that, when the original sediment was decidedly poor in quartz, the deficiency of silica showed itself in a sufficiently advanced grade by the formation of certain characteristic non-silicate minerals, viz. corundum and various spinels (p. 59). In crystalline schists of like composition these minerals are not found. The reason of this

---

[1] The attempt of Miss Elles to distinguish Barrow's zones on the Banffshire coast serves to illustrate the discrepancy ; *Geol. Mag.*, vol. lxviii (1931), pp. 24–34.

[2] *Trans. Roy. Soc. Edin.*, vol. lvii (1933), p. 590.

[3] *Quart. Journ. Geol. Soc.*, vol. xc (1934), pp. 405–7.

is to be seen partly in the deferring under stress-conditions of the muscovite-orthoclase reaction (which would set free alumina), but mainly in the absence here of the characteristic anti-stress mineral cordierite, the place of which is largely taken by garnet. We have already pointed out relations which illustrate this antagonism, and it is easy to set down others:

$$2 \text{ Garnet} + 9 \text{ Cyan.} = 3 \text{ Cordt.} + 5 \text{ Corund.}$$
$$\text{Garnet} + 2 \text{ Cyan.} = \text{Cordt.} + \text{Pleonaste.}$$
$$2 \text{ Garnet} + 2 \text{ Staur.} + \text{Cyan.} + 10 \text{ Qtz.}$$
$$= 5 \text{ Cordt.} + 2 \text{ Corund.} + H_2O.$$

In each of these equations the stress-association on the left is contrasted with the anti-stress on the right.

But, although corundum and the spinel minerals do not figure among the products of normal regional metamorphism in the argillaceous sediments proper, it is otherwise when we turn to those less common types which represent bedded deposits of aluminium, iron, and manganese ores. In these rocks, containing little or no silica, such minerals as corundum and various spinellids (including magnetite) become constituents of the first rank. From bauxitic deposits, composed largely of aluminium hydrates, come the beds of *emery* which are found associated with crystalline schists, marbles, etc., in some countries. The best-known examples are those of the Island of Naxos [1] and of Asia Minor.[2] They come from more or less ferruginous bauxites, so that the essential corundum is mingled with magnetite in varying proportions. In such a rock there can be no schistosity, but there is often a rude foliation, due to segregation of the corundum into lenticles (Fig. 111). Of other minerals, margarite and chloritoid are the most characteristic. They answer respectively to the anorthite and fayalite which form in rocks of like composition in simple thermal metamorphism (pp. 63–65), and now, under stress conditions, take up additional alumina and water:

$$CaAl_2(SiO_4)_2 + Al_2O_3 + H_2O = H_2CaAl_4Si_2O_{12};$$
$$Fe_2SiO_4 + Al_2O_3 + H_2O = H_2FeAl_2SiO_7 + FeO;$$

The FeO set free goes to make magnetite or pyrites. Pyrites is not uncommon in Naxos, as also is tourmaline (Fig. 111, C), indicating the co-operation of the pneumatolytic factor. Muscovite, biotite, cyanite, and sillimanite are occasional accessory constituents, due to a certain

---

[1] Zirkel, *Tscherm. Min. Petr. Mitt.*, vol. xiv (1895), pp. 311–42 ; Papavasiliou, *Zeits. Deuts. Geol. Ges.*, vol. lxv (1913), pp. 1–123.

[2] Krämer, *Kleinasiatische Smirgelvorkommnisse* (Inaug. Diss., Berlin, 1907).

admixture of ordinary argillaceous material, and indeed there are transitions from emery to mica-schist, as well as to marble.

The iron-ores having this mode of origin exhibit some variety.   In a low grade of metamorphism the iron is in the form of haematite.   In Brazil and in North Carolina occur thick beds of *haematite-schist* (itabirite) composed of little flakes of that mineral with a variable amount of quartz, and showing a perfect schistosity.   The bedded iron-ores of northern Norway present a different type, in a higher grade of metamorphism.   They occur as very numerous intercalations in a

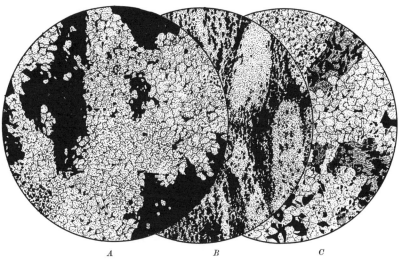

A   *B*   *C*

FIG. 111.—EMERY, Isle of Naxos, Ægean Sea ;   × 23.

A.   Koxaki : a granular aggregate of corundum and magnetite, showing a strong tendency to segregation.
B.   Mavropetri : a fine-grained variety with well-marked foliation :   corundum largely replaced by finely-divided diaspore : some chloritoid.
C.   Prosthini Skaphi : a tourmaline-bearing variety.

series of mica-schists and marbles, and were believed by Vogt [1] to represent chemical sediments (lake deposits).   Here both haematite and magnetite are found, together with quartz and various silicates, especially epidote.   Where any noteworthy quantity of manganese is present, garnet is a characteristic mineral, though as much as 3 per cent. of manganous oxide may go into the magnetite.   In an advanced grade of metamorphism haematite gives place more or less completely to magnetite, and *bedded magnetite ores* are found in various countries in circumstances which mark them as regionally metamorphosed sediments.   With the ore-mineral occur quartz and various silicates, sometimes also sulphides.   It is not always possible to decide whether

[1] *Zeits. prakt. Geol.*, vol. xi (1903), pp. 24–8, 59–65.

the original sediment has been in the state of hydrate or of carbonate.

It is to be remarked that in all these rocks, although quartz is constantly associated with the magnetite or haematite, there is no sign of chemical reaction between the iron-oxide and the free silica. It appears, however, that, under some conditions which we are not able to predicate, siliceous and otherwise impure iron-ores may undergo the same changes which are exceptionally recorded also in a contact-aureole (p. 65), giving rise to fayalite, grunerite, and other minerals.

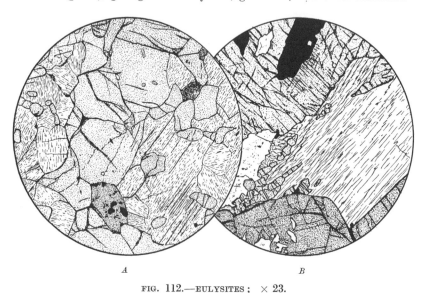

FIG. 112.—EULYSITES ; × 23.

  A. Tunaberg, Sweden : composed essentially of a yellowish pleochroic mangani-ferous fayalite and pale green hedenbergite : one crystal of red garnet.
  B. Collobrières, Dept. Var : showing fayalite, almandine garnet, fibrous grunerite, magnetite, and some felspar.

Here we may place the rare rock-type known as *eulysite*, usually massive but sometimes rudely foliated. The characteristic mineral is fayalite, which in the Swedish occurrences (Tunaberg,[1] Mansjö[2]) is a manganese-bearing variety (knebelite). With this may be associated hedenbergite, a hypersthene very rich in iron,[3] (grunerite, red garnet (almandine-spessartine), and minor constituents (Fig. 112, *A*). Some associated rocks at Tunaberg are composed almost wholly of grunerite

[1] Palmgren, *Bull. Geol. Inst. Upsala*, vol. xiv (1917), pp. 109–228.

[2] v. Eckermann, *Geol. För. Stock. Förh.*, vol. xliv (1922), pp. 253–93.

[3] Originally described as iron-anthophyllite, but see Henry, *Min. Mag.*, vol. xxiv (1935), pp. 221–6.

and garnet with little or no fayalite. According to von Eckermann,
the amphiboles result from the reaction of fayalite with the quartz of
neighbouring pegmatites ; but in some of the rocks, and notably in
a very similar type described by Lacroix [1] from Collobrières, near
Toulon, grunerite seems to play a primary part (Fig. 112, *B*). Tilley [2]
has described typical eulysites from the Loch Duich district in Ross-
shire. They are associated with hedenbergite-garnet-rocks, grunerite-
garnet-rocks, and other allied types, making part of a group of
paragneisses enclosed in the Lewisian orthogneisses. Here again the
rocks are rich in manganese, which is contained in both fayalite and
pyroxene.

Grunerite-schists are rather widely distributed in the iron-bearing
(Animikie) series of the Lake Superior region.[3] In most instances the
original rock has been a cherty ferrous carbonate, and the first stage
of metamorphism is the conversion of the chalybite to magnetite.
In a higher grade grunerite is formed, though some magnetite may
still remain, in association with the silicate and in separate bands
with quartz. With the grunerite may occur actinolite and intermediate varieties, often also garnet and other minerals. In some
districts, such as that of Mesabi in Minnesota,[4] what is regarded
as the original rock is composed largely of a ferrous silicate or hyd-
rous silicate (' greenalite '), to which the formula [5] assigned is
$3FeO . 4SiO_2 . 2H_2O$. This, however, may give rise to chalybite. As
before, progressive metamorphism gives first magnetite and then grunerite.[6] These changes, which, as it seems, develop gradually in a distance of forty or fifty miles, belong manifestly to regional metamorphism.
The Mesabi grunerite-schists, however, enter the aureole of the large
gabbro intrusion of Duluth, and have there suffered metamorphism
of the purely thermal type.[7] The resulting rock is a coarse-grained
aggregate of quartz, magnetite, olivine (which is often fayalite),
hypersthene, augite, hornblende, and occasionally some grunerite or
cummingtonite.

Of *manganiferous rocks* in regional metamorphism the most inter-

[1] *Bull. Soc. Fra. Min.*, vol. xl (1921), pp. 62–9.

[2] *Min. Mag.*, vol. xxiv (1936), pp. 331–42.

[3] Van Hise and Leith, *Monog. lii U.S.G.S.* (1911) ; Richarz, *Journ. Geol.*,
vol. xxxv (1927), pp. 690–707.

[4] Leith, *Monog. xliii U.S.G.S.* (1903), pp. 100–68.

[5] Jolliffe, *Amer. Min.*, vol. xx (1935), pp. 405–25.

[6] According to Richarz (*loc. cit.*, pp. 701–2), greenalite may also yield a crystallized ferrous silicate which has not the properties of an amphibole, and seems
to be unknown elsewhere.

[7] Grant, *Bull. Geol. Soc. Am.*, vol. xi (1900), pp. 503–10.

esting are those described by Fermor [1] from the ' Gondite Series ' of
the Central Provinces of India. They are crystalline rocks, massive
or foliated, derived from deposits of siliceous manganese-ores. The
most distinctive minerals are manganese-garnets and various manganiferous pyroxenes and amphiboles. Olivine-minerals (fayalite, knebelite, tephroite) are not found. Part of the ore often remains in the
oxide form, especially as braunite, $3Mn_2O_3 . MnSiO_3$ (Fig. 113) ; but
it is evident that the manganese-oxides in general react with free
silica more freely than do those of iron. Crystalline schists rich in

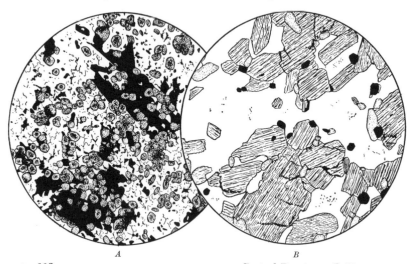

A   B

FIG. 113.—METAMORPHOSED MANGANESE-ORES, Central Provinces, India ; × 18.

A. Gondite (spessartine-quartz-rock) with braunite, Ukua, BáFighát district.
There is also much secondary manganese-oxide in the garnet.
B. Blanfordite-Quartz-rock, Kácharwáhi, Nágpur district. Blanfordite is a manganese-bearing monoclinic pyroxene with very distinctive pleochroism (rose, lavender,
and blue). The other minerals shown are braunite and apatite.

manganese and reproducing some of the Indian types have been
described from the Gold Coast.[2]

METAMORPHISM OF THE PURER ARENACEOUS SEDIMENTS

We shall consider in the first place ordinary sandstones or grits,
supposed sensibly free from calcarous matter, in which the constituent
grains are closely packed together with only a minimum of interstitial
cementing material. Owing to their superior rigidity, the behaviour

[1] *Mem. Geol. Sur. Ind.*, vol. xxxvii (1909). See also Hezner, *Neu. Jahrb.
Min.*, 1919, pp. 7–28.
[2] Kitson, *Tr. Amer. Inst. Min. Met. Eng.*, vol. lxxv (1927), pp. 372–84, and
Junner, *ibid.*, pp. 385–95.

of these rocks in the lowest grades of metamorphism is very different from that of argillaceous sediments. Any interstitial sericitic and chloritic material recrystallizes as a fine-textured aggregate, with or without evident parallel arrangement ; but the grains of quartz and felspar, which make the chief bulk of an ordinary grit, suffer at first mainly mechanical accidents. The quartz develops strain-shadows and sometimes regularly oriented trains of secondary fluid-pores (Fig. 114, *B*). Various cataclastic effects follow, such as have been described under the head of dynamic metamorphism, the normal result

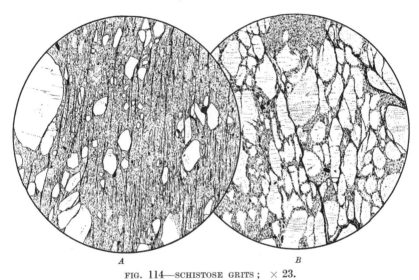

A   *B*

FIG. 114—SCHISTOSE GRITS ; × 23.

A. Inellan, Cowal, Argyllshire. A very argillaceous variety, in which the matrix behaves like a slate, showing a system of parallel flaws, the quartz-grains being merely rotated into a common orientation. Some small flakes of clastic mica still remain, with much new filmy sericite (p. 249).

B. Glen Rosa, Arran. There is only a beginning of cataclastic breaking down, but the effect of strain is seen in rows of secondary fluid-pores traversing the quartz-grains at right angles to the direction of tension. There is abundant fine sericitic matter (in other parts of the slice also epidote).

tending to a platy or lenticular fracture with the common direction imposed by the orogenic forces. With this go the mineralogical changes proper to shearing stress at low temperatures. Besides the recrystallization of any sericite and chlorite originally present, there is a production of new sericitic mica at the expense of grains of potash-felspar and sometimes also of new chlorite from detrital biotite or hornblende. These appear especially as thin films coating surfaces of discontinuous movement, and felspar within the relatively unbroken lenticles largely escapes sericitization. Total destruction of felspar, with the reduction of a felspathic sandstone to a sericite-quartz-schist, is an extreme result of dynamic metamorphism occurring only locally, as in the immediate neighbourhood of a great overthrust.

All the earlier stages of metamorphism are well seen in the belt of *schistose grits* (Ben Ledi Grits, etc.) exposed a little within the Highland Border. The first definite landmark of advance is the recrystallization of the clastic quartz. At first the shapes of the original grains are still discernible, if there is sufficient interstitial material to outline them (Fig. 115, *A*). This residual structure,

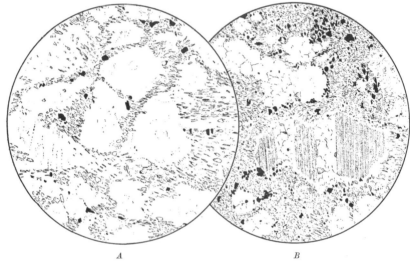

A   *B*

FIG. 115.—SCHISTOSE GRITS IN A HIGHER GRADE OF METAMORPHISM ; × 23.

A. Bull Rock, Cowal, Argyllshire. Here the quartz is recrystallized, but the outlines of the grains, with ragged edges, are indicated by the distribution of the secondary chlorite and sericite.

B. North coast of Arran. Here recrystallization has proceeded farther, and the outlines of the grains are being lost. A crystal of albite is merely broken and dragged apart, quartz filling the interspaces. The recrystallized iron-oxide figures as little octahedra of magnetite.

however, is soon obliterated owing to the relative mobility of quartz at the temperature now reached. Of the other main constituents, the iron-oxide has already recrystallized as magnetite at an earlier stage, but clastic felspar remains after the quartz is wholly recrystallized (Fig. 115, *B*), and the same is true of tourmaline, which is widely distributed in the Dalradian arenaceous sediments. With the exception of magnetite, and in carbonaceous sediments graphite, the only new mineral produced in a low grade of metamorphism is chloritoid. This seems to form as readily in arenaceous as in argillaceous sediments, and is recorded in many localities in the Alps and elsewhere.[1] It

[1] Niggli, *Ber. Math. Phys. Kl. Kön. Sachs. Ges.*, vol. lxvii (1915), pp. 238-9.

occurs in some of the schistose grits of the Highlands, e.g. in the Stonehaven district.

A special type of rock seen in the Loch Lomond and Cowal districts is worthy of notice. It has been remarked that some parts of the Ben Ledi Grits are very rich in clastic grains of a sodic felspar (p. 212). These are sometimes so abundant as to constitute nearly half of the rock. Recrystallization gives rise to pellucid grains of untwinned albite, with a strong tendency to porphyroblastic develop-

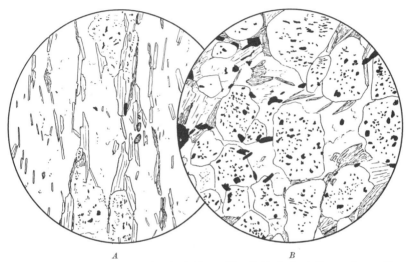

A   *B*

FIG. 116.—ALBITIC ROCKS, in Ben Ledi Grits, Glen Falloch, Loch Lomond ; × 23.

A. A common type with evident schistose structure.
B. A richly albitic variety. The albite contains numerous inclusions. The other constituents are quartz, muscovite, chlorite, and magnetite. The crystalloblastic order is : magnetite, albite, muscovite, chlorite, quartz. These rocks are in the ' biotite-zone ', but contain no biotite.

ment, and usually enclosing numerous granules of magnetite (Fig. 116).

In the more ordinary quartzose grits this conspicuous habit is not found, and any potassic felspar present, recrystallizing soon after the sodic, likewise takes its place with quartz in a general mosaic. This stage coincides with a somewhat increased mobility of the quartz in accord with the rising temperature, with the result that the transition is rather rapid from a rock which may be described as a recrystallized grit to a typical *quartzite* with its characteristic granoblastic structure. Since this structure holds, usually with some increase of grain-size, up to the highest grade, the progress of metamorphism in rocks of this kind is indicated more especially by the behaviour of those minor constituents which figured originally in the interstices of the quartz and felspar grains.

Setting aside the partly calcareous rocks, and dismissing the case of a ferruginous cement, which can yield only magnetite, the constituents to be noticed are broadly the same that make the main bulk of argillaceous sediments, including white mica and chlorite as of first importance. We must expect therefore that the same reactions will give rise to the same new minerals, but it is not to be assumed that the reactions will be equally prompt when the argillaceous

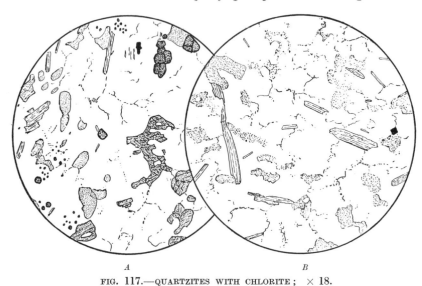

A   *B*

FIG. 117.—QUARTZITES WITH CHLORITE ; × 18.

A. Loch Lomond. This has been a slightly calcareous grit, and contains some calcite and a little of a lime-bearing garnet, which forms in a very early stage of metamorphism.
B. Glen Clunie, Aberdeenshire. Some felspar enters with the quartz. This rock is in the staurolite-zone.

material is disseminated through a mass mainly of quartz. It is necessary that the reacting bodies shall be so close together as to be included in the sphere of diffusion, and in the lower grades of metamorphism this sphere is narrowly limited. If muscovite is the only minor constituent, it naturally recrystallizes without change. So too chlorite, in the absence of micaceous material, may pass up into a high grade (Fig. 117, *B*), and the same is true of magnetite. When these various constituents are all present, but only sparsely distributed, they may still fail to react at the appropriate temperature, and so the production of biotite is necessarily deferred. Even when those constituents are present in close proximity, they are sometimes seen to have recrystallized without reaction, although biotite has begun to form in associated argillaceous beds. Here we are probably to recognize the influence of the superior rigidity of arenaceous rocks, in virtue of which they sustain a more intense shearing stress [1] (p. 187). The effect of this would be to postpone the appearance of biotite (p. 212). If this be so, it appears that this superior rigidity is lost as the temperature rises higher, owing to the enhanced solubility of quartz, for the higher index-minerals, when they are developed, come in at about the same stage in arenaceous as in argillaceous rocks.

More or less *micaceous quartzites*, with light or dark mica or both

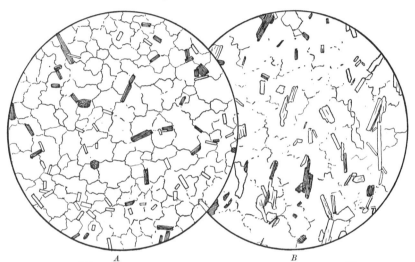

A   B

FIG. 118.—QUARTZITES WITH MUSCOVITE AND BIOTITE; × 18.
A. Pass of Killiecrankie, Perthshire.   *B.* Loch Lee, Forfarshire.

together, are well represented in the Highland Quartzite group and in the psammitic parts of the Moine Series. Set in a granoblastic matrix of quartz, usually with some felspar, the flakes may or may not show a pronounced orientation. In general the parallel arrangement is more regular in proportion as the mica bulks more largely (compare Figs. 118 and 85, *B*). The most highly micaceous type may be named a *quartz-mica-schist*, and in this the parallel orientation is always strongly marked, though often modified by a puckering or fine plication like that of many phyllites (Fig. 119). It should be remarked, however, that rocks of this type may originate in more than one way. Greenly [2] has described in Anglesey the conversion of a felsite into a typical quartz-mica-schist (Fig. 119, *B*). In this case much of the alkalies must have been removed in solution.

[1] Compare Turner, *Geol. Mag.*, vol. lxxv (1938), p. 164.
[2] *Geology of Anglesey* (*Mem. Geol. Sur. G.B.*, 1919), pp. 122-3.

In the higher grades of metamorphism a *garnetiferous quartzite* is a common type. The garnet is doubtless of the same kind as in the associated argillaceous rocks, and it shows the same tendency to build rather large crystals with inclusions of quartz (Fig. 81, *A*, above). Staurolite, cyanite, and sillimanite are of less frequent occurrence in the highly quartzose rocks now considered.

METAMORPHISM OF MORE FELSPATHIC TYPES

Distinct from the quartzites proper, we may recognize a parallel

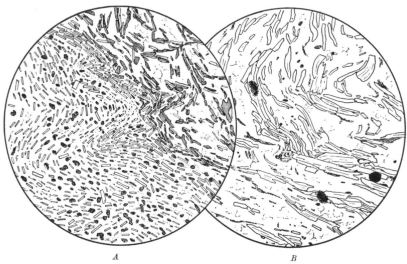

A   B

FIG. 119.—HIGHLY QUARTZOSE MICA-SCHISTS, SHOWING PLICATION ; × 18.
A. Quartz-Muscovite-Biotite-schist, Ardvasar, Skye.
B. Quartz-Muscovite-schist, Holland Arms, Anglesey.

series of types which, while still rich in quartz, contain also a large amount of felspar. For such rocks there is no well-established name. By some writers they would be called gneisses, but many of them have neither a notably coarse grain nor any gneissic structure. Following the Geological Survey, we use the name ' granulite '.[1] Rocks of this kind make up most of the psammitic part of the Moine Series, which occupies such large tracts in the western and northern parts of the Highlands, including also the Struan Flags, to the south of the Grampians,[2] and the Eilde Flags of Argyllshire and Lochaber. The

---

[1] This term likewise is open to objection, since it is in use for igneous rocks of more than one kind. The nomenclature of metamorphosed rocks in general is sorely in need of authoritative revision.

[2] See various Memoirs of the Geological Survey, and Barrow, *Quart. Journ. Geol. Soc.*, vol. lx (1904), pp. 400–46.

original sediments must have been exceptionally pure felspathic sandstones, containing abundant fresh felspars,[1] and these have for the
most part passed through the earlier stages of metamorphism without
change other than recrystallization.   The actual grade is usually that
of the garnet- or some higher zone, as indicated by the associated pelitic
beds ; but in the granulites themselves, owing to their relatively
pure nature, the higher-grade index-minerals are very often wanting.
The common types are *biotite-granulite* and *garnet-granulite* (Fig. 120),
and there is also a variety rather rich in magnetite (Fig. 88, *C*, above).

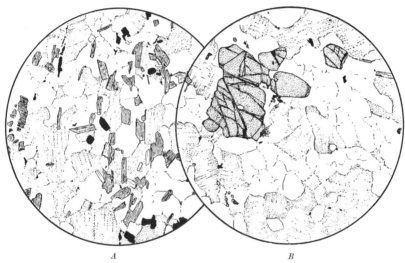

A   *B*

FIG. 120.—PSAMMITIC TYPES IN THE MOINE SERIES, near Glen Urquhart, Inverness-
shire ;   × 23.

The main bulk of these rocks is made by a granoblastic aggregate of quartz and
felspars.
A. Biotite-Granulite, with some muscovite : the abundant felspar includes both
oligoclase and microcline.
B. Garnet-Granulite, with little biotite : here the felspar is largely oligoclase.

Two felspars are constantly present, often in such quantity as
jointly to outweigh the quartz.   Sometimes, as in the Struan Flags,
the potash-felspar is microcline ; in other districts orthoclase is found,
and microcline is rare.   The sodic felspars include albite (usually
untwinned) and oligoclase (often with twin-lamellation).   Felspars and
quartz together make a granoblastic aggregate, through which the
flakes of mica are scattered.   Biotite predominates over muscovite,
and shows a deep brown colour with intense pleochroism, from yellow-
brown to nearly black.   The common accessory minerals are magnetite,
apatite, and zircon, all doubtless minor constituents of the original

[1] One hypothesis correlates these rocks with the Torridonian sandstones.

sands. In beds which have had a slight calcareous admixture epidote is common. When garnet enters, it may be in numerous small scattered crystals or in larger individuals with the usual inclusions of quartz.

In the granulites generally a parallel orientation of the mica-flakes does not suffice to impart any effective fissility. There is often, however, a flaggy parting, due to parallel laminae or films of biotite, which represent thin chloritic seams in the original sediment. In these the schistosity coincides with the local direction of stratification, and not necessarily with that dictated by the regional system of stress. The growing flakes, narrowly confined between relatively rigid boundaries, found least resistance in the plane of bedding. The case is somewhat analogous to that of a mica-schist in an aureole of thermal metamorphism (p. 36).

METAMORPHISM OF IMPURE ARENACEOUS SEDIMENTS

We have still to examine the metamorphism of those sandstones and grits which, in their original state, contained a considerable amount of finely divided material in addition to the grains of quartz and felspar. Setting aside ferruginous matter, which by itself can yield only magnetite, this extraneous element, in non-calcareous sediments, is broadly of an argillaceous nature. Metamorphism accordingly follows the same general lines that have already been sufficiently discussed, with differences which it is not difficult to anticipate. The due reactions are not here suspended owing to a too sparse distribution of the reacting substances ; so that the characteristic new minerals make their appearance at the appropriate stages. As in the granulites, potash-felspar, as well as an acid plagioclase, may figure in all grades. Notwithstanding the abundance of quartz, and possibly of felspars, there is enough of dark and coloured minerals present to give a general aspect different from that of quartzites or of granulites.

As leading types we may recognize *quartzose mica-schists* and *quartzose garnet-mica-schists*, in which the schistose structure is often well developed, though not in the degree shown by the phyllites. The representatives of these rocks in the more advanced grades include *staurolite-*, *cyanite-*, and *sillimanite-gneisses*. These last in particular are often very coarse-textured rocks with conspicuous foliation. They may have as main constituents quartz, two felspars, two micas, garnet, and sillimanite. In rocks which were originally conglomerates, quartz-pebbles, losing their outlines and changing their shape, become a factor in the coarse foliation. Here too, as among more purely argillaceous rocks, are types which were not rich enough in alumina to yield the most distinctive minerals of the highest grades. They usually contain garnet, and have the same coarse texture and rude foliation as the sillimanite-gneisses themselves (Fig. 121, *A*). Here belong the ' kinzigites ' of German petrologists ; a name applied to gneisses of sedimentary origin—paragneisses in Rosenbusch's convenient nomenclature—in which garnet is a prominent constituent. Such rocks are found associated with sillimanite-gneisses in the Highlands of Forfarshire and elsewhere. With them occur also micaceous

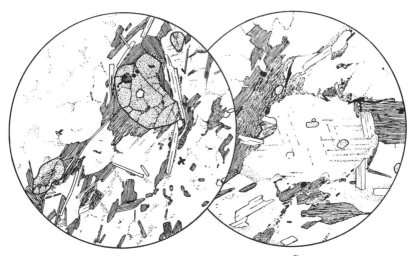

FIG. 121.—HIGHLY METAMORPHOSED GRITS, Glenesk, Forfarshire ; × 18.

A. Garnet-Gneiss ; with red garnet, muscovite and biotite, orthoclase, plagioclase and predominant quartz.
B. Muscovite-Biotite-Gneiss ; showing two micas, felspar (mostly an acid plagioclase), and abundant quartz, with some apatite and magnetite.
These rocks are in the sillimanite zone.

paragneisses devoid even of garnet (Fig. 121, *B*). The coming in of cordierite in rocks of this kind has already been discussed.

When in the original sediment interstitial argillaceous matter was in such quantity as to isolate the detrital grains from one another, we have to do with a gritty slate-rock rather than an argillaceous sandstone. It is enough to observe that, in the earliest stages of metamorphism, the grains, being largely relieved from stress, are little affected ; except that, if of elongated or flattened shape, they may become rotated into approximate parallelism (Fig. 114, *A*, above). At the same time the matrix acquires, according to circumstances, either a true schistosity or some form of false cleavage. With advancing metamorphism the quartz grains recrystallize, and the further course follows lines already sufficiently discussed.

PNEUMATOLYSIS AND METASOMATISM IN REGIONAL METAMORPHISM

Igneous intrusion being so frequent a concomitant of regional metamorphism, *pneumatolysis* often enters, to complicate in some measure the results of metamorphism proper. Its effects may be manifested at a considerable distance from any visible igneous rock, as compared with what is seen in an ordinary thermal aureole. Doubtless the mechanical disturbance which goes with regional metamorphism facilitates the penetration of gaseous emanations into the country-rocks. There is no need to recapitulate the various mineralogical transformations due to this cause, as already discussed in Chapter IX. It is perhaps in the calcareous rocks that pneumatolytic effects are most often noticeable, such minerals as scapolite, phlogopite, apatite, pyrrhotite, etc., having sometimes a rather wide distribution. In the non-calcareous sediments an abundance of tourmaline is often a significant feature.

Grubenmann [1] proposed to account for the richness in alumina of some Swiss sillimanite-gneisses by an introduction of aluminium-fluoride, giving the reaction:

$$Al_2F_6 + 3H_2O = Al_2O_3 + 6HF.$$

This seems to be a gratuitous hypothesis, and if alumina can indeed be introduced in this fashion, we should look for topaz rather than sillimanite as the resulting product.

As already remarked, there is a school of geologists in whose conception of metamorphism *metasomatism* of a far-reaching kind holds an important place (p. 133). We may take as typical the question of ' feldspathisation ', i.e. the supposed impregnation on an extensive scale of metamorphosed rocks with felspar derived from a magmatic source. This hypothesis dates from a time when the formation of various felspars as normal products of metamorphism was not adequately appreciated. The proof of addition of material to a rock should naturally be sought in a series of comparative chemical analyses, but such evidence is seldom forthcoming. A good instance is that of the Stavanger district, studied by Goldschmidt. [2] Here the metamorphosed sediments are found to become decidedly richer in albite as the trondhjemite (soda-granite) boundary is approached. A series of analyses, however, makes it clear that the main part of the albite has been derived from the substance of the rocks themselves. There

---

[1] *Viert. Zür. Nat. Ges.*, vol. lii (1907), and *Kristallinen Schiefer* (2nd ed., 1910), p. 165.

[2] *Vidensk. Skr.*, 1920, No. 10 (1921), pp. 108–21.

has been some accession, not of felspar, but of soda and silica.[1]   The metasomatic effects are perceptible to a considerable distance from the main plutonic mass ; but there are many small intrusions among the albitic schists, and it appears that the actual diffusion of material does not surpass a very moderate range.

[1] Goldschmidt infers also a certain addition of lime, which could not well come from a magmatic source, but here the figures of the analyses are less convincing.

CHAPTER XVI

REGIONAL METAMORPHISM OF CALCAREOUS SEDIMENTS

Metamorphism of Impure Non-Magnesian Limestones—Metamorphism of Impure Magnesian Limestones—Metamorphism of Calcareous Shales and Slates —Metamorphism of Highly Chloritic Types—Metamorphism of Calcareous Grits.

METAMORPHISM OF IMPURE NON-MAGNESIAN LIMESTONES

WE have seen that, in sediments in which calcareous and non-calcareous material is closely commingled, simple thermal metamorphism has certain well-marked characteristics. Reactions between the two discordant elements take place with great promptitude, in so much that they are not only initiated, but proceed to completion, at a comparatively early stage of metamorphism. Now reactions of this class, involving always the liberation of a volatile phase, are especially subject to the control of pressure as an adverse condition. We have to remark at the outset that, under the great pressures which ordinarily rule in regional metamorphism, these characteristic reactions may be incomplete, or may be wholly suspended. Moreover, this may be found even in a high grade of metamorphism, for in general high temperature and high pressure go together. These considerations, in addition to the influence of shearing stress, will be seen to import a profound difference between the two types of metamorphism in the class of rocks with which we have now to deal.

Of the pure carbonate-rocks little need be said. They give rise to *simple marbles* (Fig. 122). In a calcite-marble the grains often show a certain amount of interlocking ; in a dolomite-marble they are of simpler shape ; where the two minerals are associated, the dolomite is always idioblastic. Rocks of such simple constitution naturally show little change in different grades of metamorphism, nor do they differ essentially from the corresponding rocks in a thermal aureole. Any noticeable parallel elongation of grains is not very common in calcite (Fig. 85, *A*), and is not found in the stronger dolomite. The pencatite type is not represented in regional metamorphism, pressure being too great to permit even so much dissociation of a carbonate in the absence of silica.

252

Consider now the metamorphism of a non-dolomitic limestone, originally containing some admixture of foreign material. We may disregard for the moment those substances which can have no reaction with carbonates—muscovite, albite, magnetite, pyrites, graphite, etc. These will at first recrystallize without change at the appropriate temperatures. The noteworthy fact is that quartz also recrystallizes side by side with calcite without any mutual reaction (Figs. 92, 123, *A*). This holds good up to the highest grade (Fig. 125, *B*), and accordingly

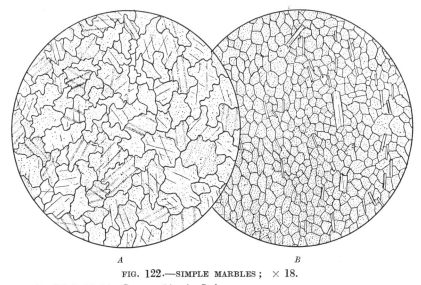

A   *B*

FIG. 122.—SIMPLE MARBLES ;   × 18.

A.   Calcite-Marble, Carrara, Liguria, Italy.
B.   Dolomite-Marble, Binnental, Valais, Switzerland, with a few flakes of white mica.

wollastonite does not normally figure as a mineral of regional metamorphism.[1]

It is otherwise with the aluminosilicates, some of which form readily in the metamorphism of an earthy limestone ; but here other principles find their application. In the absence of such substances as kaolin and gibbsite, the source of alumina must be found first in sericitic and chloritic material. In a low grade of metamorphism (chlorite-zone) muscovite and chlorite recrystallize separately, and indeed, if only sparingly dispersed in the rock, they may continue to do so in a somewhat higher grade. Otherwise they yield biotite, which generally appears at about the same stage in an impure limestone as in associated argillaceous rocks. Alumina being released in the reaction, an aluminosilicate is normally produced at the same

[1] Compare Goldschmidt, *Vidensk. Skr.*, 1912, No. 22, p. 6.

time, and this is constantly *zoisite*. Its associates at this stage are quartz, muscovite, biotite, and often albite. In a more advanced grade (garnet-zone) other lime-bearing minerals make their appearance. The Loch Tay Limestone, which is usually poor in magnesia, affords good material for study.[1] In addition to the minerals already mentioned, it often contains abundant *grossularite* and *idocrase*, and in one locality even wollastonite is found. At the same time biotite gives place gradually to *diopside* and muscovite to *microcline* (Fig. 124).

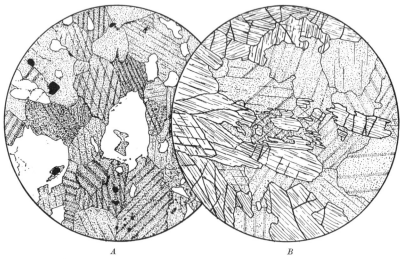

A   *B*

FIG. 123.—METAMORPHOSED LIMESTONES (MARBLES), Perthshire ; × 23.

A. Loch Tay Limestone, Glen Ogle. A siliceous limestone, in which quartz and calcite have recrystallized together without reaction. This is in the garnet-zone.
B. Blair Atholl Limestone, Glen Tilt. A dolomitic (and partly siliceous) limestone, which has given rise to abundant diopside.

Sphene is of less common occurrence. All these minerals may be found dispersed through the crystalline limestone or, in the more impure parts, making lime-silicate-rocks with quartz. The idocrase tends often to a porphyroblastic habit, while the garnet is usually in an aggregate of small crystals with diopside, microcline, etc.

The assemblage here recorded is a remarkable one. Zoisite is recognized as a stress-mineral, though not exclusively so, but grossularite and idocrase are highly characteristic minerals of simple thermal metamorphism, and the formation of diopside in preference to tremolite is significant in the same sense. It is clear that the influence of

[1] *Geol. of Cowal* (*Mem. Geol. Sur. Scot.*, 1897), pp. 47–8 ; Tilley, *Geol. Mag.*, vol. lxiv (1927), pp. 372–6. The Blair Atholl Limestone is usually more magnesian, but grossularite and zoisite are locally abundant near Blair Atholl itself.

the stress-factor is here much reduced. The metamorphism of a rock of this kind, i.e. a somewhat impure limestone, is in strong contrast, as we shall see, with that of calcareous shale or slate, in which, at a corresponding stage, only stress-minerals are formed. Clearly pressure cannot be the determining factor, though indeed zoisite, garnet, idocrase, and diopside are all minerals favoured by high pressure. The explanation is to be found in a consideration already put forward (p. 187). The relatively yielding nature of a

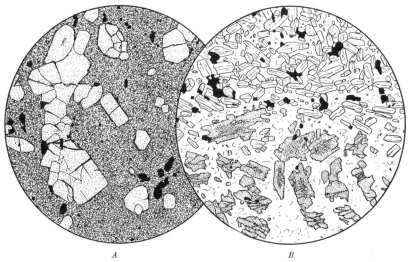

A   *B*

FIG. 124.—LIME-SILICATE-ROCKS IN THE LOCH TAY LIMESTONE, Perthshire ; × 23.

A. A silicate nodule in limestone, Tirarthur, Loch Tay ; showing porphyroblasts of idocrase and magnetite in a fine aggregate mainly of diopside and grossularite.
B. Band in limestone, Keltnie valley ; showing zoisite (below) and diopside (above) with quartz, microcline, plagioclase, pyrrhotite, and sphene.

rock composed mainly of calcite makes any very high measure of shearing stress impossible.

The formation of a potash-felspar at the expense of mica in a medium grade of metamorphism is a consequence of the same principle. We have seen that in crystalline schists of argillaceous composition this reaction is inhibited by shearing stress except in the highest grade. In a limestone the stress is much less intense, and in addition the reaction is promoted by the presence of abundant lime, which affords a ready way of disposing of the excess of alumina in zoisite. Since silica alone does not react with calcite, the production of new silicate-minerals in a non-magnesian or poorly magnesian limestone depends upon a provision of alumina from some source. Here it comes from the dissociation of muscovite and biotite, which

give place to a potash-felspar and a pyroxene. In this medium grade
the second felspar is always of a sodic variety, albite to oligoclase.

The micro-structure of these crystalline limestones is granoblastic
as regards the calcite matrix, with modifications arising from the
idioblastic and sometimes porphyroblastic development of the silicate
minerals. In the absence of mica there is no pronounced schistosity.
A marble rich in white mica (cipollino) is an exceptional type.

The highest grades of metamorphism in the rocks under consideration are marked especially by the disappearance of the epidote-zoisite
minerals, which go to make anorthite. If any micas have survived
to so late a stage, the conversion of these to potash-felspar and
pyroxene furnishes additional anorthite. The particular variety of
plagioclase finally produced depends then upon the ratio of the anorthite thus set free to the original content of albite ; and it follows from
what has been said that the more calcic felspars are found in rocks
which contained originally a notable amount of micaceous material.
The ferro-magnesian mineral is diopside, hornblende being at most a
subordinate constituent. Of more calcic silicates, grossularite may
still be present, but not wollastonite. Recalling the relation :

$$\text{Grossularite} + \text{Quartz} = 2\,\text{Wollastonite} + \text{Anorthite},$$

we may infer that, under the high pressures which prevail, the
association on the left side is the stable one.

The Deeside Limestone of the Aboyne and Banchory districts,
studied by Hutchison,[1] was a very impure rock, and often contains
but little residual calcite. Here grossularite is almost unknown.
The characteristic type is a diopside-plagioclase-limestone, which
may also be rich in orthoclase (seldom microcline). The plagioclase
is usually andesine or labradorite, but anorthite is also found. Some
hornblende may accompany the diopside ; and it is observed that
this occurs in bands rich in plagioclase but not in those where orthoclase is abundant. These rocks are in the sillimanite-zone of Barrow,
but in a region where there are grounds for believing that the shearing
stress was not always at its maximum value (p. 186). The plagioclase
in the metamorphosed limestones of Deeside is often partly replaced
by scapolite.[2] This widespread scapolitization, distinct from that
which is found near contact with Newer Granite intrusions, is doubtless related to the mechanical conditions proper to regional meta-

[1] *Trans. Roy. Soc. Edin.*, vol. lvii (1933), pp. 557–92.
[2] On the scapolite-bearing rocks of Scotland see also Flett, *Summ. Progr.
Geol. Sur.* for 1906 (1907), pp. 116–31.

morphism, which facilitate the permeation of the rocks by volatile bodies.

METAMORPHISM OF IMPURE MAGNESIAN LIMESTONES

In limestones which were partly or completely dolomitized, metamorphism has followed a different course ; for dolomite, with its relatively low dissociation-pressure as compared with calcite, has not normally been prevented from reacting with any free silica present. Here, however, the first product is not forsterite, as in simple thermal

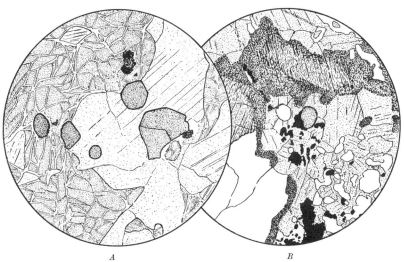

A   B

FIG. 125.—HIGHLY METAMORPHOSED DOLOMITIC LIMESTONES, Glenelg, Inverness-shire ; × 18.

A. The minerals shown are forsterite, with incipient serpentinization, spinel, and calcite.
B. This has been only a partly dolomitized rock. The dolomite has given rise to diopside, which is seen as a fringe of granules bordering quartz and also as distinct crystals. The remaining carbonate is calcite, and contains inclusions of quartz. The opaque mineral is graphite.

metamorphism, but the stress-mineral amphibole. It appears as numerous needles or slender crystals of *tremolite*, or less commonly a pale green actinolite, being at this early stage non-aluminous. With it may be associated more or less muscovite, biotite, and magnetite, representing micaceous, chloritic, and ferruginous impurities in the original rock. By reaction with some of these minerals, tremolite gives place with advancing metamorphism to a *green aluminous hornblende*. If, however, the original magnesian limestone contained none but siliceous impurity, tremolite persists, in crystals of increasing dimensions, far into the garnet-zone (as determined in argillaceous sediments), being then replaced by diopside and forsterite. This is well illustrated, for example, in the purer parts of the Blair Atholl Limestone in Glen Tilt (Fig. 123, *B*). The aluminous hornblendes generated in more impure rocks include some variety of characters, and doubtless of composition, though information on this point is scanty. The colourless edenite, which is sometimes found, may be distinguished from tremolite by its rather higher extinction-angle, and positive optical character.

In the highest grades of metamorphism the aluminous hornblendes,

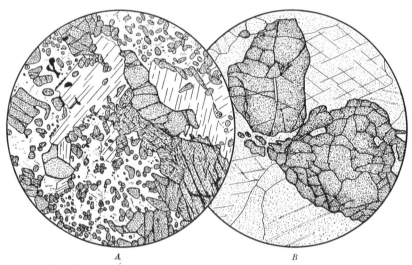

FIG. 126.—FLUOSILICATES IN METAMORPHOSED DOLOMITES ; × 25.

A. Silicate-rock, Hakgala, Ceylon. The minerals, in order of idioblastic develop-ment, are diopside, colourless phlogopite, blue fluo-apatite, and anorthite, the last enclos-ing granules of diopside in poeciloblastic fashion.
B. Chondrodite-Marble, Ferme Park Reservation, New York. Large crystal-grains of yellow chondrodite are embedded in a matrix of calcite.

like the non-aluminous, normally give place to *diopside* and *forsterite*, the alumina released going to make a *spinel* (Fig. 125, *A*). It appears, however, that pyroxenes rich in alumina may be stable in association with spinellid minerals and calcite, though not in presence of free silica.[1] If hornblende is still found in a high grade, it appears to be always in the immediate vicinity of a plutonic intrusion, and may be attributed to pneumatolytic influence, but analyses of such hornblendes are a desideratum. The characteristic fluorine-bearing mineral in these rocks, however, is *phlogopite* (Fig. 126, *A*). This is derived by reaction of dolomite either with potash-felspar, formed at an earlier stage, or

[1] Tilley, *Geol. Mag.*, vol. lxxv (1938), pp. 81–5.

with muscovite.   In the latter case there is a surplus of alumina to be disposed of in spinel or some other mineral :

$$3CaMg(CO_3)_2 + KAlSi_3O_8 + H_2O$$
$$= H_2KMg_3Al(SiO_4)_3 + 3CaCO_3 + 3CO_2 ;$$
$$3CaMg(CO_3)_2 + H_2KAl_3(SiO_4)_3$$
$$= H_2KMg_3Al(SiO_4)_3 + Al_2O_3 + 3CaCO_3 + 3CO_2.$$

These reactions involve another illustration of dedolomitization.

Again, the place of forsterite is sometimes taken by *chondrodite*

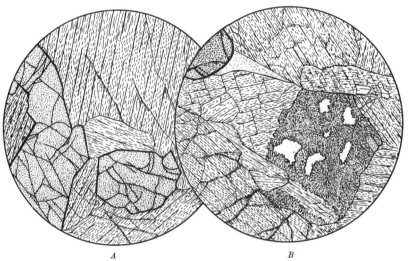

FIG. 127.—PYROXENE-PLEONASTE-ROCKS, Adhekanwela, Ceylon ;   × 18.

The pyroxene here is an aluminous variety comparable with fassaite ; Tilley, *Geol. Mag.*, vol. lxxv (1938), p. 82.   In *B* there is some phlogopite and a little residual calcite ; also a pseudomorph of zoisite and clinozoisite after grossularite.

(Fig. 126, *B*).   Scapolite and apatite, like grossularite, anorthite, and sphene, belong rather to the less magnesian rocks than to dolomites proper.   The common sulphide mineral in this high grade is *pyrrhotite*. It appears that pyrites is converted to pyrrhotite at about the same stage that tremolite gives place to diopside.[1]   Good examples of highly metamorphosed impure dolomites are afforded by the masses intercalated in the Lewisian gneiss complex, especially in Tiree and in the Glenelg district (Fig. 125).   A prominent feature is the strong tendency to segregation, giving rise to nodular aggregates of various silicates—diopside, phlogopite, etc.—embedded in crystalline lime-stone, often of coarse grain.   Differential stress being almost negligible, there is no tendency of segregation to spread in a prescribed direction

[1] Laitakari, *Bull.* 54 *Comm. Géol. Finl.* (1920), p. 22 (Pargas).

imposed from without, and any banding that may be developed follows the original bedding.

With a sufficient original admixture of siliceous and aluminous material a limestone or dolomite will suffer complete de-carbonation, and we find accordingly *silicate-rocks*, not merely as nodules or inconstant bands in a silicate-bearing limestone, but occurring in mass. Most of the characteristic minerals mentioned above may figure in rocks of this kind. Shearing stress has doubtless more influence upon the mineralogical constitution here than in rocks with a calcite

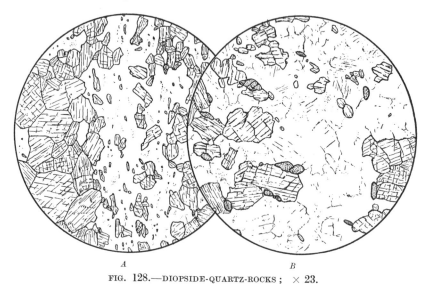

A   B

FIG. 128.—DIOPSIDE-QUARTZ-ROCKS ;   × 23.

A.   Eyre Peninsula, South Australia (see Tilley, *Geol. Mag.*, vol. lvii (1920), pp. 492–3).
B.   Bellenwald, Black Forest.   Here is a considerable amount of plagioclase with the quartz and some sphene associated with the diopside.

matrix.   So, for example, an almost pure tremolite-rock, a member of the Blair Atholl group, retains that character into a very high grade of metamorphism (cyanite-zone), merely becoming coarser in grain.   Since magnesia may enter in chlorite as well as in dolomite, the rocks are almost always more or less magnesian, and the most abundant mineral is usually a pyroxene.   It is commonly spoken of as diopside, but, as already remarked, aluminous varieties are also found in rocks without free silica.   With pyroxene are associated in different instances tremolite, spinel or pleonaste, phlogopite, plagioclase, apatite, scapolite, pyrrhotite, graphite, etc. (Figs. 126, *A* ; 127). In general there has been sufficient silica present to satisfy all demands, so that forsterite and chondrodite are absent.   There may be an excess of silica, and so we have diopside-quartz-rocks with or without other significant minerals (Fig. 128). Another mineral which may figure is potash-felspar, which here as before is commonly a characteristic microcline. It is usually a subordinate constituent, and, since it can come only from original muscovite, it must always be accompanied by some more aluminous mineral.

METAMORPHISM OF CALCAREOUS SHALES AND SLATES

We pass now to the case in which the carbonate element enters only in subordinate quantity in sediments which are in the main argillaceous. The difference of composition, as compared with the former case, imports a decided difference in the chemistry of metamorphism. Many of the characteristic minerals met with before are still found ; but of the more extreme calcic and magnesiam silicates some are absent from this association, and others have now a more restricted occurrence. Such minerals as lime-garnet and idocrase may indeed be considered as ruled out by the stress-conditions, for there is not here that relaxation of shearing stress which we had to note in the case of rocks rich in calcite. For the same reason potash-felspars are found only in a very advanced grade of metamorphism. The more limited distribution of tremolite and the absence of forsterite are to be explained by the fact that the type of sediment in question, owing to its relatively impermeable constitution, has not often suffered dolomitization. If there is a notable richness in magnesia, it comes rather from an abundance of chlorite, a case to be discussed later. Even with these limitations ordinary calcareous shales exhibit in their metamorphism a rather wide range of diversity.

In non-chloritic shales calcite usually plays a merely passive part, entering into no reaction with the quartz, sericite, albite, magnetite, etc. There are accordingly calc-phyllites and calc-muscovite-schists, showing often an incipient foliation due to the segregation of calcite and quartz. Even chlorite, if present only in small amount, may go wholly to the making of biotite. We find thus, parallel to the ordinary argillaceous series, a series of calc-mica-schists and calc-garnet-mica-schists containing no lime-bearing silicate (Fig. 92, *A*, above).

Most argillaceous sediments, however, have chlorite as an important constituent. We have then in the lowest grade such types as *calc-chlorite-sericite-schist* and *calc-chlorite-albite-sericite-schist* (Fig. 129, *A*). Here the part of the calcite is still a passive one ; but, as the production of biotite, etc., proceeds with advancing metamorphism, alumina is liberated, and calcite enters into reactions. The characteristic product is epidote. Epidote may appear exceptionally before the biotite stage is reached, and must then be attributed to the presence of kaolin in the original sediment :

$$3H_4Al_2Si_2O_9 + 4CaCO_3$$
$$= 2Ca_2(AlOH)Al_2(SiO_4)_3 + 5H_2O + 4CO_2.$$

More generally epidote makes its first appearance concurrently with biotite, and we may recognize as a distinctive type *calc-epidote-mica-schist* (Fig. 129, *B*) or sometimes *calc-zoisite-mica-schist* (Fig. 136, *A*, below). Garnet comes in at about the same stage in these calcareous

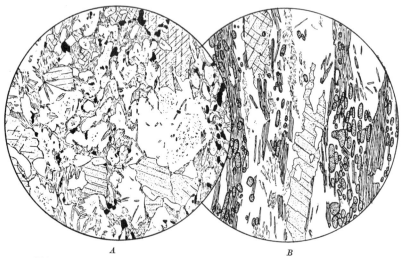

A   *B*

FIG. 129.—CALCAREOUS SCHISTS, Ardrishaig Phyllites, Dalmally, Argyllshire ; × 23.

A. Calc-Chlorite-Albite-schist ; a low-grade type, composed of chlorite, albite, calcite, magnetite, and quartz.
B. Calc-Epidote-Biotite-schist ; a more advanced grade of metamorphism (biotite-zone). Calcite and quartz are intimately associated in streaks following the direction of schistosity. Abundant epidote is associated with the biotite.

schists as in purely argillaceous rocks, and there is nothing to suggest that it is of different composition. This is what should be expected in view of the undoubted status of lime-garnet as an anti-stress mineral. Zoisite often takes the place of epidote at this stage. Light and dark micas are usually well represented, with some chlorite in addition. Sphene is a common accessory constituent. Here as elsewhere, its amount tends to vary inversely with that of biotite which is always titaniferous.

The influence of calcite as modifying the course of metamorphism now becomes increasingly evident. About the middle of the garnet-zone, as laid down in ordinary argillaceous rocks, garnet in these semi-calcareous sediments fails, the remaining chlorite disappearing at the same time. Their place is taken by a green hornblende, formed by reaction of these and perhaps other minerals with the calcite. There results a *calc-hornblende-schist*, containing in addition biotite and often muscovite (Fig. 92, *B*). This is a distinctive and widely distributed type. Zoisite or epidote is often present; but these minerals dwindle as metamorphism proceeds, and are represented in the highest grades by a lime-bearing felspar.

The various types which have been noticed all contained recrystal-

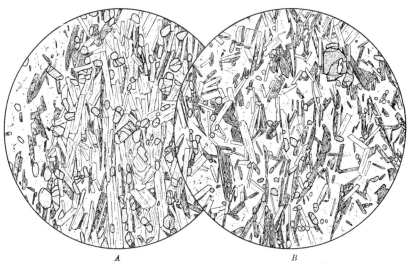

A   *B*

FIG. 130.—EPIDOTE-MICA-SCHISTS, Moine Series, Loch Morar, Inverness-shire ;
× 23.

A. A variety rich in muscovite.
B. A garnetiferous type : there is here some plagioclase with the quartz.

lized calcite in addition to lime-bearing silicates. Where, however, the carbonate in the original sediment was less abundant, it has been wholly consumed in the reactions already discussed. It is needless to describe in detail the resulting rock-types. With a small amount of calcite evenly disseminated, there is usually enough kaolin and iron-oxide present to convert it to epidote. This mineral is accordingly a common accessory constituent of many mica-schists, figuring as numerous small stout prisms with rounded ends. We have accordingly a series of *epidote-mica-schists* (including epidote- and zoisite-phyllites) and *epidote-garnet-mica-schists* (Fig. 130). When the lime-silicate is abundant, it is more often zoisite. The finest examples of *zoisite-mica-schist* in the Highlands are found in Glencoe (Fig. 131). Here the crystals, up to 1½ inch in length and with regular parallel arrangement, often make up the greater part of the rock. Biotite is abundant, but garnet is lacking, although the locality is in the interior of the garnet-zone. The original sediment must have contained abundant kaolin or aluminium-hydrate with only a moderate amount of chlorite and sericite. In the same grade a more chloritic rock is represented by a hornblende-schist, usually with some calcite.

Although, as already remarked, the carbonate element in argillaceous sediments is commonly calcite, *dolomitic shales* and slates are found, and their behaviour is necessarily different from that of the

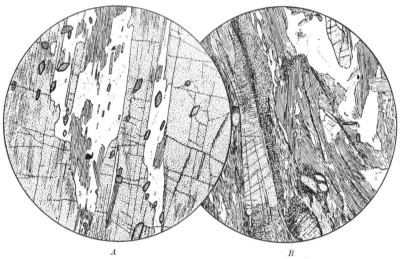

A   *B*

FIG. 131.—ZOISITE-MICA-SCHISTS (Ballachulish Limestone), Glencoe, Argyilshire ;
× 18.

   A. Large crystals of zoisite, with parallel arrangement, make the greater part of the rock. The rest is of biotite, sphene, and quartz.
   B. Here biotite is a more abundant constituent.

ordinary calcareous types. Recrystallized dolomite takes on idioblastic shape, contrasting with the interstitial and lenticular habit of the weaker calcite. As metamorphism proceeds, there is formed, in what may be considered the normal course, an amphibole of the tremolite type, in place of the epidote or zoisite of the non-magnesian sediments. The characteristic rock is then a *tremolite-mica-schist*, showing well-marked schistosity. In addition to tremolite, or less frequently a pale actinolite, it contains commonly light and dark micas with quartz, often some albite, and sometimes a little calcite (Fig. 132, *A*). Some of the best British examples are found on the Banffshire coast. These are often highly phyllitic, with corrugation on a microscopic scale. Sometimes the tremolite porphyroblasts have the spiral shape due to rotation during the growth of the crystals (p. 221).

Although dolomite normally disappears at a somewhat early stage, there are instances in which it seems to have persisted through the biotite grade into that of almandine. We may then see the apparent anomaly of dolomite and epidote in close association (Fig. 132, *B*). Possibly the carbonate as a distinct and later origin. Examples are found in the Ben Lawers Schists of Perthshire.

In partly calcareous rocks in general the advance of metamorphism

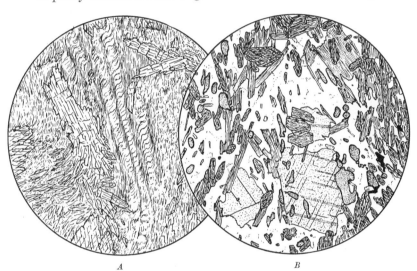

A   *B*

FIG. 132.—METAMORPHOSED DOLOMITIC SHALES ; × 23.

A. Tremolite-Mica-schist, Crathie Point, west of Portsoy, Banffshire. Porphyroblasts of tremolite lie in a phyllitic matrix, partly corrugated.
B. Biotite-Hornblende-schist with dolomite and epidote, upper part of Pass of St. Gotthard. The dolomite touches and encloses grains of epidote.

is often marked by a gradual change in the composition of plagioclase felspars, due to reaction between original albite and calcite or epidote. In the Sulitelma district Vogt [1] recognizes a distinct oligoclase-zone, following that of almandine. In New Zealand, where the rocks studied are mainly of more psammitic nature, Turner [2] has also made an oligoclase-zone, but regards it as equivalent to the almandine zone and perhaps also that of cyanite. It is evident that the changing composition of the plagioclase can be used as a guide only with some caution, and only so long as both plagioclase and epidote remain in the rocks.

[1] *Norges Geol. Und.*, No. 121 (1927), pp. 209–19, 485–6.
[2] *Trans. N.Z. Inst.*, vol. lxiii (1933), pp. 238, 244–6.

METAMORPHISM OF HIGHLY CHLORITIC TYPES

Sediments which have been richly chloritic and also more or less calcareous present in their metamorphism some features of special interest. Excellent examples are furnished by the rocks known to Highland geologists as the ' Green Beds ', which make a well-characterized group in the Dalradian sequence, and can be followed through every grade of metamorphism.[1] Their bulk-composition is that of material derived directly from the waste of basic igneous rocks. There

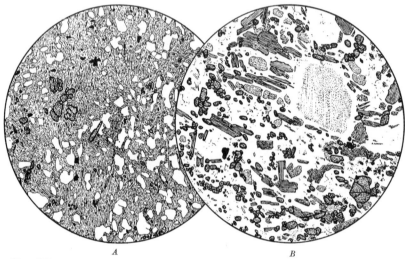

A   *B*

FIG. 133.—EARLY STAGES OF METAMORPHISM IN THE ' GREEN BEDS ', Perthshire ; × 23.

A. Albite-Chlorite-schist, Loch Lubnaig. This shows mainly a dense mass of little scales of chlorite with grains of clear albite and some crystals of epidote. There are also smaller amounts of magnetite, sericite, and quartz, and at scattered points in the rock biotite is beginning to form.
B. Epidote-Albite-Biotite-schist, foot of Loch Katrine. Besides the minerals named there is some chlorite remaining and a little sericite and quartz. The porphyroblastic development of the albite is very noticeable.

is, however, a variable admixture of detrital quartz, and, when this becomes abundant, the rocks may pass rather into chloritic grits, sometimes pebbly.

The lowest grade is represented by an *albite-chlorite-schist* (Fig. 133, *A*) or a *calc-albite-chlorite-schist* with a little sericite and magnetite, chlorite being always preponderant. The proportion of calcite varies. Some part may survive to a fairly advanced grade of metamorphism,

[1] See various *Memoirs of the Geological Survey of Scotland* ; also Phillips, *Min. Mag.*, vol. xxii (1930), pp. 239–56. ' On Green Beds in Co. Antrim,' see Bailey and McCallien, *Trans. Roy. Soc. Edin.*, vol. lviii, pp. 171–5.

but in the main it is consumed in reactions yielding epidote, and the next term in the series is accordingly an *epidote-albite-chlorite-schist.* Very often, however, there appear among the chlorite numerous flakes of a green or green-brown mica, the true nature of which is at present problematical. It is quite distinct from ordinary biotite, and is probably poorer in potash. Beginning well within the chlorite zone, as laid down in simple pelitic sediments, it persists into a higher grade, but apparently gives place gradually to normal biotite (Fig. 133, *B*). Another noteworthy point, which finds a parallel also in

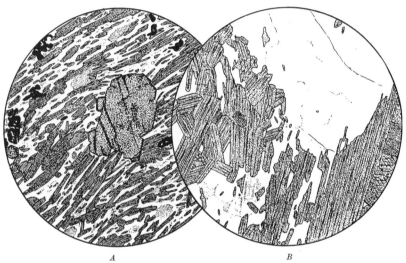

FIG. 134.—HORNBLENDE-SCHISTS, ' Green Beds ' ;   × 18.

A. Garnetiferous Hornblende-schist, Aberfeldy, Perthshire.   The minerals shown are garnet, hornblende, magnetite, quartz, and some plagioclase.
B. Pebbly Hornblende-schist, Clova, Forfarshire : composed of green hornblende, quartz, and labradorite, and showing part of a pebble of quartz. This rock is in the cyanite-zone.

the basic igneous rocks to be noticed later, is the early production of hornblende, presumably from residual granules of augite. In the Green Beds of the Highlands it sometimes makes its appearance in the chlorite zone, figuring at first as slender needles embedded in the chlorite and then as somewhat larger crystals. All the lower grades of metamorphism in this group of rocks may be studied in the Lower Cowal district in Argyllshire and about Loch Katrine and Lock Lubnaig in Perthshire.

In a higher grade hornblende becomes the most prominent constituent, while chlorite and epidote dwindle and disappear. The green mica, too, has gone, and instead appears ordinary brown biotite. The amount of this is of course dependent upon the original content of sericitic material, usually scanty ; so the proportion of biotite is variable, and muscovite is normally absent. At this stage then the rocks are *hornblende-schists*, or sometimes *hornblende-biotite-schists*, usually with well-marked parallel structure. Red garnet comes in at about the same stage as in ordinary argillaceous rocks, and a *garnetiferous hornblende-schist* is a characteristic type through a wide range. The mineral is, however, by no means constant, and garnetiferous and non-garnetiferous bands sometimes alternate. Good examples are seen in the Aberfeldy district of Perthshire (Fig. 134, *A*). Some epidote or zoisite may still be present, but here (in the garnet zone) these minerals soon disappear, while the plagioclase is no longer albite but albite-oligoclase or oligoclase. Representatives of the Green Beds in the highest grades (cyanite and sillimanite zones) are seen in the Clova district of Forfarshire. They are relatively coarse-textured hornblende-schists, with or without garnet (Fig. 134, *B*). The plagioclase now is mostly andesine. Biotite is usually absent, and, since a potash-felspar is not often to be detected, it is probable that part of the potash has been taken up by the hornblende.[1]

METAMORPHISM OF CALCAREOUS GRITS

The metamorphism of arenaceous rocks containing some calcareous admixture, though somewhat simpler, reproduces many of the features noticed already in the calcareous shales. If a carbonate is the only impurity in a sediment composed essentially of quartz, or of quartz with fresh felspar, there can be no chemical reaction, and accordingly calcite may be found as an occasional constituent of quartzites. More usually there has been a certain amount of impurities, including kaolin and hydrated ferric oxide, to give rise at an early stage to some epidote. In higher grades this mineral is likely to be produced by other reactions, and it is found in some Highland quartzites and more often in the granulites of the Moine Series. An epidote-quartz-rock or ' *epidosite* ', in which the distinctive mineral bulks largely, is not a common type [2] (Fig. 135, *A*). Usually, when the lime-silicate is abundant, it is not epidote but zoisite. Thin bands of *zoisite-granulite* are intercalated in the psammitic parts of the Moine Series in many parts of the Highlands.[3] Such a rock

---

[1] Analyses of hornblendes from metamorphosed semi-calcareous rocks often show a notable content of potash : Glen Urquhart, Inverness, 2·20 per cent. ; Längban, Sweden, 1·89 ; Pargas, Finland, 2·70.

[2] Most rocks of this composition have a different origin, not related to metamorphism proper.

[3] See *Memoirs of the Geological Survey*, especially *Geology of Ben Wyvis, etc.* (1912), pp. 42–5, with a chemical analysis.

consists mainly of quartz, felspars (mostly plagioclase), biotite, and zoisite, which is sometimes very abundant. In many cases garnet and sphene are also present (Fig. 136, *B*). The garnet, as seen on a hand-specimen, is apparently in most cases the common red variety, but sometimes shows a lighter red or orange colour. It is isotropic in section. We have no data concerning its chemical composition.

Many of the rocks just mentioned carry also a green hornblende, and *zoisite-hornblende-granulite* with or without garnet, may be reckoned as a distinct variety. Speaking generally, however, there

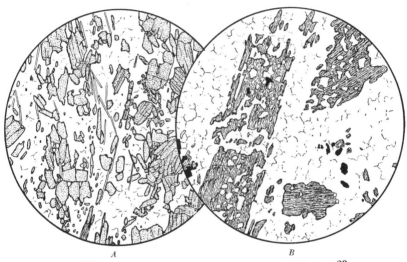

A   *B*

FIG. 135.—EPIDOTIC AND HORNBLENDIC QUARTZITES ; × 23.

A. Foliated Epidosite, Roxburgh, New Zealand : composed of epidote and quartz with a few flakes of muscovite.
B. 'Feather-Amphibolite', Amable, near Bancroft, Ontario : showing large poeciloblastic crystals of hornblende, set in a granoblastic ground-mass of quartz and felspar.

is an inverse relation between the zoisite and hornblende, and particular bands are characterized by one or the other mineral. Hornblende by itself, arising mainly from reaction between calcite and chlorite, is not uncommon as an accessory mineral in some of the Highland quartzites. It makes porphyroblasts of some size, always with very numerous inclusions. We may regard it as taking the place of garnet wherever a little calcareous matter was present (Fig. 81, *B*). In some more impure arenaceous rocks large poeciloblastic crystals of hornblende come to constitute the major part of the bulk. This is the very striking type known to German petrologists as ' *Hornblendegarbenschiefer* '. The hornblende porphyroblasts, often accompanied by biotite and epidote, are enclosed in a quartzose or quartzo-felspathic ground-mass with granoblastic structure (Fig. 90).

Such rocks have been described from the Binnental,[1] from Val Piora [2], from the Tremola Series of St. Gotthard,[3] and elsewhere in the Alps. Here too comes the 'feather-amphibolite' of the Grenville Series in Ontario [4] (Fig. 135, *B*). The blades of hornblende, about an inch in length, are spread out along the bedding-planes with a tendency to stellate grouping. Rocks comparable with these occur in the Ben Lawers schists at Ben Vrackie, Perthshire, and in close association with the zoisite-mica-schists of Glencoe, and there are good examples in the Moine Series near Kinlochewe, Ross-shire (Fig. 137).

FIG. 136.—CALCITE-BEARING CRYSTALLINE SCHISTS ; × 23.

A. Calc-Zoisite-Mica-schist (Ben Lawers Schists), Ben Vrackie, Perthshire.
B. Calc-Zoisite-Garnet-Granulite (a calcareous band in the Moine granulites), Malaig, Inverness-shire ; showing zoisite, poeciloblastic garnet of pale colour, rather abundant sphene, calcite, quartz, and felspars.

It is worthy of remark that the typical Hornblendegarbenschiefer differ little in total chemical composition from some igneous rocks. In Grubenmann's chemical classification they are grouped with the plagioclase-gneisses, the equivalents of the diorites. As a rule, the bulk-analysis of any crystalline schist enables us to assign it with some confidence to either an igneous or a sedimentary origin.[5] Exceptions must obviously be found, when a sediment has been derived

[1] Bonney, *Quart. Journ. Geol. Soc.*, vol. xlix (1893), pp. 107–10.

[2] Krige, *Ecl. Geol. Helv.*, vol. xiv (1918), pp. 565–6, with analysis.

[3] Hezner, *Neu. Jb. Min.*, Beil. Bd. xxvii (1908), pp. 161–176, with analyses.

[4] Adams and Barlow, *Geol. of Haliburton and Bancroft, Mem. 6 Geol. Sur. Can.* (1910), pp. 166–9.

[5] Rosenbusch, *Tsch. Min. Pet. Mitt.*, vol. xii (1891), pp. 49–61 ; Bastin, *Journ. Geol.*, vol. xvii (1909), pp. 445–72.

directly from some particular igneous rock without important chemical change—e.g. an arkose of the ' Green Beds ' described above. Setting these aside, the high alumina content of an argillaceous and the high silica of an arenaceous rock (relatively to alkalies) are very significant, and especially in both the predominance of potash over soda. The calcareous sediments are characteristically rich and the non-calcareous poor in lime. It is easy to see, however, that for *partly* calcareous rocks these several criteria may sometimes fail, and it is especially here that ambiguous cases present themselves.

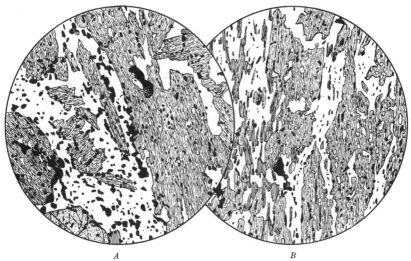

A   *B*

FIG. 137.—PORPHYROBLASTIC HORNBLENDE-SCHIST (Hornblendegarbenschiefer), near Kinlochewe, Ross-shire ; × 18.

With biotite, garnet, and magnetite, and interspaces occupied by quartz with some felspar. *A* is cut parallel and *B* perpendicular to the schistosity.

Reviewing what has now been set down, we are warranted in concluding that the zonal method of investigation, first devised by Barrow for the most common type of argillaceous sediment, is no less applicable, with the appropriate modifications, to those more special types which are characterized by richness in iron-oxide, in manganese, in lime, or in magnesia. The choice and due definition of the index-minerals to be employed, and the correlation of the resulting scheme with that already laid down for the simple pelitic series, will be more satisfactorily effected when our knowledge is more complete. It is especially desirable that such observations as have been recorded should be confirmed, and if necessary corrected, by study in other countries.

CHAPTER XVII

REGIONAL METAMORPHISM OF IGNEOUS ROCKS

Twofold Relation of Igneous Rocks to Regional Metamorphism—Regional Metamorphism of Ultrabasic Rocks—Regional Metamorphism of Basic Rocks—Regional Metamorphism of Intermediate and Acid Rocks—Regional Metamorphism of Alkaline Rocks.

TWO-FOLD RELATION OF IGNEOUS ROCKS TO REGIONAL METAMORPHISM

A COMPREHENSIVE survey of any extensive metamorphic tract will recognize a wide diversity of rock-types, all presumably related in one way or another to regional metamorphism ; and among them rocks which by their mineral and chemical composition are plainly marked as of igneous origin often bulk largely. In respect of composition they may include representatives of all the various families recognized among ordinary igneous rocks. Their structures vary from typically schistose to coarsely gneissic or simply massive. The more or less distinctly banded crystalline rocks which are conveniently styled orthogneisses (in contradistinction to paragneisses, which are highly metamorphosed sediments) attain in some countries a vast development, but their precise relation to the regional metamorphism is a matter for inquiry. A very moderate acquaintance with ordinary igneous intrusions in the field is enough to teach us that neither a banded disposition nor a parallel orientation of crystals is in itself evidence of a metamorphic origin.

It is to be observed at the outset that the regional metamorphism of igneous rocks involves complications which do not arise in relation to other classes of rocks, and that in consequence a study on genetic lines encounters here peculiar difficulties. Our mode of procedure in other cases has been, theoretically at least, quite simple. Starting from a rock supposed to be at ordinary temperature and sensibly free from stress, we endeavoured to follow the successive changes induced in it in response to progressively rising temperature, with the further condition of powerful shearing stress. Evidently a granite or a basalt, equally with a grit or a limestone, may be subjected to the same conditions, and undergo progressive metamorphism after its kind. Such rock belonged, in its origin, to some earlier period of igneous activity, in no wise related to the metamorphism. It is certain, however, that a large part of the igneous rocks to which the name 'metamorphic' is applied have a different history. Igneous activity, being closely bound up with crustal stress and crustal displacement, is a very usual concomitant of regional metamorphism. It takes the form of the intrusion of molten magmas, sometimes upon a large scale, among the rocks of various kinds which are in process of being metamorphosed. In this way originate igneous rocks which are in a general sense of the same age as the metamorphism, though their relation in time to the gradual progress and culmination of metamorphism in the associated sediments admits of considerable latitude. The special mechanical conditions under which intrusion is here effected are often reflected in certain peculiarities of micro-structure in the resulting igneous rocks, as well as in fluxional effects of various kinds. In so far as these peculiarities have been impressed upon the rocks from their birth, they do not strictly fall within the definition of metamorphism, but they may closely simulate structures which are developed in the course of metamorphism proper. Further, these features which belong to primary igneous gneisses are very likely to be complicated by truly metamorphic effects superposed upon them, and this applies to mineralogical as well as structural characters. It is evident that such quasi-contemporaneous igneous rocks must share in so much of the regional metamorphism as post-dates their consolidation.

In general this succeeding metamorphism will be controlled by *falling temperature*, with the stress-factor still more or less operative. Rocks of sedimentary origin, and also older igneous rocks,which have undergone metamorphism at elevated temperatures, have subsequently passed through a period of cooling, during which they were liable to experience some 'retrograde' metamorphism—a subject to be more particularly discussed in a later chapter. Often the effects are not very noticeable. The quasi-contemporaneous igneous rocks, however, must cool down from considerably higher temperatures and in presence of a somewhat richer content of water and other solvents. Since, moreover, their initial mineralogical constitution often included forms which lose their stability on cooling, changes consequent upon a decline of temperature are here more general.

We have then to recognize that crystalline schists, granulites, and gneisses of igneous origin fall into *two distinct categories*, which in a merely descriptive treatment of the subject are confused together. On the one hand are those which either belong to some earlier system of igneous activity or at least were intruded and completely consolidated before the climax of regional metamorphism was reached ; so that they have been subjected to practically the same conditions as the associated sediments.   On the other hand are igneous rocks intruded near the climax of metamorphism or during the earlier waning stages ; so that the special characters impressed on them have been acquired in the main under the condition of declining temperature.   In the absence of characteristic residual characters, mineralogical or structural, a petrographical study may sometimes be insufficient to discriminate between the two cases, and we must then have recourse to the geological relations of the rock and to considerations of a general kind.   None the less the distinction is of fundamental import, and we shall endeavour to carry it out.

We shall be concerned in the present chapter with the former of the two cases so discriminated ; viz. that of igneous rocks of earlier date which are brought under the influence of regional metamorphism of lower or higher grade.   Such rocks may comprise not only abyssal but hypabyssal types, as well as volcanic products, both lavas and tuffs; but it will not be necessary to insist upon these distinctions.   Initial differences of texture may indeed affect the behaviour of the rocks in the lowest grades, but cease to be of importance when recrystallization has set in.   To avoid needless expansion of a range already sufficiently wide, we shall also disregard generally those impure tuffs ('tuffites' of some authors) which contain an admixture of detrital or calcareous material.

Of Grubenmann's twelve groups of crystalline schists, founded upon chemical composition,[1] six correspond roughly (allowing for some exceptions) with sedimentary and six with igneous rocks, as shown below ; and the same broad chemical grouping of the igneous rocks will serve our present purpose.

Sedimentary

II.   Tonerdesilikat-Gneise (argillaceous).
VIII. Quartzitgesteine (arenaceous).
X.   Marmore (calcareous).
IX.   Kalksilikatgesteine (semi-calcareous).
XII.   Aluminiumoxydische Gesteine (bauxitic).
XI.   Eisenoxydische Gesteine (ferruginous).

Igneous

V.   Magnesiumsilikatschiefer (ultrabasic).
IV.   Eklogite und Amphibolite (basic).
III.   Kalknatronfeldspatgneise (intermediate).
I.   Alkalifeldspatgneise (acid).
VI.   Jadeitgesteine   }
VII.   Chloromelanitgesteine   } (alkaline).

[1] See also Niggli, *Schw. Min. Petr. Mitt.*, vol. xiv (1934), pp. 464–72.

REGIONAL METAMORPHISM OF ULTRABASIC ROCKS

The ultrabasic rocks are represented mainly by plutonic types, including the peridotites and the pyroxenites. Metamorphosed examples may be studied in the Archaean of Scotland, and especially in the Shetland Isles.[1] Other illustrations are afforded by the ultrabasic dykes which intersect the Lewisian complex of Sutherland.[2] These latter have been modified by a metamorphism, of no high grade, which is of regional type but local distribution, as discussed in a later chapter.

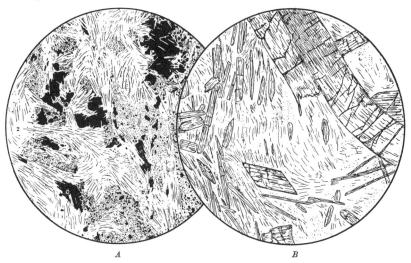

A   B

FIG. 138.—ULTRABASIC SCHISTS, Mainland, Shetland Isles; × 23.

A. Antigorite-Talc-Magnetite-schist, Scousburgh. The flakes of antigorite show a plumose grouping. The talc is in a confused aggregate, speckled with magnetite dust. *B*. Tremolite-Talc-schist, Fethaland.

The simplest low-grade representative of a peridotite is a *serpentine-schist*. Interspersed among the fibres of serpentine, which tend to a pronounced parallel arrangement, is usually some residual spinellid mineral (picotite or chromite)—sometimes also relics of olivine—while part of the iron comes out as secondary magnetite. In any considerable development of such rocks, however, the serpentine commonly has the scaly habit of antigorite, and *antigorite-schist* is a very char-

[1] Phillips, *Quart. Journ. Geol. Soc.*, vol. lxxxiii (1927), pp. 622–51; Read and Dixon, *Min. Mag.*, vol. xxiii (1933), pp. 309–16 (stichtite in serpentine); Read, *Quart. Journ. Geol. Soc.*, vol. xc (1934), pp. 662–6, 669–72 (Unst). Some of the minerals were described and analysed by Heddle, *Min. Mag.*, vol. ii (1879) and *Mineralogy of Scotland* (1901).

[2] *Geological Structure of the North-West Highlands* (*Mem. Geol. Sur. G.B.*, 1907), pp. 85–9.

acteristic type.[1]   The little flakes often have an interlaced arrangement, or tend to fan-like and feathery groupings, and schistosity is then little apparent (Fig. 138, *A*).   Doubtless many antigorite-schists come from the metamorphism of common serpentine-rocks, but antigorite may also be formed directly, both from olivine and from pyroxene, and relics of these minerals sometimes remain.   It is evident, however, that the conversion of a peridotite to a serpentine- or antigorite-schist demands a very considerable accession of water, and under normal conditions talc is produced instead.[2]   So *antigorite-talc-schist* and

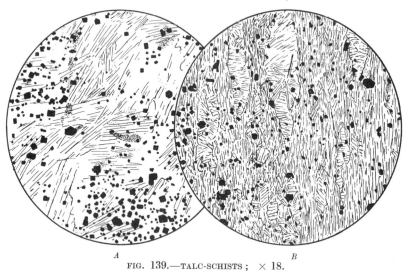

A   B
FIG. 139.—TALC-SCHISTS ;   × 18.

A.   Zillertal, Tirol.   *B*.   Zöptau, Moravia.
Both rocks are composed essentially of talc, with numerous little octahedra of magnetite.

simple *talc-schist* are to be recognized as common types. The geological evidence makes it apparent that talc, moreover, represents a more advanced metamorphism in the purely dynamical sense ; or, in other words, talc is more emphatically a stress mineral than the serpentines.   It is clear too that talc-schists, as judged by their associations, persist into a higher grade of metamorphism.   Here the flakes become larger, and the finely divided magnetite, which is always present, collects into little octahedra (Fig. 139).   Schistosity is usually well marked, coming from a parallel arrangement, partly of the individual flakes, partly of lenticles in which the flakes are set transversely.

[1] On Alpine antigorite-schists see Bonney, *Geol. Mag.*, 1890, pp. 533–42 ; *Quart. Journ. Geol. Soc.*, vol. lii (1896), pp. 452–9 ; vol. lxi (1905), pp. 690–714 ; vol. lxiv (1908), pp. 152–70.

[2] Pure magnesian serpentine contains 13·04 per cent. of water ; talc 4·76.

If we have regard to the molecular ratio $RO : SiO_2$ in the several minerals concerned :

Olivine, R_2SiO_4	. . . . . .	$2 : 1$
Enstatite, $RSiO_3$	. . . . . .	$1 : 1$
Serpentine, $H_4R_3Si_2O_9$	. . . . .	$3 : 2$
Talc, $H_2R_3(SiO_3)_4$	. . . . . .	$3 : 4$

it is evident that in the making of an antigorite-schist from an olivine-rock a certain part of the bases, magnesia and ferrous oxide, must be set free, and for a talc-schist a larger amount. From a pyroxene-bearing peridotite the surplus will be less. This excess of bases comes out largely in the form of magnetite, but under appropriate conditions there is a production of magnesian and ferrous carbonates. These may be subsequently removed in solution, as is not infrequently shown by veins traversing the rock ; but some part often remains, giving an *antigorite-carbonate-schist* or *talc-carbonate-schist*.

Certain *chlorite-schists* and *chlorite-magnetite-schists* are the meta-morphosed representatives of ultrabasic igneous rocks. Chemical analyses show that the characteristic mineral is a variety very poor in alumina, contrasting with that found in chlorite-schists of sedi-mentary origin. Like the talc-schists, this chloritic type may pass up into a fairly advanced grade of metamorphism.

Another series of ultrabasic schists is characterized by members of the amphibole group, viz, the monoclinic tremolite and the rhombic anthophyllite. These may be associated with talc (not commonly with antigorite), giving such transitional types as *tremolite-talc-schist* (Fig. 138, *B*), and *anthophyllite-talc-schist* ; but there are also rocks consisting essentially of the amphibole minerals, separate or together. A *tremolite-schist* having this origin may be indistinguishable from one derived from a siliceous magnesian limestone. The texture may be finer or coarser, and may show either a plexus of interlacing fibres or a regular parallel disposition (compare Figs. 140, *A* and 88, *B*). A pure *anthophyllite-schist* is not a common type. Good examples are known in the Shetland Isles [1] (Fig. 140, *B*). If we set aside the quite gratuitous hypothesis of an accession of silica from without, it is evident that the composition of these amphibole-schists allies them with the pyroxenite rather than the peridotite family. Sometimes too the coming in of some zoisite indicates a certain original content of lime-felspar. Magnetite and rutile are common accessory con-stituents, and spinellid minerals may be present as residual elements. With advancing metamorphism these various minerals come to be

---

[1] For an analysis of this rock see Heddle, *Min. Mag.*, vol. iii (1880), p. 21.

merged in a green hornblende of complex constitution, adapted to the bulk-composition of the rock. We have then a relatively coarse-textured *hornblende-schist*, or a more massive hornblende-rock, often composed almost wholly of the one mineral, though a red garnet may enter in addition.

Of the highest possible grades of metamorphism of ultrabasic rocks we possess little or no definite knowledge. What we have seen elsewhere would lead us to expect a change from amphiboles to pyroxenes, and presumably, if a sufficiently high temperature can be

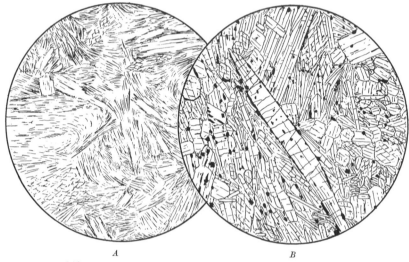

FIG. 140.—ULTRABASIC AMPHIBOLE-SCHISTS, Shetland Isles ;   × 23.

A. Tremolite-schist, Muckle Head Geo, Balta : showing an interlacing or felted arrangement of fibrous crystals.
B. Anthophyllite-schist, Hillswickness, Mainland. Finer-textured varieties show a stronger schistosity.

realized, the various magnesian minerals observed in the lower grades should yield olivine. Both peridotites and pyroxenites, sometimes garnetiferous, are indeed found in tracts of high-grade metamorphism ; but the evidence goes to show that they are normal plutonic rocks, not greatly modified from their pristine state. In rocks of such simple mineralogical constitution it is scarcely possible to discriminate between a crystalloblastic and a primary igneous structure ; so that geological relations may afford the only criterion.

REGIONAL METAMORPHISM OF BASIC ROCKS

Basic igneous rocks, as compared with ultrabasic, contain less magnesia and more lime, a higher percentage of silica, and considerable amounts of alumina and soda.   With these differences in chemical composition, metamorphism necessarily follows a different course as regards mineralogical reconstruction.   Since the transformations to be observed are instructive and typical of what is found also in some other families of rocks, they will be examined in some detail.   As before, we shall take our illustrations, so far as is possible, from British areas.   The earlier stages of metamorphism are well illustrated in the Start district of Devonshire [1] by a group of rocks representing old basic lavas and tuffs, possibly with intercalated sill-intrusions.   Rocks

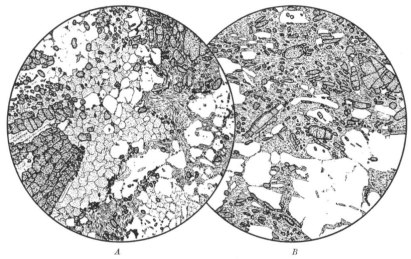

FIG. 141.—ALBITE-EPIDOTE-CHLORITE-SCHISTS, Start district, South Devon ;   × 18.

A.   Prawle : rich in calcite and containing a little quartz and sericite.
B.   Southpool Creek : a variety devoid of calcite.   The porphyroblastic habit of the albite is very noticeable.

generally similar to these figure in the Gwna Group of Anglesey,[2] where, however, there is a tendency to more sodic types.   The higher grades may be studied in the Lizard district of Cornwall.[3]   Most instructive, however, is the important group known to Highland geologists as ' epidiorites '.[4]   These represent dolerite sills intruded at numerous horizons in the Dalradian and Moine series and subsequently metamorphosed in common with those strata.   Their grade of meta-

[1] Tilley, *Quart. Journ. Geol. Soc.*, vol. lxxix (1923), pp. 180–8, with analyses, *Geol. Mag.*, 1938, pp. 501–11, with analyses.

[2] Greenly, *Geology of Anglesey* (*Mem. Geol. Sur.*, 1919), pp. 76–8, with analyses.

[3] Flett, *Geology of the Lizard* (*Mem. Geol. Sur.*, 1912), pp. 44–51.

[4] See various *Memoirs of the Geological Survey of Scotland* ; also Wiseman, *Quart. Journ. Geol. Soc.*, vol. xc (1934), pp. 354–416 (a comprehensive study of progressive metamorphism).

morphism can therefore be directly correlated with the scheme of zones already established for the pelitic sediments, and representatives are found in every grade from the lowest to the highest. Excepting the general absence or scarcity of sericitic material, and of biotite derived from it, there is a fairly close parallel between these intrusive rocks and the sedimentary ' Green Beds ' already noticed (pp. 266-8).

In the lowest grade we have a *calc-albite-chlorite-schist*, and next an *albite-epidote-chlorite-schist*, which may still carry calcite in greater or less amount (Fig. 141). The minerals named are the essential

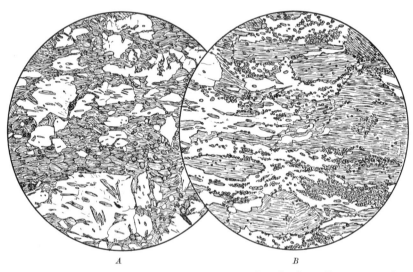

FIG. 142.—ALBITE-EPIDOTE-HORNBLENDE-SCHISTS, Prawle, Start district ; × 23.

A. With porphyroblastic albite. Some chlorite is mingled with the hornblende.
B. With porphyroblastic hornblende, often fringed by granules of epidote with a little sphene.

constituents, with little or no quartz. The albite tends, as usual, to a porphyroblastic development. A not infrequent accessory constituent is hornblende, as slender needles embedded in the chlorite and sometimes enclosed in the albite. Its appearance well within the chlorite zone is at first sight anomalous. The probable explanation, as put forward by Wiseman and also by Tilley, is that, at this early stage, it is produced at the expense of residual granules of augite present in the original rock. With progressive development of hornblende in this and the succeeding (biotite) zone, coming now from reaction of chlorite with calcite or epidote, the chlorite rapidly dwindles, and the epidote is somewhat reduced in amount. Little granules of sphene appear about the same time. We have thus a distinctive type of *albite-epidote-hornblende-schist* (Figs. 142, 143), in which horn-blende gradually becomes the principal constituent.   It figures either as a plexus of little crystals with grains of epidote interspersed or as distinct porphyroblasts, often bordered by epidote granules.   Accord-ing to Tilley's analyses, it is an actinolitic variety.   At this stage the felspar, though still of a highly sodic kind, is no longer a pure albite.

Hornblende continues to be the characteristic mineral in most higher grades, and the rocks may accordingly be named amphibolites.[1]

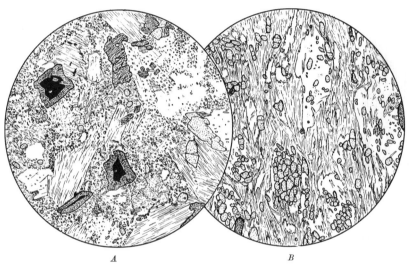

A   B

FIG. 143.—ALBITE-EPIDOTE-HORNBLENDE-SCHISTS from the 'epidiorite' group
of the Highlands ; × 23.

A.   Taycreggan, Loch Awe, Argyllshire (chlorite zone).   The chief constituents are a pale fibrous hornblende, clear albite studded with granules of epidote, and aggregates of chlorite flakes.   Epidote occurs also in distinct grains, and abundant sphene has been formed round grains of iron-ore.
B.   Calder Bridge, Balquhidder, Perthshire (biotite zone).   This shows a strong schistosity with a tendency to porphyroblastic development of the felspar, which is no longer a pure albite.

Metamorphism, however, has not always followed quite the same lines, as appears on comparison of the Lizard rocks with those of the Scottish Highlands.   The green schists of the 'Old Lizard Head Series', fine-textured, highly schistose, and often corrugated, consist of a dense mass of hornblende needles with some quartz as well as felspar, grains

---

[1] More or less schistose hornblende-plagioclase-rocks have very commonly been styled by British geologists 'hornblende-schists', but in the systematic nomenclature of Grubenmann and other Continental writers they are 'plagio-clase-amphibolites'.   The name hornblende-schist is more conveniently reserved for quartz-bearing rocks with little or no felspar.

of epidote, and sometimes flakes of chlorite. The associated sediments are in the garnet-zone. The ' Landewednack Hornblende-schists ' of the same district are of the same and higher grades, reaching in places the sillimanite stage. They are *plagioclase-amphibolites*. The simplest and most characteristic type is composed essentially of a green hornblende and a felspar near andesine with only a small amount of other minerals—epidote, sphene, magnetite, and apatite, seldom quartz. There are indeed varieties containing more epidote, and often well-

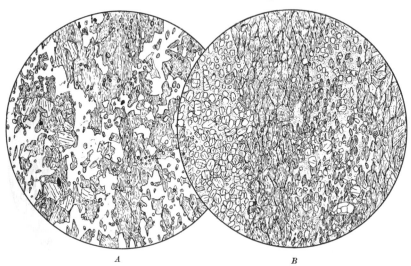

A   B

FIG. 144.—PLAGIOCLASE-AMPHIBOLITES WITH DIOPSIDE in the ' Landewednack Hornblende-schists ' of the Lizard Cornwall, ; × 23.

    A. Carnbarrow, near Cadgwith. Essentially of green hornblende and a plagioclase of an intermediate variety ; also (towards the left) grains of a colourless diopside and granules of epidote.
    B. Church Cove, near Landewednack. Here diopside is abundant, characterizing particular bands, which alternate with hornblendic bands. The felspar is largely saussuritized.

marked bands which are highly epidotic ; but these are probably due to weathering and calcification of the rocks prior to metamorphism. The parallel arrangement of crystals may be more or less pronounced, and is best developed in the most richly hornblendic rocks. One variety is composed essentially of hornblende, almost to the exclusion of felspar. It is clearly suggested here that felspar may, in some measure, pass into a potential state [1] as part of the composition of a complex amphibole.

    [1] The expression is that of Lacroix, who has shown that a rock consisting mainly of pyroxene and plagioclase may be converted, without change of total composition, into a non-felspathic amphibolite ; *Comptes Rendus*, vol. clxiv (1917), pp. 969–74.

This absorption of felspar-substance into the hornblende is found especially in the highest grade of metamorphism. At the same stage, in bands which were presumably richer in lime, the place of hornblende is partly or wholly taken by a colourless diopside (Fig. 144). These bands contain abundant felspar, and it is clear that the pyroxene is not readily capable of taking the highly complex constitution of the amphibole. This diopside-bearing type is found especially in close association with intrusions of peridotite (now serpentine),[1] but there

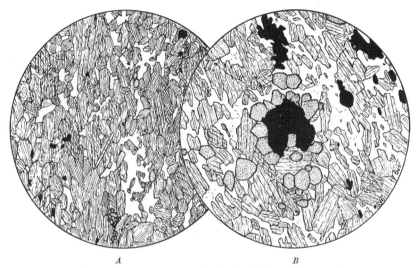

A   B

FIG. 145.—PLAGIOCLASE-AMPHIBOLITES in the 'epidiorite' group, Cluny Bridge, Pitlochry, Perthshire; × 25.

A. The orientation of the hornblende crystals gives a pronounced schistosity. Magnetite, sphene, and apatite are present as accessory constituents.
B. A garnetiferous variety, the garnet tending to cluster about grains of iron-ore. Both rocks are in the garnet-zone, and occur in close proximity.

is no reason to doubt that it belongs to the general regional system of metamorphism.

Much of the foregoing account is equally applicable to the 'epidiorite' group of the Highlands. Here, when we enter the garnet-zone, chlorite soon disappears, and the amount of epidote diminishes. Concurrently a green hornblende becomes the dominant mineral, and the felspar passes through oligoclase to andesine. The epidote minerals may, however, survive in part well into the garnet-zone, giving as a recognized type an *epidote-* or *zoisite-amphibolite* [2] (Figs. 87, 146, *A*),

[1] Fox and Teall, *Quart. Journ. Geol. Soc.*, vol. xlix (1893), p. 201.

[2] Compare Teall, *British Petrography* (1888), plate XXVIII, Fig. 1; but the 'epidote-amphibolite' of Fig. 2 is an epidote-hornblende-mica-schist of sedimentary origin.

Failing this, we have a simple *plagioclase-amphibolite*, a type which has a wide distribution (Fig. 145, *A*). Very often, however, there enters in addition another important constituent, viz. a red garnet. It makes its first appearance in these basic igneous rocks at about the same stage as in the associated argillaceous sediments ; but then and thereafter its occurrence is inconstant, so that a *garnet-plagioclase-amphibolite* and a non-garnetiferous type are often found in close proximity (Fig. 145). This can be due in general only to differences in the bulk-composition of the rocks ; but the abundance of garnet

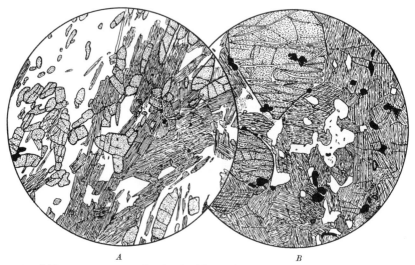

A   B

FIG. 146.—AMPHIBOLITES in the 'epidiorite' group of the Highlands ; × 25.

A. Epidote-Amphibolite, Ben Vrackie, Perthshire (garnet-zone). Essentially of epidote, green hornblende, and clear plagioclase.
B. Garnet-Hornblende-Plagioclase-Gneiss, Glen Doll, Forfarshire (sillimanite-zone). Of coarse texture and devoid of schistosity. It is composed of red garnet, hornblende, and andesine, with a little magnetite, sphene, and quartz.

in the Highland rocks, in contrast with its almost complete absence in the Lizard district, seems to point rather to the influence of pressure. Indeed there is evidence in the epidiorite group itself that dynamical conditions, rather than chemical composition, may be the determining factor. The interior of a thick sill may carry garnet, while the marginal part is devoid of that mineral, and is at the same time richer in hornblende and much more decidedly schistose.[1]

In the highest grades of metamorphism (cyanite and sillimanite zones) the mineralogical constitution is generally the same, but the rocks are of coarser texture, and often show little or nothing of the

---

[1] A good example is seen at Craig an Eunaich, three miles west of Dunkeld.

schistose character. From amphibolites in the ordinary sense they pass to *hornblende-plagioclase-gneiss* and *garnet-hornblende-plagioclase-gneiss* (Fig. 146, *B*). The felspar is commonly a medium andesine.

The prominent part taken in all these rocks by hornblende, to the exclusion of pyroxene, emphasizes the influence of the dynamic factor in regional metamorphism. In the simple thermal metamorphism of basic igneous rocks we saw that the conversion of augite to hornblende in a medium grade was reversed in a higher grade (p. 110). Shearing stress has the effect of promoting the amphibolization of any original augite, and postponing the reverse change, which in fact is attained only rather exceptionally. A colourless diopside, which we have noted in some of the most highly metamorphosed of the Lizard rocks, is found under like conditions in the Highland ' epidiorites ', but only in certain districts, such as Ardgour in Argyllshire, Glenelg in Inverness-shire, and the Banchory district of Deeside. It occurs too in what is probably a corresponding group of basic sills in Connemara, Co. Galway.

Whether in the highest grade hornblende does or does not give place to pyroxene is not determined merely by the ratio CaO : MgO + FeO in the bulk-composition, for a non-calcic (rhombic) pyroxene may come in. Another relevant condition presumably is the proportion of silica, which in amphiboles of the kind in question is constantly below the metasilicate ratio ; but it is probable that great hydrostatic pressure is often the decisive factor. In the rocks known as *pyroxene-granulites* this is further emphasized by the frequent occurrence of garnet. The best-known pyroxene-granulites are found in the so-called Granulitgebirge of Saxony, where they are associated with other ' granulites ' of intermediate and acid composition. The principal constituents are hypersthene, augite, frequently garnet, labradorite, and magnetite. The structure is granoblastic, but sometimes modified by a radiate arrangement of the pyroxene about crystals of garnet (Fig. 89, *B*, above). Pyroxene-granulites, with and without garnet, are found locally in the Lewisian complex of Sutherland (Scourie and Lochinver). With their simple mineralogical constitution and microstructure, devoid of any schistosity or foliation, such rocks have a considerable resemblance to the most highly metamorphosed basalts and dolerites from thermal aureoles, and they can also be matched in many particulars among high-grade representatives of impure calcareous sediments.

Of other pyroxenic rocks which figure among the ' Kata-Gesteine ' of Grubenmann, and more especially of the eclogites, it is safe to assert that they have never passed through lower to higher grades of metamorphism but represent plutonic rocks intruded under special conditions. Their place is therefore in the following chapter.

In most of the examples which have been cited, even in the lowest grades of metamorphism, the original fabric of the igneous rock (gabbro, dolerite, or basalt) had been completely broken down. In certain circumstances, however, it is possible for *residual structures* to be preserved, at least in the case of large plutonic masses, which possessed, even during metamorphism, a certain degree of rigidity. In the Lizard district the ' Traboe Hornblende-schists ' [1] are metamorphosed

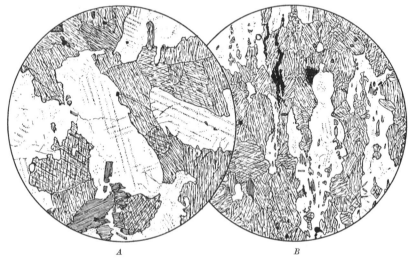

A   *B*

FIG. 147.—CONVERSION OF GABBRO TO AMPHIBOLITE Portsoy, Banffshire ; × 23.

A.   Interior of mass.   The augite is converted wholly to hornblende, but the original structure of the plutonic rock is retained.
B.   Margin of mass.   A schistose plagioclase-amphibolite with a little biotite, sphene, magnetite, and apatite.

gabbros which often retain something of their original structure, modified by the usual cataclastic accidents.   Exceptionally even some primary augite remains unchanged.   In the Highlands, related to the ' epidiorite ' group and involved in the same regional metamorphism, there are coarse-grained bodies of more massive habit (gabbros) which were able to offer a more effective resistance.   As seen, for instance, in the Glen Shee district of Perthshire, they have undergone the appropriate mineralogical changes, while still preserving much of the structure of plutonic rocks.   The same thing is exhibited more strikingly in that tract of Aberdeenshire and Banffshire where, as we have seen reason for believing, shearing stress fell short of its normal

[1] *Geology of the Lizard*, pp. 49–52.

intensity (p. 186). Here, even in an intrusive mass of quite moderate dimensions, we find the original relations of the minerals largely preserved in the interior, while the marginal part has the structure, as well as the mineralogical constitution, of a crystalline schist (Fig. 147).[1] The epidiorites themselves in the same area often preserve relics of their original igneous structures and of original minerals, and have been recognized by Wiseman [2] as abnormal from the zonal point of view.

REGIONAL METAMORPHISM OF INTERMEDIATE AND ACID ROCKS

After the somewhat full consideration which has been devoted to the metamorphism of basic igneous rocks, those of intermediate and acid composition will be dismissed more briefly ; the more so because British examples are few. Metamorphism follows indeed the same general lines as before, but with mineralogical differences such as are easily predicable from the different bulk-composition of the rocks. As we pass from more basic to more acid, there is a falling off in lime and magnesia with a rise in silica and alkalies, especially in potash. How these differences translate themselves into mineralogical terms is readily appreciated.

Consider first rocks in which the original structure has been quite obliterated. This is the case, not only of lavas and tuffs, but of plutonic rocks which have been completely broken down at an early stage. In the lowest grade a rock of mean acidity is represented by a *calc-albite-sericite-chlorite-schist*, and this is succeeded by *albite-epidote-sericite-chlorite-schist*. These differ from the corresponding basic types by the coming in of a certain amount of white mica and quartz and usually by a higher content of albite as compared with chlorite and the lime minerals. The early formation of acicular hornblende, which was a noteworthy feature in the former case, has not been recorded here, but hornblende makes its appearance with advancing metamorphism, as the chlorite and epidote dwindle. Biotite also becomes a prominent constituent, with or without some muscovite, and such rocks may be styled *biotite-hornblende-schists*. The felspar at this stage is oligoclase, and has not the porphyroblastic habit often noticeable in the albite of the lower grades. Of the still higher grades of metamorphism in rocks of this kind we possess less certain knowledge. Hornblende- and biotite-plagioclase-gneisses, with andesine or oligoclase and some potash-felspar, are well known, but doubtless most of these rocks represent diorites or intermediate pyroxenic rocks intruded

---

[1] Compare Read, *Geol. of the Country around Banff* (*Mem. Ord. Sur. Scot.*, 1923), pp. 99–101.   [2] *Loc. cit.*, pp. 405–7.

under stress-conditions.   When, however, such a rock makes part of a
gneissic complex like the Lewisian, it may suffer metamorphism at a
later stage.   It is interesting to note that a rock so reconstructed
sometimes reproduces many of the features of simple thermal metamorphism (Fig. 148).   This is probably due to its being enveloped by
a volume of acid magma, sufficient to relieve it from any serious
shearing stress during its metamorphism.

This brief treatment is necessarily of a generalized kind.   What
we have named ' intermediate ' igneous rocks comprise a wide range

A   B

Fig. 148.—METAMORPHOSED DIORITIC ROCKS, in the Lewisian complex, Sutherland ;   × 25.

A.   Near Scourie.   The minerals shown are pale green fibrous hornblende, in places
honeycombed with quartz, biotite with a tendency to radiate grouping, and fresh andesine,
with magnetite and some apatite.
B.   Lochinver.   A more acid type.   Here the hornblende is of a deeper green ;
biotite is abundant and is sometimes intergrown with magnetite ; the felspar is oligoclase,
and shows incipient saussuritization ; and a noteworthy amount of quartz is present.

of bulk-analysis, and this is reflected in the relative proportions of
the minerals enumerated.   As we pass from sub-basic to sub-acid,
quartz, potash-felspar, and micas increase in amount, while chlorite,
epidote, and hornblende decrease.   The plagioclase too becomes more
restricted to the sodic end of the series, even in the most highly
metamorphosed rocks.

In rocks of definitely acid composition the proportion of chlorite
and epidote is further reduced, and hornblende does not form.   Quartz
is more abundant, and white mica is the next constituent in importance.
In rocks completely broken down the lowest grade is a *sericite-schist*,

often with little porphyroblasts of albite. There is usually sufficient chlorite and iron-oxide present to give rise in due course to biotite ; but, failing this, we have a *muscovite-quartz-schist*, not distinguishable from a certain sedimentary type (Fig. 119, *B*, above), and this may persist into a high grade with no other change than an increasing coarseness of texture.

It is to be remarked, however, that for rocks in which a potash-felspar is an important constituent, metamorphism may follow two alternative courses. Sericitization may begin and be complete at an

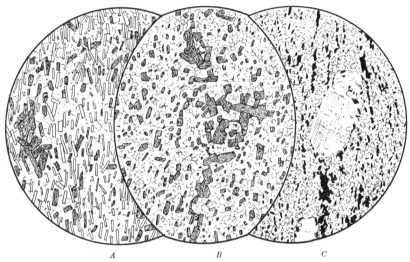

A   B   C

FIG. 149.—METAMORPHOSED RHYOLITES ;   × 23.

   A.   Muscovite-Biotite-Granulite.
   B.   Biotite-Granulite.   These two rocks are from the ' hälleflinta ' formation of Utö, Sweden.   The flakes of mica are set in a granoblastic aggregate of quartz and felspar. Both rocks have had scattered blasto-porphyritic felspars, not shown in the figures.
   C.   Riebeckite-Granulite, Berlin, Wisconsin.   This is the ' rhyolite-gneiss ' of Weid-man (*Bull. III, Wis. Geol. Nat. Hist. Sur.*, 1898).   It shows a quartzo-felspathic ground with granoblastic structure enclosing very numerous little crystals of riebeckite (deep blue but represented by black in the figure).   There are also phenocrysts of a sodic felspar, partly crushed and modified.

early stage ; but this is dependent upon the conditions and in particular upon a sufficient quantity of water. The reverse change, from mica to felspar, belongs to an advanced grade, the higher in proportion as the stress-factor enters. It is clear that there is a wide intervening range of temperature in which potash-felspar and mica may be present in practically stable association, and this has already been exemplified in the case of ' granulites ' derived from felspathic sandstones (p. 246). In like manner igneous rocks of appropriate composition come to be represented in various grades by quartzo-felspathic rocks, carrying

more or less mica, to which the same name *granulite* is applicable.
A parallel orientation of mica-flakes, light or dark, may impart a more
or less pronounced schistosity (Fig. 149, *A*, *B*).   In sodic types, with a
sufficiency of iron, the place of mica is taken by riebeckite (Fig. 149, *C*).
Only in a very advanced grade of metamorphism does the muscovite
of potassic rocks disappear by conversion to felspar.   Here most
probably we may include the acid members of the well-known Saxon
granulite-group, though their true relations have been the subject of
some debate.   They are clearly in a very high grade of metamorphism,

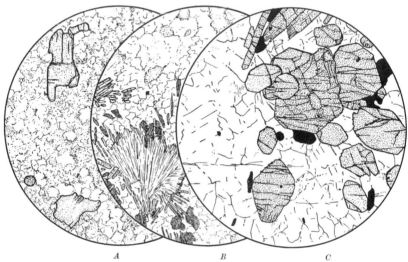

FIG. 150.—ACID AND ALKALINE GRANULITES, from the Mittelgebirge or Granulitgebirge of Saxony ;   × 23.

    A.   Cyanite-Granulite, Röhrsdorf ; with cyanite and small garnets in a granoblastic
matrix of felspar and quartz.
    B.   Sillimanite-Granulite, Markersdorf ; with sillimanite and biotite (garnet else-
where in the slice) ; the felspar is microperthite.
    C.   Prismatine-Albite-Granulite, Waldheim ; essentially of prismatine and a felspar
near albite, with crystal-grains of a dark spinellid mineral.

for, not only muscovite, but often biotite too has suffered dissociation,
its place being taken partly or wholly by a red garnet (Fig. 89, *A*,
above).   The frequent occurrence of such minerals as cyanite, sillimanite, and hercynite [1] is due to the liberation of alumina in this
process of dissociation, for the granulites contain no more alumina in
their bulk-analysis than a granite of like acidity (Fig. 150, *A*, *B*).
These rocks often show cataclastic structures due to subsequent
mechanical disturbance.
    In the most usual case a granitic rock involved in regional meta-

---

[1] The iron-spinel, unlike the magnesian, is stable in presence of free silica.

morphism has acquired a more or less strongly foliated character, often taking the form of wavy lenticular and 'augen' structures. The general fabric is typically granoblastic or 'granulitic', but with possible modifications, becoming schistose where there has been a copious production of new mica, and often being affected by cataclastic accidents of later date. Examples are afforded by some of the granite-gneisses of the Highlands, such as that which forms Ben Vuroch, in Perthshire.[1] On its western or north-western side, which was the lee-side with reference to the direction of thrust, it retains, with only

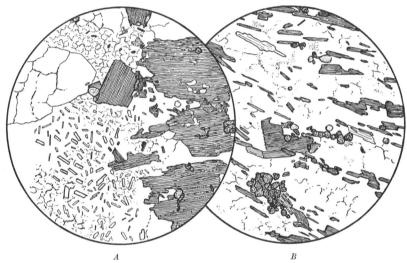

A   *B*

FIG. 151.—BIOTITE-GRANITE AND GNEISS, Ben Vuroch, Perthshire ; × 23.

A. Granite from the western (protected) border. Besides incipient cataclastic effects there is a production of flakes of muscovite in the microcline ; also grains of epidote associated with the biotite.

B. The same rock transformed to a gneiss. There is a parallel orientation of mica, muscovite as well as biotite, and garnet has become an important constituent. In other parts the gneiss shows 'flaser' and 'augen' structures.

minor accidents, its original character of a biotite-microcline-granite (Fig. 151, *A*). This passes into an augen-gneiss, which makes the greater part of the mountain.[2] To the original minerals are added garnet, muscovite, and granular sphene derived from ilmenite. The 'eyes' of microcline taper off into streaks consisting of fragments of the felspar and films of mica formed from it, and inconstant ribbons of mica, dark and light, wrap round the 'eyes'. Where the augen-structure is not developed, we see merely parallel flakes of mica and

[1] *Geology of Blair Atholl* (*Mem. Geol. Sur. Scot.*, 1905), pp. 87–91.

[2] See Teall, *British Petrography* (1888), plate XLIII, Fig. 1, for photograph of a cut surface.

clusters of garnet set in a quartzo-felspathic aggregate (Fig. 151, *B*). On the south-eastern side of the mass the rock becomes highly schistose and rich in white mica. It is probable that metamorphism has been effected not all at one time or at one temperature. Another granite which was intruded at an early epoch, and has undergone regional metamorphism is that of Càrn Chuinneag in eastern Ross.[1]

REGIONAL METAMORPHISM OF ALKALINE ROCKS

The various igneous rocks which have been noticed above belonged all to the calcic or ' Kalk-Alkali ' branch. There remain those which are characterized chemically by a greater relative richness in alkalies, and usually in soda. The only rocks of this category which call for any detailed examination are those of the *spilitic series*, of which the distinguishing mark is a special richness in albite. The salient fact to be observed is that in metamorphism much of the soda present is taken up into a soda-amphibole. Glaucophane is the common form, but with it may be included gastaldite and crossite, the one richer and the other poorer in alumina. With glaucophane substituted for actinolite and common hornblende, there is to be recognized a series of types parallel with that of the chlorite-schists and amphibolites of ordinary basic composition. It should be remarked that the sodic part of glaucophane is equivalent to jadeite (with some acmite) ; but, while the amphibole, as a stress-mineral, figures iu the lower and medium grades of metamorphism, the pyroxene, as a high-temperature and high-pressure mineral, is to be looked for only in a very high grade.

Although most glaucophane-bearing rocks included here [2] come from spilites and spilite-dolerites, it appears that dolerites and gabbros not specially rich in soda have sometimes followed the same line of metamorphism.[3]

The lowest representative is the type known as *prasinite*, composed essentially of albite, chlorite, epidote, glaucophane, and sometimes calcite. In the relative proportions of these minerals and in the degree of development of a schistose structure the rocks show considerable variety. Analyses indicate a spilitic composition, and the manner of occurrence is suggestive of metamorphosed lavas and tuffs. Some of the best Alpine examples come from the Val de Bagne, in Valais,[4]

---

[1] *Geology of Ben Wyvis* (*Mem. Geol. Sur. Scot.*, 1912), pp. 89–93.

[2] It has already been shown that the acid glaucophane-schists have a different origin (p. 134).

[3] Joplin, *Min. Mag.*, vol. xxiv (1937), pp. 536–7 (New Caledonia).

[4] Grubenmann, *Fest. Rosenbusch* (1906), pp. 1–24 ; Woyno, *Neu. Jb. Min.*, B. Bd. xxxiii (1911), pp. 141–56.

where also are varieties with muscovite and quartz, indicating some admixture of sedimentary material. The prasinite type is represented among the glaucophane-bearing rocks of Anglesey.[1]

In Anglesey, too, a more advanced grade of metamorphism is represented by a *glaucophane-epidote-schist*.[2] This is a well-marked type, composed essentially of glaucophane and epidote with a variable amount of albite and scattered grains of sphene (Fig. 152, *A*). To derive such a rock from a prasinite, the chlorite, calcite, and part of the albite must react to produce more glaucophane. These Anglesey

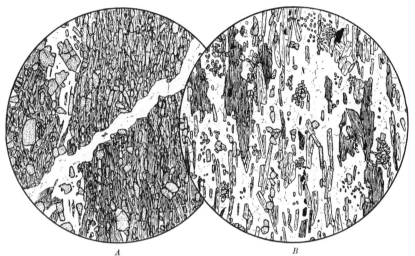

A   *B*

FIG. 152.—CRYSTALLINE SCHISTS RICH IN ALKALIES ; × 23.

A. Glaucophane-Epidote-schist, Anglesey Monument, Menai Straits. Some albite and quartz are present, and a veinlet of clear albite traverses the slice.
B. Biotite-Albite-schist, Fibbia, St. Gotthard : containing also hornblende, epidote, and a little magnetite and apatite. This type, representing a lamprophyre dyke, will be noticed below (p. 296).

rocks, however, come from spilitic lavas. Of the two chemical types $RSiO_3$ and $NaAl(SiO_3)_2$, which go in fairly equal proportions to the making of glaucophane, the former represents roughly the pyroxene of the original rock and the latter the sodic felspar. It is possible that some other glaucophane-schists may be derived from aegirine-bearing rocks, but evidence of such origin is lacking.,

Sometimes, e.g. in the Monte Viso district New Caledonia,[3] and elsewhere, the place of epidote may be taken by lawsonite. This

[1] Greenly, *Geology of Anglesey* (*Mem. Geol. Sur. G.B.*, 1919), pp. 119-20.

[2] Blake, *Geol. Mag.*, 1888, pp. 125-7 ; Greenly, *loc. cit.*, pp. 115-18, with analysis ; for a coloured figure, see Teall, *British Petrography* (1888), plate XLVII, Fig. 1.

[3] Joplin, *loc. cit.*, pp. 534-7.

mineral,[1] with the composition $H_2Ca(AlOH)_2(SiO_4)_2$, is equivalent to anorthite with two molecules of water, and it perhaps represents a first step to the formation of zoisite or epidote. Glaucophane-schists with lawsonite were first described from California,[2] where rocks of this series are found in some abundance and variety (Fig. 153, A). Among them is a *garnet-glaucophane-schist*, composed essentially of the minerals named with epidote but without albite. This represents a somewhat more advanced grade of metamorphism, and a still higher grade is indicated by the disappearance of epidote. Similar rocks

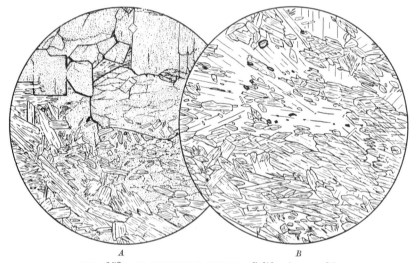

A B

FIG. 153.—GLAUCOPHANE-SCHISTS, California ;   × 25.

A. Lawsonite-Glaucophane-schist, Tiburon Peninsula. The lawsonite here is in large crystals. The rest is of glaucophane, chlorite, and a little albite.
B. Muscovite-Glaucophane-schist, San José.

occur in some force in the Ile de Groix, off the coast of Brittany.[3] Besides typical glaucophane-epidote-schists, there are rocks of which garnet and glaucophane are the essential constituents. To this type the name glaucophane-eclogite has often been applied ; but in view of its geological relations, as well as its pronounced schistosity, *garnet-glaucophane-amphibolite* seems a more suitable designation. The glaucophane-bearing rocks of the Ile de Groix, occurring as numerous sheets, alternating with mica-schists and chloritoid-schists, doubtless represent spilitic lavas and possibly tuffs. Some at least of the Alpine

[1] Lawsonite is exceptional among the minerals of metamorphism in having a prismatic habit with a pinacoidal cleavage.

[2] Ransome, *Bull. Geol. Univ. Cal.*, vol. i (1895), pp. 301–12.

[3] Barrois, *Ann. Soc. Géol. Nord.*, vol. xi (1884), pp. 45–63.

rocks styled glaucophane-eclogites have a different significance, having been intruded, like ordinary eclogites, in direct relation with regional metamorphism.

To avoid possible confusion, it should be recalled here that, while the better-known glaucophane-bearing schists represent rocks of igneous origin, there are others which certainly come from the metamorphism—not necessarily under stress conditions—of bedded sediments relatively rich in soda. A *muscovite-glaucophane-schist*, such as is recorded from the Greek island of Syria,[1] and is found also in California (Fig. 153, *B*), has probably been formed by the regional metamorphism of a tuffaceous sediment.   The quartz-glaucophane-schists, however, are to be interpreted as metamorphosed adinoles.[2] It is significant that these glaucophane-bearing rocks of sedimentary origin are often associated with intrusive igneous rocks which, as now metamorphosed, also carry glaucophane.

The metamorphosed equivalents of other families of sodic igneous rocks, viz. *soda-syenites* and *nepheline-syenites* (including soda-trachytes and phonolites) are little known.   The Saxon granulite group, already mentioned, comprises a type [3] consisting mainly of a felspar near albite, with which are associated such minerals as cyanite, sillimanite, and tourmaline, but especially corundum and the rare silicate prismatine (Fig. 150, *C*).   The richness in alumina here indicated is in no wise peculiar, for corundum occurs abundantly as a normal pyrogenetic mineral of albite-syenites in Ontario and elsewhere.

A nepheline-syenite-gneiss was discovered by Osann [4] at Cevadaes, near Campo Maior in Portugal, where it occurs as a sill intruded among bedded sediments and metamorphosed in common with them.   Other examples have been described by Lacroix [5] from Madagascar (Fig. 154, *A*).   These rocks have the mineralogical composition of ordinary nepheline-syenites but with a parallel orientation of some of their constituents.

Of a different type is a rock described as an aegirine-granulite, found intercalated in the Moine Series in Glen Lui, Aberdeenshire.[6] This yields the analysis of a phonolitic trachyte, and has perhaps been a tuff.   The felspar is anorthoclase, and the coloured silicates aegirine

---

[1] Foullon and Goldschmidt, *Jb. k. k. Geol. Reichs.*, vol. xxxvii (1887), pp. 1–34.

[2] Compare Rosenbusch, *Sitz. k. preus. Ak. Wis.*, vol. xlv (1898), p. 712.

[3] Sauer, *Zeits. Deuts. Geol. Ges.*, vol. xxxviii (1886), pp. 704–5 ; Kalkowsky, *Abh. Nat. Ges. Isis*, 1907, pp. 47–65.

[4] *Neu. Jahrb. Min.*, 1897, vol. ii, pp. 109–28.

[5] *Comptes Rendus*, vol. clv (1912), pp. 1123–7 ; *Minéralogie de Madagascar*, , vol. iii (1923), pp. 170–2.

[6] *Geology of Braemar* (*Mem. Geol. Sur Scot.*, 1912), pp. 52–4.

and a yellow garnet of the melanite kind.   In other bands epidote takes the place of aegirine, suggesting a banded tuffaceous sediment.

Representatives of the *alkali-granites* again are little known in a metamorphosed state.   Keyserling [1] has described a riebeckite-gneiss from Gloggnitz, in Lower Austria, where it occurs as an intrusive sill in the schists of the Semmering district.   The rock is composed of potash-felspars, a sodic plagioclase, quartz, riebeckite, and aegirine, with accessory constituents, and shows a strongly marked parallel structure (Fig. 154, *B*).   In chemical composition it is comparable

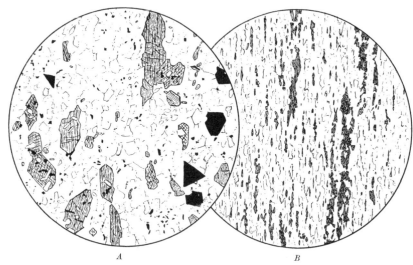

FIG. 154.—METAMORPHOSED ALKALINE IGNEOUS ROCKS ;   × 23.

A.   Nepheline-Syenite-Gneiss, Makaraingo, Madagascar.   The essential constituents are microcline, albite, nepheline, magnetite, and an alkali-hornblende near hastingsite. Parallel structure is little apparent, except in the orientation of the amphibole crystals.
B.   Riebeckite-Gneiss, Gloggnitz, Lower Austria, showing pronounced schistosity. Alkali-felspars and riebeckite are the chief constituent minerals.

with some known riebeckite-granites and with comendites : whether the rock originally contained a pyroxene or an amphibole, it is not possible to pronounce.   As already remarked for glaucophane, the presence in a rock of riebeckite, even in fair abundance, is not in itself enough to relegate it to the alkaline category (p. 290, and Fig. 149, *C*).

Those basic igneous rocks which are distinguished chemically by richness in potash may be dismissed more briefly.   Of metamorphosed leucitic lavas we possess no knowledge, and it will be sufficient to notice the *mica-lamprophyres*, in which the alkali is contained partly in felspar but largely also in biotite.   The usual occurrence of these

[1] *Tsch. Min. Pet. Mitt.*, vol. xxii (1903), pp. 109–58.

in the form of small dykes and sills makes them liable to be overlooked in a tract of advanced metamorphism, and our information concerning them is scanty. The numerous lamprophyre dykes of the Scottish Highlands are of Mid-Palaeozoic Age, and so have no part in the general metamorphism of the region. In some districts, however, they have been modified by a later metamorphism, apparently of rather local distribution and rising only to a moderate grade.[1] Reconstruction, beginning with merely cataclastic effects, may culminate in strongly marked schistosity, and the concurrent mineralogical changes indicate a decided elevation of temperature, due probably to heat developed by the crushing of the rocks themselves. A characteristic transformation is the replacement of olivine by fibrous tremolite (' pilite ') instead of the usual serpentine and carbonate. The common brown biotite, with idiomorphic habit, is converted to a green variety in shapeless flakes, and hornblende gives place to green mica and epidote. Sphene is formed from the titaniferous iron-ore. At the same time the alkali-felspar and quartz are recrystallized to a mosaic, in which calcite, epidote, and sphene also take part, as well as small flakes of mica. These changes are observed in dykes which show the effect of shearing in various degrees. In the more highly metamorphosed rocks (' lamproschists ') the biotite has recovered its brown colour, and a green hornblende also becomes a prominent constituent. These minerals make fairly defined folia separating lenticles mainly of felspar and quartz.

The metamorphosed equivalents of minettes and kersantites are known from various Alpine districts, where they figure as *biotite-plagioclase-schists* [2] (Fig. 152, *B*, above). They are composed usually of biotite and albite with more or less hornblende and some epidote or zoisite, sometimes also quartz. In other examples the felspar is an oligoclase.[3]

[1] See *Memoirs of the Geological Survey of Scotland*, especially *Geology of Braemar* (1905), pp. 125–30, and *Geology of Ben Wyvis* (1912), pp. 121–6.

[2] Grubenmann, *Die Kristallinenschiefer* (2nd ed., 1910), pp. 241–2.

[3] Becke, *Denk. Wien. Ak.*, 1903, p. 29.

CHAPTER XVIII

PLUTONIC INTRUSION IN RELATION WITH REGIONAL
METAMORPHISM

*Intrusion under Orogenic Stress—Composite and Hybrid Gneisses and Injection-
Gneisses—Ultrabasic Rocks in Regional Metamorphism—Basic Rocks in Regional
Metamorphism—Intermediate and Acid Rocks in Regional Metamorphism—
Localized Analogues of Regional Metamorphism.*

INTRUSION UNDER OROGENIC STRESS

THE rocks now to be considered include doubtless the great majority
of those evidently plutonic rocks which are styled igneous gneisses or
orthogneisses. Occurring in a tract of regional metamorphism, and
associated often with highly metamorphosed sediments, they are the
result of igneous intrusion genetically bound up with the metamorphism
itself. Such peculiarities as they may show, distinguishing them from
ordinary granites, diorites, and the like, are due largely to the special
conditions attending their intrusion; and they are therefore not
primarily and essentially metamorphosed rocks, though subsequent
metamorphism has usually set its mark upon them. The fact that
such highly crystalline rocks are in many countries developed upon a
vast scale, without any representatives of lower grades to suggest
progressively advancing metamorphism, would itself suffice to raise
a strong presumption that their intrusion and the metamorphism of
the region were parts of one connected sequence of events. But, while
a general connexion may be assumed, the precise relation in time of
the intrusion to the culminating epoch of metamorphism must vary,
and becomes therefore a subject of inquiry in each case. This comes
out clearly when we have to do with a series of intrusions following
one another at greater or less intervals.

The ' Older ' (presumably Archaean) intrusions of the Scottish
Highlands illustrate this well.[1] They comprise a wide range of types,
and, as exposed in the country between the Highland Border and the
Great Glen, they are not in such force as to confuse their relations.

[1] See Barrow in Hatch's *Textbook of Petrology*, 5th ed. (1909), p. 292, and
Geol. of Braemar (*Mem. Geol. Sur Scot.*, 1912), pp. 66–72.

298

The sequence was the usual one of decreasing basicity. The early basic magmas were intruded among well-bedded Dalradian sediments at a time anterior to the folding and metamorphism, and made a group of regular sills, together with some more massive gabbroitic cores. The metamorphism of this 'epidiorite' group has been described in the preceding chapter. The acid intrusions which followed fall, as Barrow has shown, into several distinct groups. Some of the earlier granites date from a time anterior to the general metamorphism, and have felt the full effects of it, the Ben Vuroch rock being a

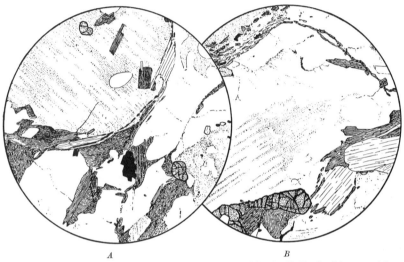

A   B

FIG. 155.—BIOTITE-OLIGOCLASE-GRANITE-GNEISSES, Glen Doll, Forfarshire ; × 25.

    A. Round-grained Gneiss : showing one of the large crystals of oligoclase. The other minerals are garnet, biotite, muscovite, and quartz, with some magnetite.
    B. Augen-Gneiss : in which the oligoclase crystals have the lenticular shape, and the fluxional arrangement of the mica is well shown.

prominent instance (p. 290). A very different case is presented when the dynamical factor in the general metamorphism has come powerfully into action during the progressive crystallization of an intruded magma. Under these conditions it is possible for much of the still liquid part to be strained off and carried forward to make pegmatites elsewhere. This has been well described by Barrow [1] in the biotite-oligoclase-gneiss of Glen Doll, near Clova. At the time when the crustal stress culminated, biotite had already separated out, and large crystals of oligoclase were well advanced in the process of crystallization, while the liquid part contained especially the constituents of

[1] *Geol. Mag.*, 1892, pp. 64–5 ; *Geology of Braemar (Mem. Geol. Sur. Scot.,* 1912), pp. 67–9.

potash-mica and potash-felspar with free silica. Much of this residual magma was forcibly squeezed out, and the consequences of this are seen, not only in the altered total composition of the rock, but in the peculiar structure impressed upon it (Fig. 155). The unfinished oligoclase crystals give it a ' round-grained ' appearance which is very characteristic. At the same time the removal of much of the interstitial magma has forced the remaining minerals closer together, and trains of mica-flakes are seen winding round the larger crystals and squeezed between them. As in so many other granitic rocks either crystallized or recrystallized under high pressure and stress, garnet is a characteristic constituent.

Parallel and lenticular structures, whether of a larger or a smaller order, as congenital features of igneous rocks are essentially fluxional effects, and the degree in which they are developed must vary greatly. It must vary, not only according to the character and intensity of the incident stress, but with the physical status of the body on which the stress is brought to bear. In an environment of highly-heated country-rocks the process of crystallization certainly covers a prolonged time, and the mechanical conditions which cause movement in a partially crystallized magma may culminate at an earlier or later stage of the process, or may be repeated. If consolidation has reached a point at which crystals of tabular or columnar habit make a considerable part of the whole, enforced flow is likely to set up a pronounced parallel orientation of these. It will be still more effective if the forcing out of part of the liquid residuum causes a closer packing of the crystals. Narrowness of the channel of flow will obviously be a favouring condition, and in fact intrusion under great stress tends always to excessive subdivision.

When by continued crystallization, aided perhaps by a straining-off process, the interstitial liquid has been greatly reduced, and the mass is now composed mainly of crystals, it may still be forced into flowing movement under sufficiently great differential stress. A new factor now enters ; for the crystals, violently ground together, become strained or bent, and often suffer fracture according to their individual properties. Actual fractures may be repaired by renewed crystallization, but many of the resulting structures remain indelible. Broadly speaking, they correspond with the cataclastic structures set up in crystalline rocks by dynamic causes at low temperatures ; but for these effects, produced before the completion of magmatic crystallization, the term *protoclastic* [1] is conveniently employed. There may be a more or less advanced granulation of the rock, or of its more brittle minerals,

---

[1] Brögger, *Zeits. Kryst.*, vol. xvi (1890), p. 105.

usually with relics of larger crystals. Such relict crystals tend naturally to lenticular shapes, and many 'eyed' gneisses (Augengneise) have originated in this way. Here, as in simple dynamic metamorphism, these purely mechanical effects are complicated by others which involve solution, diffusion, and mineralogical changes, such as the production of new white mica from the alkali-felspars and epidote from oligoclase or hornblende. Especially in these rocks long maintained at high temperatures, segregation-foliation has been active. This is well seen in the way in which flaky minerals such as micas fall into trains which border lenticles of felspar and quartz, or wind in the fashion of stream-lines about 'eyes' made by larger crystals.

It is perhaps needless to remark that the characters of such rocks, as now observed, have not always been acquired at one time and at one temperature. Superposed upon those mineralogical rearrangements which date from the birth of a rock, there have often been others brought about during the subsequent cooling ; and these changes, in a general sense of a retrograde kind, have doubtless been precipitated in many cases by some belated revival of stress-conditions. They are likely to be accompanied by cataclastic effects, which can often be verified as belonging to a late stage in the decline of temperature.

COMPOSITE AND HYBRID GNEISSES AND INJECTION-GNEISSES

Foliation in igneous gneisses arises, as we have seen, partly from internal movement, which may be discontinuous, under the influence of great differential stress, partly by segregation due to molecular diffusion, the latter factor emphasizing and exaggerating the effects of the former. *Gneissic banding*, i.e. the alternation of distinct litho-logical types in parallel bands, streaks, and lenticles, has a different origin. It results from the drawing out in fluxional fashion of a mass which was *already of a heterogenous nature*. It is in no wise confined to tracts of regional metamorphism, and is in fact a common incident of stratiform intrusion. In a magma free from complications, however, the requisite initial heterogeneity is seldom very pronounced except in the basic and ultrabasic families. The Tertiary plutonic rocks of the Hebrides afford some beautiful illustrations [1] ; and here the strongly marked banding is often accompanied by a parallel orientation of crystals sufficient to impart a certain degree of schistosity.

The necessary conditions for the setting up of a prolonged gneissic banding are, however, realized much more fully in a *plutonic complex*. Here heterogeneity arises, not as in the former case from incomplete

---

[1] Geikie and Teall, *Quart. Journ. Geol. Soc.*, vol. 1 (1894), pp. 645–59 ; Harker, *Tertiary Igneous Rocks of Skye* (*Mem. Geol. Sur. U.K.*, 1904), pp. 75–7.

differentiation in a single body of magma, but from partial intermingling of distinct rock-types. An earlier (normally a more basic) rock, already solidified, is invaded by a later (more acid) magma, which sends veins into it. The veins branch and anastomose, and in this way fragments of the older rock become detached and, enveloped by the new magma, may be further broken up. This mechanical process is complicated both by simple metamorphism of the older rock and by chemical reaction between the two rock-types, giving rise to hybrid products of intermediate composition. Cooling being gradual, a considerable time must elapse before the complex mass becomes too rigid to yield to deformation ; and, if during this time it is forced into flowing motion, it will take on the character of a group of banded gneisses.

The phenomena thus summarily described may be observed locally even in ordinary plutonic intrusions, such as those of Tertiary age in the Inner Hebrides. In the Isle of Rum [1] it is possible to follow in detail every stage of the passage from a complex of eucrite and granite into well-banded gneisses. The process is, however, eminently promoted under the physical conditions which rule in a large tract of regional metamorphism. A magma intruded under powerful crustal stress tends, as has been said, to extreme subdivision and intimate penetration of the rocks invaded. Under the same differential stress, maintained or renewed, the resulting complex yields in the manner of fluxion, and the very slow rate of cooling allows ample time for the elaboration of a pattern of regular banding. In a great system like the Lewisian, built up of numerous units and having an involved history, this is, of course, only one factor among others, but it is an important one. That differential movement here took place in response to powerful orogenic forces does not remove this type of gneissic banding from the category of primary fluxion structures.

So far we have been speaking of intermingling only among successive members of a plutonic sequence. There are, however, composite gneisses of another kind. A granitic magma, with its liberal content of water and perhaps other fluxes, intruded among rocks already heated, is freely fluid, and, when forced in under great pressure, often penetrates the country-rock in a remarkably intimate fashion. If the rock be one of sedimentary origin, it may have a pronounced fissile character, whether due to original bedding or to slaty cleavage or schistosity. The magma is then likely to invade it in the form of numerous thin parallel sheets or leaves, giving the effect described by Michel-Lévy as ' *intrusion lit par lit* '. In this way there arises a composite rock, partly ortho- and partly para-gneiss, consisting of

[1] Harker, *Geology of the Small Isles* (*Mem. Geol. Sur. Scot.*, 1908), pp. 105–7.

closely alternating thin bands of granite-gneiss and highly meta-morphosed sediment. Injection of so intimate a kind demands suitable conditions, including high pressure as well as high temperature. It may be found locally as part of an aureole of purely thermal meta-morphism bordering a granite batholite [1] but such effects are possible upon an extensive scale only when the country-rocks invaded had already been raised to a high temperature prior to the intrusion. It is then an incident of regional metamorphism ; and, as already remarked, the igneous intrusion, while closely related to the meta-morphism, is not to be regarded as its sole and sufficient cause.

When the spacing is wide enough to exhibit the two distinct rocks in their individuality, the dual origin of the complex is sufficiently evident. Often dark bands, rich in biotite, are seen alternating with pale bands, mainly quartzo-felspathic. In many of the rocks styled *injection-gneisses*, however, the blending of the two elements is more complete. The crystals of the different bands interlock, and the parallel orientation of crystals and coarseness of grain are common to both. Moreover, there has been a certain amount of chemical reaction between them at the junction. The chemistry of this has been fully discussed by Bowen [2] though not in this connexion. It is not to be conceived as a melting in, but is correctly described as assimilation, in the etymological sense of the word. The effect is to make over the sedimentary material in contact with the magma into new minerals which are in chemical equilibrium with the magma itself. In so far therefore as its bulk-composition permits, the part of the sediment so reconstituted consists of the same minerals as the igneous part of the complex, though in different relative proportions. The material in contact with the magma having been brought into chemical equilibrium with it, there can be no further reaction ; so that assimila-tion is normally restricted within very narrow limits. If, however, the injections are very close together, the intervening thin leaves of sedimentary material may be entirely assimilated, and the true nature of the resulting composite rock is then to be perceived only by careful scrutiny. The presence of felspars, unless in considerable amount, is not to be taken as evidence of igneous injection ; nor is microcline of diagnostic value in the same sense, as has sometimes been claimed. Perthitic, granophyric, and myrmecitic structures, however, are always

---

[1] Good examples are described by Bosworth, *Quart. Journ. Geol. Soc.*, vol. lxvi (1910), pp. 380–5 (Ross of Mull), and Sugi, *Jap. Journ. Geol.*, vol. viii (1930), pp. 29–112 (Tsukuba district).

[2] *Journ. Geol.*, vol. xxx (1922), pp. 513–70, and *The Evolution of the Igneous Rocks* (1928), pp. 175–220.

significant. On the other hand, a pelitic element in a rock mainly igneous is often to be detected by the presence of parallel seams or mere films of richly micaceous nature (Fig. 156). Even when this fails, the evidence is sometimes supplied by the presence of sillimanite, a mineral normally foreign to igneous rocks (Fig. 157).

Some of the best-known injection-gneisses are those of the Locarno district [1] and the Stavanger district of Norway.[2] Among British areas in which the phenomena can be studied with advantage are Eastern and Central Sutherland,[3] Middle Deeside,[4] and Donegal.[5]

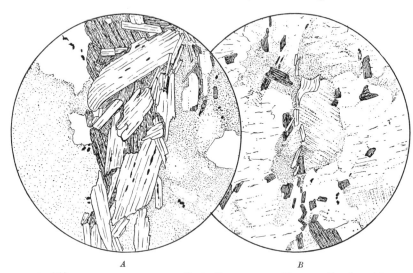

A   B

FIG. 156.—INJECTION-GNEISSES, Craig Ferrar, near Aboyne, Aberdeenshire ; × 23.

The igneous part is a granite composed of oligoclase, orthoclase, and quartz, with a little biotite. In this are intercalated streaks made up essentially of mica, light and dark, representing the pelitic element. One of these is shown in *A*. There are also a few garnets, not seen in the figure. In *B* we have a thinner pelitic intercalation, now reduced to a mere train of mica-flakes.

ULTRABASIC ROCKS IN REGIONAL METAMORPHISM

The matters dealt with in the last section, lying on the borderland of our main theme, have been discussed only briefly. It remains

[1] Klemm, *Sitz. k. preuss. Akad. Wis.*, 1904, pp. 46–65 ; 1905, pp. 442–53 ; 1906, pp. 420–31 ; 1907, pp. 245–58 ; Gutzwiller, *Ecl. Geol. Helv.*, vol. xii (1912), pp. 5–64.

[2] Goldschmidt, *Vidensk. Skr.*, 1920, No. 10 (1921).

[3] Horne and Greenly, *Quart. Journ. Geol. Soc.*, vol. lii (1896), pp. 633–48 ; Read, *Geology of Central Sutherland* (*Mem. Geol. Sur. Scot.*, 1921).

[4] Read, *Trans. Roy. Soc. Edin.*, vol. lv (1927), pp. 317–53.

[5] Cole, *Proc. Roy. Ir. Acad.* (B), vol. xxiv (1902), pp. 203–30.

to examine more particularly the leading petrographical characters of plutonic rocks of various kinds which have been intruded and consolidated under intense stress. At the high temperatures which rule at the time of crystallization, the principal element of stress is hydrostatic pressure. In certain circumstances, at which we have already glanced, shearing *movement* may set up special structures ; but any considerable measure of shearing *stress* becomes possible only at a somewhat later stage, when declining temperature has brought about a more effective resistance. The influence of pressure is seen in the

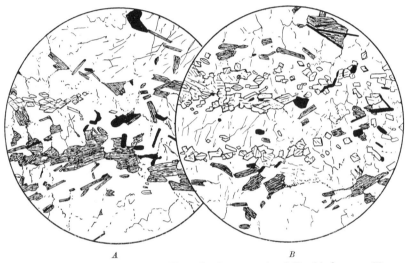

A   B

FIG. 157.—INJECTION-GNEISS, Västerhaninge, south of Stockholm ;   × 23.

The igneous portion is a garnetiferous granite-gneiss. Besides garnet (unevenly distributed and not included in the figure), the constituents are biotite, magnetite, plagioclase, microcline, and quartz. The sedimentary (argillaceous) intercalations have been almost completely absorbed, but are still indicated by seams rich in biotite and especially by strings of little well-shaped crystals of sillimanite.

widespread occurrence of minerals of minimum volume (garnets, rutile, pyroxenes, etc.) : stress-minerals have more often been formed later by reactions which are, in fact, of the retrograde kind. For reasons easily understood, the rocks vary much in structure. Not a few of them are of thoroughly massive habit, with little or nothing of any parallel arrangement. The current use of the term ' gneiss ', as applied to such rocks in such an environment, is a legacy from obsolete theoretical conceptions and dates from a time when, whether banded or not, they were believed to represent highly metamorphosed sediments.

The ultrabasic rocks related to regional metamorphism usually show little indication of any special conditions attending their intrusion.

Of simple constitution, they are composed entirely of minerals chemically stable at high temperatures and pressures and relatively strong, both in the crystalloblastic and in the mechanical sense. *Peridotites* are found, never in great force, associated with basic types in various regions, and typical examples have been described from Norway,[1] the Ticino,[2] and other countries. They are for the most part indistinguishable from peridotites intruded under more ordinary conditions, a certain tendency to a parallel disposition of the crystal elements being common in both (Fig. 158, *A*). The not infrequent occurrence

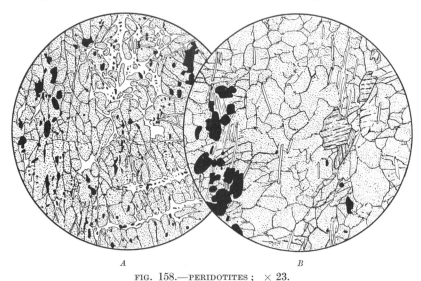

A   B

FIG. 158.—PERIDOTITES ; × 23.

A. Dunite, Loderio, Ticino ; with partial serpentinization.
B. Peridotite, Almeklovdalen, Söndmore, Norway ; with a few crystals of green diallage. Fibres of anthophyllite, with parallel orientation, have been formed from the olivine.

of large crystals of pyrope garnet is, however, significant. The common ' celyphite ' border, composed of fibrous amphibole and spinel, is to be ascribed to a magmatic, not a metamorphic, reaction. Again, in some Norwegian examples there has been a development of amphibole (cummingtonite, anthophyllite, tremolite, actinolite) at the expense of the olivine (Fig. 158, *B*). The little fibres and shreds of amphibole have a regular parallel arrangement. At the margin of a body of peridotite retrograde metamorphism has sometimes gone much farther,

[1] Brögger, *Neu. Jahrb.*, 1880, vol. ii, pp. 187–92 ; Eskola, *Vidensk. Skr.*, 1921, No. 8, pp. 19–23.
[2] Grubenmann, *Viert. Nat. Ges. Zür.*, vol. liii (1908), pp. 129–56 ; Hezner, *ibid.*, vol. liv (1909).

and there is a passage into actinolite-schist or further into talc- or chlorite-schist. Here there is evidence that shearing stress has been operative at more than one stage of declining temperature.

Peridotites are represented sparingly in the Lewisian complex of the North-West Highlands and Western Isles of Scotland. In the same region occur coarse-textured rocks composed essentially of pyroxene, or of hornblende, or of those minerals in varying proportions.[1] The *pyroxenites* commonly have both a rhombic and a monoclinic pyroxene, often with some hornblende in addition and as

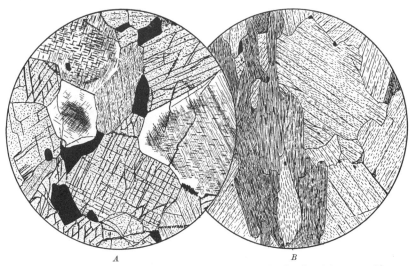

FIG. 159.—PYROXENIC AND HORNBLENDIC ROCKS in the Lewisian ; × 23.

A. Pyroxenite, Scourie, Sutherland : composed of pleochroic hypersthene, colourless augite with strong ' schiller ' structure, pale green hornblende, and magnetite.
B. Hornblende-Biotite-rock, An Accarsoid Thioram, South Rona.

accessories magnetite and pleonaste (Fig. 159, *A*). There are transitions to rocks in which hornblende occurs to the exclusion of pyroxene, and here biotite may also be present (Fig. 159, *B*). It is not easy to decide how far the hornblende is a primary constituent and how far derivative after pyroxene. There is sometimes, however, a close interbanding of more pyroxenic and more hornblendic types, and here the proportions of the two minerals must have been determined by chemical composition. All these rocks are normally of massive habit.

BASIC ROCKS IN REGIONAL METAMORPHISM

Among basic rocks the *eclogites*, though of restricted occurrence, have an importance all their own. They are plutonic rocks which,

[1] Teall, *Geol. Struct. of the N.W. Highlands* (*Mem. Geol. Sur.*, 1907), pp. 44-7.

intruded under especially deep-seated conditions, have acquired a peculiar mineralogical constitution in consequence of the great pressure under which they crystallized. It has been held that some eclogites may represent the extreme result of metamorphism of ordinary basic igneous rocks, amphibolite and garnet-amphibolite representing intermediate grades, but decisive evidence of such origin is lacking.

Typical eclogites occur in some abundance in the Lewisian of the Glenelg district,[1] on the borders of counties Inverness and Ross (Fig. 160, *A*), and allied types are recorded from other parts of the Highlands.[2]

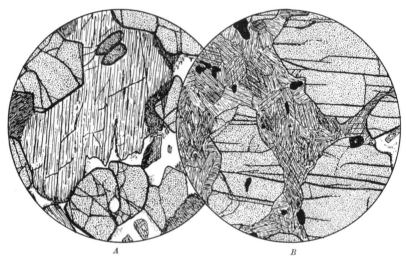

A   *B*

FIG. 160.—ECLOGITES in the Lewisian ; × 23.

A. Eclogite, N. of Glenelg, Inverness-shire : composed of garnet, light green pyroxene (omphacite) with some deeper green hornblende, rutile, plagioclase, and quartz.
B. Hornblende-Eclogite, S. of Loch Laxford, Sutherland : garnet and green hornblende, with a little magnetite, plagioclase, and quartz. The parallel cracks in the garnet, perpendicular to the direction of tension, belong to a relatively late epoch.

The eclogites of the Fichtelgebirge in Bavaria (Fig. 161, *A*) and other European districts are well known ; and we have important memoirs on the rocks of the Tirol,[3] France,[4] and Norway.[5] These rocks are

[1] Teall, *Min. Mag.*, vol. ix (1891), pp. 217–18 ; *Geol. of Glenelg* (*Mem. Geol. Sur. Scot.*, 1910), pp. 32–5 ; Alderman, *Quart. Journ. Geol. Soc.*, vol. xcii (1936), pp. 488–528.

[2] Harker, *Geol. Mag.*, 1891, p. 171 (Fig. 160, *B*) ; Peach, Geol. of C. Ross (*Mem. Geol. Sur. Scot.* (1913), p. 48 ; Read, *Geol. of C. Sutherland* (1931), pp. 81–83.

[3] Laura Hezner, *Tscherm. Min. Petr. Mitt.*, vol. xxii (1903), pp. 437–71, 505–80.

[4] Y. Brière, *Bull. Soc. Fra. Min.*, vol. xliii (1920), pp. 72–222.

[5] Eskola, *Vidensk. Skr.*, 1921, No. 8.

of special interest as illustrating how, in a particular case, the principle of minimum volume may become paramount. Its influence is seen not only in the essential minerals, garnet and omphacite, but also in the characteristic accessory constituents, rutile, zoisite, and cyanite.

The garnet of the eclogites is of a mixed variety—almandine, pyrope, and grossularite, sometimes with a notable content also of andradite. The omphacite is an aluminous augite, and it appears that the alumina enters largely in the compounds jadeite, $NaAl(SiO_3)_2$, and ' pseudojadeite ' or lime-jadeite, $CaAl_2(SiO_3)_4$. These two silicates answer respectively to albite and anorthite, the one with a deficiency and the other with an excess of silica. It seems therefore that under a sufficiently great pressure [1] diopside is capable of taking up ' potential ' felspar in the fashion which is otherwise peculiar to hornblende (p. 282). Since the felspar of gabbroitic rocks is more calcic than sodic, there is likely to be some excess of silica on the balance, and a little quartz is often present. The not infrequent occurrence of plagioclase indicates that the limit of solid solution, even under high pressure, has been reached. With relief of pressure the constrained solid solution is, theoretically at least, no longer stable. Usually the omphacite persists as a metastable form. In some instances described, however, it has broken down, yielding an intricate intergrowth of diopside and plagioclase. [2]

Another and more frequent change of the retrograde kind which may affect eclogites is amphibolization, and here the determining condition is to be sought, not merely in relief of pressure, but in decline of temperature, with the increased measure of shearing stress which this makes possible. Whether hornblende in rocks of this kind is ever a primary constituent, crystallized perhaps under the influence of volatile ' mineralizers ', is a debatable question. There are coarse-textured *hornblende-eclogites*, in which this mineral occurs to the exclusion of any pyroxene, and with no indication of a derivative origin (Fig. 160, *B*). True eclogites, however, may be seen in all stages of amphibolization. Amphibolization is perhaps a more appropriate description ; for it is often not a mere replacement of pyroxene by amphibole but a reaction between garnet and pyroxene,

---

[1] Compare Eskola, *Norsk Geol. Tidsskr.*, vol. vi (1920), pp. 173–5. Jadeite is presumably a stable form only under high pressure. Under laboratory conditions it has no place in the crystallization of Na_2O—Al_2O_3—SiO_2 mixtures ; Greig and Barth, *Amer. J. Sci.* (5), vol. 35 A (1938), pp. 93–112.

[2] Eskola, *Vidensk. Skr.*, 1921, No. 8, pp. 70–4 : see also Barviř, *Sitz. k. Böhm. Ges. Wis.*, 1893.

M.—11

yielding an aggregate of hornblende and plagioclase.[1] The process
starts from the ' celyphite ' borders which so commonly invest the
garnet crystals, and may spread until the pyroxene disappears. Sometimes the garnet too is exhausted, its place being indicated merely by
knots of green hornblende with same magnetite. The amount of
visible plagioclase varies, since some part of its substance may be
incorporated in the hornblende. An incidental change is the replacement of rutile by sphene, having at first the ' leucoxene ' character.
A *garnet-amphibolite*, originating in this way from the degradation of

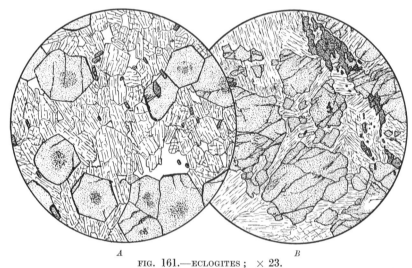

A   B
FIG. 161.—ECLOGITES ; × 23.

A. Eclogite, Fichtelgebirge, Bavaria : composed of garnet and omphacite with some
rutile and quartz.
B. Glaucophane-Eclogite, Val d'Aosta, Piedmont. The minerals shown are garnet,
glaucophane (with a little green hornblende), and deep brown rutile. Elsewhere in
the slice are more abundant green hornblende and rutile, epidote, and a little pale mica.
The figure is selected to show glaucophane wrapping about the fragments of shattered
garnet.

an eclogite, shows often a more or less evident schistosity, and may
be indistinguishable from a rock formed by the direct metamorphism
of a basalt or dolerite.[2]

If the omphacite was rich in the jadeite component, the derived
amphibole includes glaucophane as well as green hornblende. When
the soda-amphibole is abundant, it is accompanied by epidote. Among
Alpine occurrences a good example has been described by Bonney[3]

[1] See especially Hezner, *loc. cit.*
[2] On these retrograde changes see Alderman, *loc. cit.*, pp. 513–23.
[3] *Min. Mag.*, vol. vii (1886), pp. 1–8, with coloured plate.

from the Val d'Aosta (Fig. 161, *B*). The late origin of the amphibole is shown by its relation to cataclastic structures in the rock. Other examples come from Monte Viso [1] and from the Allalin district,[2] near Zermatt. The last-named locality furnishes a great diversity of rock-types derived from the degradation of eclogites and gabbros and containing various amphiboles, besides epidote, zoisite, talc, chlorite, etc. The emerald-green smaragdite, with 2–3 per cent of soda, is derived from diallage.

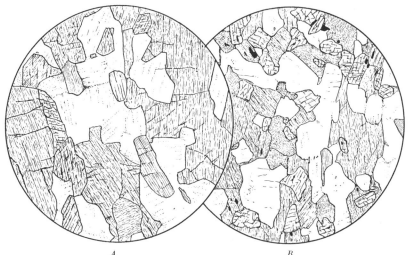

FIG. 162.—HORNBLENDE-PLAGIOCLASE-GNEISSES, in the Lewisian ; × 23.

A. Loch Maddy, North Uist : composed of light green hornblende and labradorite.
B. Tarbert, Harris : here, in addition to a brownish green hornblende, is a colourless augite, the two minerals being evidently independently crystallized.

The eclogite type constitutes a special facies, and the more usual representatives of the basic plutonic rocks are *hornblende-plagioclase-gneisses*, massive or with flow-banding. They occur in force in many parts of the Lewisian tract (Fig. 162, *A*). Probably most of them have been originally pyroxenic rocks [3] gabbros and norites—and, indeed, the hornblende sometimes contains a core of residual augite. Less commonly the two minerals are found associated in a manner which proves the hornblende to be of primary crystallization (Fig.

[1] Rutley, *Quart. Journ. Geol. Soc.*, vol. xlv (1889), pp. 60–2.
[2] Bonney, *Phil. Mag.* (5), vol. xxxiii (1892), pp. 237–50 ; Schäfer, *Tsch. Min. Pet. Mitt.*, vol. xv (1895), pp. 1–48.
[3] The rocks, mainly of augite and plagioclase, styled ' Erlanfels ', ' Augit-fels ', and ' Augitschiefer ', are lime-silicate-rocks of sedimentary origin,

162, *B*). Garnet is usually absent, and titanium goes into ilmenite rather than rutile. In the conversion of pyroxenes to hornblende there is a liberation of silica, and accordingly a little quartz may enter in rocks of thoroughly basic composition. Hornblende often preponderates greatly over plagioclase, and has probably incorporated in its complex constitution part of the felspar substance and perhaps also of the iron-ores.

The normal structure of the rocks is rather coarsely granoblastic and with little approach to parallel orientation of the crystals. These, like other gneisses, however, are liable to suffer ' *granulitization* '. This results from crushing at a stage when recrystallization was still possible, and the process is therefore in some sense intermediate between the protoclastic and the cataclastic. All the minerals are broken down ; but while the felspar (with quartz if present) makes a simple mosaic, the hornblende, with a superior force of crystallization, forms little imperfect prisms or fibrous patches (compare Fig. 164, *A* below).

These hornblendic gneisses have often passed, as temperature declined and shearing stress asserted itself, into amphibolites with pronounced schistosity. Garnet often enters at the same time, figuring now definitely as a stress-mineral, and sphene may be a noticeable constituent. Except by their coarser average grain-size, such *plagioclase-amphibolites* and *garnet-amphibolites*, related to the declining phase of regional metamorphism, do not differ from the corresponding types produced in advancing metamorphism. The very slow rate of cooling allowed time for the establishment of equilibrium ; so that the characters of a rock were determined by the actual conditions of temperature and stress, without any survival of residual minerals or of other than large-scale structures. Further change in the direction of retrograde metamorphism, viz. a passage into chloritic schists, etc., is found only as a local incident. At the lower temperatures mineralogical transformations would be controlled mainly by the continued or renewed activity of shearing stress, and the low-grade stress-minerals in general demand a supply of water which the rock cannot provide.

INTERMEDIATE AND ACID ROCKS IN REGIONAL METAMORPHISM

Much of what has been said concerning the basic igneous gneisses is applicable, *mutatis mutandis*, to those of intermediate and acid composition, and need not be repeated. Rocks of medium acidity, as compared with basic, are poorer in lime and magnesia but richer in silica and alkalies, including a noteworthy proportion of potash.

This last goes mostly into a brown mica, but a certain amount of potash-felspar is usually present in addition. Plagioclase is more abundant relatively to hornblende, and is now andesine or oligoclase. The characteristic rocks are accordingly *hornblende-biotite-plagioclase-gneisses*, often containing, besides quartz, some orthoclase or microcline. Sphene is less common, the titanium being taken up in the biotite. As in ordinary quartz-diorites, there is some range of variety, the more acid examples being richer in biotite and quartz. Rocks of this kind, as well as the more schistose (amphibolic) type, are represented in

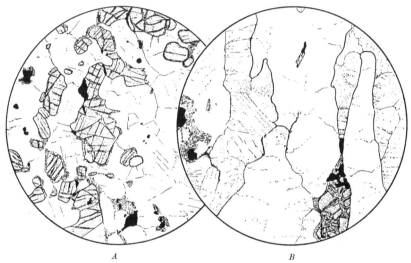

A   *B*

FIG. 163.—HYPERSTHENE-PLAGIOCLASE-GNEISS, Loch Glencoul, Sutherland ; × 23.

A. The chief constituents are hypersthene and andesine, the former having a narrow border of fibrous green hornblende. The other minerals shown are magnetite with a few flakes of biotite clinging about it, a little quartz, and a few rather large crystals of apatite (above).

B (another section from the same specimen) shows the coming in of elongated streaks of quartz with parallel arrangement. The hypersthene and its border of hornblende are being replaced by biotite.

the fundamental complex of the Lewisian. Gneissic banding and foliation are common, owing probably in many cases to a hybrid origin.

Another type, very prevalent in the Assynt and Scourie districts of Sutherland is a *pyroxene-plagioclase-gneiss* (Fig. 163, *A*). The pyroxenes may include hypersthene as well as a pale green augite, often with ' schiller ' inclusions. These and a felspar near andesine are the chief constituents, and garnet sometimes occurs in addition. The pyroxenes are usually bordered by a narrow ' corona ' of green fibrous hornblende. Quartz is present only in small amount in the normal type, but there has often been a later introduction of quartz, which then becomes a prominent constituent. It is conspicuous on a specimen owing to a bluish opalescence, caused by numerous minute inclusions. Instead of making veins, as it would do at a low temperature, it has taken the form of parallel elongated ovals and tongues with blunt or rounded extremities. This disposition, conforming also with any other parallel structure which the rock may possess, constitutes a special and well-marked type of foliation (Fig. 163, *B*). In the rocks so affected there has often been a partial replacement of the pyroxenes and their fringe of hornblende by biotite. The same type

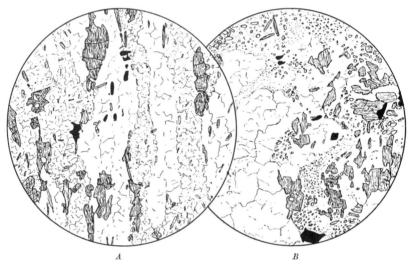

A   *B*

FIG. 164.—GRANULITIC HORNBLENDE-PLAGIOCLASE-GNEISSES WITH QUARTZ, near Lochinver, Sutherland ; × 23.

A. The granular aggregate of felspar consists of oligoclase (with a fine dust of secondary zoisite) and orthoclase. The hornblende, recrystallized at the same time and probably from pyroxene, shows a parallel orientation, and with it are scattered flakes of biotite. Quartz is abundant in parallel streaks, and much of it has probably been introduced at the time of the granulitization.

B. Here there is little of any parallel structure. The production of zoisite and epidote in the plagioclase is more advanced.

of foliation, due to an injection of quartz under stress, is seen in other rocks, now hornblendic, which have probably been derived from pyroxenic gneisses, and here it is evidently connected with a granulitization of the general mass of the rock (Fig. 164).

In rocks of definitely acid composition the dominant felspar is typically microcline, but usually accompanied by a sodic plagioclase. Quartz is always an important constituent. In the less acid of these *granite-gneisses* hornblende may still be the chief coloured silicate (Fig. 165, *A*) ; but it gives place increasingly to *biotite*, which is the characteristic mineral in all the more acid types. Almandine garnet

is often found in addition.   Pyroxene-gneisses of acid composition are
not represented in the British area.   The well-known charnockite,[1]
widely developed in Southern India, Ceylon, and Burma, is a hyper-
sthene-bearing type.

If we have regard only to rocks of massive habit, it is very noticeable
that *muscovite-bearing gneisses* are scarcely represented in such a complex
as the Lewisian.   Probably the chief reason of this is the squeezing
out of the residual fluid magma to make pegmatites rich in muscovite
and microcline (p. 299).   In the south-eastern division of the High-

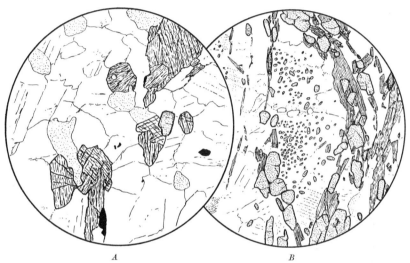

A   B

FIG. 165.—HORNBLENDE- AND BIOTITE-GNEISSES, in the Lewisian ;   × 23.

    A. Hornblende-Gneiss, near Laxford, Sutherland : a granitoid rock with only a
slight approach to parallelism of the hornblende crystals.   The fresh felspar, making
the bulk of the rock, is microcline ; the quartz is dotted with numerous fluid-inclusions
apatite and magnetite are the only other minerals.
    B. Biotite-Epidote-Augen-gneiss, between Gairloch and Poolewe, Ross.   Biotite
and epidote, in close association, make trains which wind about a large eye of oligoclase,
enclosing smaller granules of epidote.   Microcline is here subordinate to oligoclase : the
other minerals are quartz and a little sphene.

lands, on the other hand, granite-gneisses with both light and dark
mica have a wide distribution, though seldom in continuous bodies
of large size.   Probably the magmas which furnished them are themselves to be regarded as residual magmas, derived by a straining-off
process on a large scale.   If so, the process has often been repeated
after the emplacement of the magma and partial crystallization, and
the rocks may then show protoclastic foliation and eyed structures
(p. 300).   The extruded pegmatites are usually non-foliated.   They
remained liquid until after the culmination of stress in the region,

    [1] Holland, *Mem. Geol. Sur. Ind.*, vol. xxviii (1900), pp. 134–41.

and any renewal of stress-conditions after their crystallization has left only cataclastic and strain effects. Many granite-gneisses have little or nothing of any banded structure. Acid magmas, as intruded, are much more homogeneous than basic ; and, when primary gneissic banding is a conspicuous feature, we may suspect hybridization by digested basic inclusions. Orthogneisses devoid of any parallel structure can often be distinguished from ordinary plutonic rocks only by their crystalloblastic structures. In such a rock as the well-known ' Aberdeen granite ', for instance, muscovite is constantly idioblastic

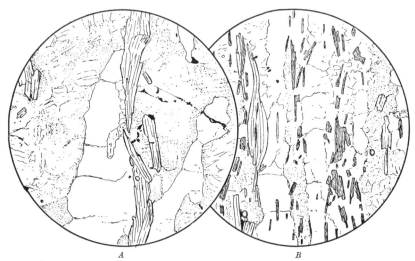

A   B

FIG. 166.—MUSCOVITE-BEARING GNEISSES, Eastern Highlands ; × 23.

A. Grantown, Elgin. The parallel arrangement of flakes of muscovite and lenticles of quartz determines a pronounced foliation. The felspar is microcline : elsewhere in the slice are biotite, garnet, and oligoclase.

B. Keith, Banffshire. Here the foliation is closer and more regular, and the felspar is largely granulitized. The minerals are muscovite, biotite, oligoclase, microcline, and quartz, with a little apatite and magnetite.

against biotite, whereas a normal granite shows biotite idiomorphic against muscovite.

While *foliated and schistose structures* in acid gneisses may date from the epoch of crystallization, they have doubtless been acquired more often at a later stage, rocks originally granitoid yielding under intense differential stress while still at a high, though declining, temperature. The attendant mineralogical changes, being essentially of the nature of degradation, belong in strictness to retrograde metamorphism. The breaking down of hornblende, in these rocks rich in potash, yields biotite together with epidote, which may also come from a plagioclase felspar. Biotite-epidote-gneisses, in which the two minerals are intimately asociated, are found in the Lewisian complex, and have probably originated from hornblende-bearing rocks (Fig. 165, *B*). In the more acid types muscovite, derived from potash-felspar, now figures prominently, and the parallel arrangement of the flakes of mica, light and dark, is mainly responsible for the schistose structure [1] (Fig. 166). The foliation is determined largely by the easily mobile quartz which, as recrystallized (not here introduced from without), tends to segregate into lenticles and streaks. Garnet is a very common constituent (Fig. 183, below). Doubtless it is often of primary origin, but it may be produced also as a stress-mineral in the

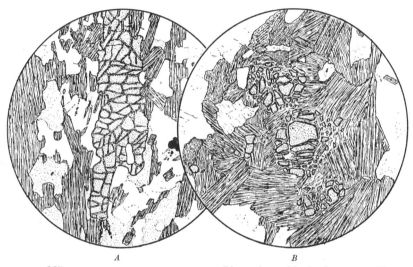

A   *B*

FIG. 167.—GARNET-BIOTITE-GNEISS, near Llanerchymedd, Anglesey; × 23.

A. The garnet is seen shattered and drawn out. Biotite is very abundant; the other minerals are oligoclase, orthoclase, quartz, and some magnetite.
B. This shows how the biotite has been largely derived from the destruction of garnet, reacting with the potash-felspar.

solid rock, still at a high temperature. If there is a revival of stress-conditions at a later time, when with declining temperature the mineral has lost its power of rejuvenation, the crystals may be shattered and their fragments separated. Often they have broken down chemically as well as mechanically, reacting with potash-felspar to produce additional biotite (Fig. 167):

$$R_3Al_2(SiO_4)_3 + 2KAlSi_3O_8 + 2H_2O$$
$$= H_2KR_3Al(SiO_4)_3 . H_2KAl_3(SiO_4)_3 + 3SiO_2.$$

[1] In Grubenmann's classification muscovite is wanting in acid gneisses of the lowest zone of depth, which are '*massig bis schieferig*', but occurs in the middle zone, where the structure is '*kristallisationsschieferig*'. The connexion of muscovite with schistosity, and so with shearing stress, rests on more assured ground than estimates of depth, which must often be conjectural.

It is to be borne in mind that, although acid rocks were intruded at considerably lower temperatures than basic, they have passed through a wide range of cooling. In the earlier stages there was ample freedom of recrystallization and rearrangement under the influence of differential stress, but this freedom was progressively abridged in accordance with the decline of temperature and the specific properties of the minerals concerned. Of the new minerals some, such as epidote and muscovite, may form at very different stages, even in one and the same rock. Only when they take their place as part of the essential fabric of the rock, are they marked as belonging to an early stage of the gradual cooling (compare the two different occurrences of epidote in Fig. 165, *B*, and of muscovite in Fig. 183, below.

LOCALIZED ANALOGUES OF REGIONAL METAMORPHISM

We have laid it down as a characteristic of metamorphism of the most general kind that it has, as normally developed, a wide areal extension with gradually rising intensity. For this reason the term ' regional ', as applied in this connexion, has a manifest propriety. None the less it is possible for the essential condition—viz. a conjunction of high temperature and intense stress—to be realized upon a strictly local scale. The phenomena which we are accustomed to associate with regional metamorphism then present themselves as a local incident interpolated in an area which may be otherwise quite unaffected. For the study of metamorphism such cases possess a special interest, since it is often possible, within the limits of a single exposure in the field, to compare the metamorphosed rock with the original type and to follow the transformation step by step. This localized rendering of the effects of regional metamorphism may come about in more than one way.

In the first place, it is to be remembered that, while there exists a certain causal connexion between the thermal and dynamic factors in regional metamorphism, it is of an indirect kind, and does not imply any precise coincidence in time. Even after the final cooling down there may be intermittent recrudescence of the orogenic forces. Rocks at low temperature may be presumed to have sufficient rigidity to offer an effective resistance or to yield only in cataclastic fashion ; but, if igneous intrusion should take place within the area during this time, a different case is presented. The intruded rocks will be liable to experience either fluxional and protoclastic effects while in process of consolidation or metamorphism (of regional type) when consolidated but still at a high temperature.

Admirable illustrations are afforded by the intrusive complex of the Lizard district, in Cornwall, as described by Flett.[1] It is situated within the area of regional metamorphism, of unknown but possibly large extent, which includes the ' Lizard Hornblende-schists ', etc., and stands to this in the relation of a sequel or appendix. The earliest intrusion of the series is represented by a boss of peridotite, since serpentinized. After an interval came gabbro, partly as a boss, partly in the form of dykes, and after another interval a numerous group of dykes of olivine-dolerite. Between this latter and the final group, of acid composition, there was a partial overlapping in time, so that hybrid intermediate varieties occur, and only the latest intrusions are of pure granite. All these various rocks are found massive in one place, schistose and foliated in another, and with a distribution which at first sight is curiously capricious. Of two members of the same group at the same place one may be massive and the other schistose ; and, when one is intersected by the other, it may be either the older or the younger that exhibits schistosity. Moreover, if both are schistose, the structure has not a common direction in the two, but is parallel to the bounding walls of each separate intrusion.

The facts here very briefly summarized make it apparent that there was not one epoch of metamorphism for the whole Lizard complex but numerous epochs, pointing to repeated temporary renewals of crustal stress during the prolonged regional decline of temperature. If the intrusion of a particular rock coincided nearly with one of these times of stress, this is marked by various fluxional and protoclastic effects, and we have such examples produced as the ' augen-gabbros ', ' flaser-gabbros ', and even ' gabbro-schists ' seen at Carrick Luz and other localities. If the intrusion came somewhat earlier, the incidence of stress found a rock already solid, but at a high temperature and therefore of much weakened rigidity. In greater or less degree according to its actual temperature the rock yielded and suffered recrystallization and the mineralogical changes proper to regional metamorphism. If, on the other hand, intrusion took place during an interval of relaxation, and there was time both for consolidation and for cooling before any new disturbance supervened, the rock was either proof against change, or incurred only accidents of the cataclastic kind (Fig. 168, *B*). The basic rocks were intruded largely in the form of dykes of no great width, and these were specially vulnerable. It is often possible to observe the passage of a dolerite into an amphibolite with strongly marked schistosity parallel to the walls of the dyke.

[1] *The Geology of the Lizard and Meneage* (*Mem. Geol. Sur.*, 1912).

Another way in which the effects proper to regional metamorphism may be localized is in connexion with faulting. Some of the most interesting illustrations of this relation are found where the rocks concerned belong to highly alkaline types. This is not the place to discuss the way in which a succession of various rock-types can be derived by the process of magmatic differentiation. It is enough to remark that in the normal course of evolution alkaline magmas are produced only, if at all, at the very latest stage. Such a residual fluid magma possesses an exceptional mobility in consequence of the

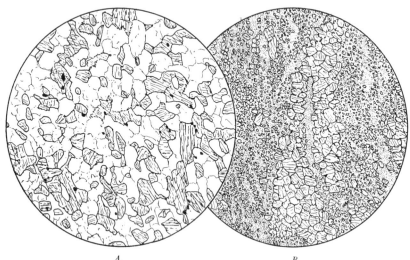

A   *B*

FIG. 168.—PROTOCLASTIC AND CATACLASTIC STRUCTURES IN GABBRO, Lizard district, Cornwall ; × 23.

A. Granulitic Gabbro, Downas Cove. The structure is the result of enforced differential movement during the time of magmatic crystallization.

B. Granulitized Gabbro, Spernic Cove : showing crush-effects produced in a coarsely crystalline rock at an advanced stage of its cooling.   The large augite crystals have been broken down, while the felspar has yielded by suffering saussuritization.

These examples represent the two extremes of conditions. Under a like stress developed at some intermediate stage of cooling the Lizard gabbros and dolerites have often been transformed to amphibolites.

concentration in it of water and other fluxes. A study of the actual distribution of different rocks shows that, in the case of igneous action related to orogenic forces, any residual alkaline magma that may be produced is invariably driven out from the disturbed area into some adjacent tract free from acute stress.[1] It is for this reason that the ' alkaline rocks ' are not represented in the class of orthogneisses,

[1] The present writer has elsewhere illustrated this principle from the history of igneous action in the British area ; *Quart. Journ. Geol. Soc.*, vol. lxxiii (1918), pp. lxvii–xcvi.

ranging from ultrabasic to acid, treated in preceding sections of the present chapter. Even a highly alkaline rock, such as a nepheline-syenite, belonging to some older series, may happen to be included later in a tract of regional metamorphism (p. 295) ; but among igneous rocks which stand in a real relation to the metamorphism alkaline types find no place. The poverty-stricken appearance of Gruben-mann's groups VI and VII is easily explained.

We have to remark, however, that, although intrusion of alkaline magmas is not found in connexion with orogenic movements, it may

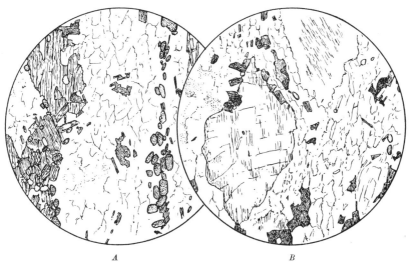

A   B

FIG. 169.—FOLIATED AND PROTOCLASTIC STRUCTURES IN NEPHELINE-SYENITES,
Langesundsfjord, Norway ;   × 23.

  A. 'Ditroitschiefer' of Brögger, Kjörtingholmen. A pronounced foliation is indicated by the distribution of the coloured silicates, viz. aegirine, lepidomelane, and sphene, which show also a parallel orientation of crystals.
  B. 'Augenditroit', Stor Arö. A large 'eye' of nepheline is conspicuous, and above is one of microperthite. The dark silicate here is lepidomelane.
  See Brögger, *loc. cit.*, pp. 110–113.

be an accompaniment of crustal displacement of another kind.   The classical example is the sunken tract of the Oslo Fjord, with its suite of igneous rocks characterized by a general richness in soda. Brögger [1] has shown that, along the boundary-fault which traverses the Langesundsfjord, intrusion has in places been effected in the line of fault and contemporaneously with continued differential movement. Consolidation taking place under these conditions has given rise to a variety of foliated and protoclastic structures (Fig. 169) ; and

[1] *Zeits. f. Kryst.*, vol. xvi (1890), pp. 104–20.

shearing stress continued or renewed after consolidation was completed has sometimes brought about a strong schistosity.

Still another way in which metamorphism of the regional type may be produced as a local incident is by heat generated in the crushing of rocks.   This type of metamorphism, which may be found affecting sedimentary as well as igneous rocks, will be discussed in the next chapter.

CHAPTER XIX

REPEATED METAMORPHISM

Cases of Multiple Metamorphism—Regional followed by Simple Thermal Meta-morphism—Mechanical Generation of Heat as a Factor in Metamorphism—Simple Thermal followed by Regional Metamorphism.

CASES OF MULTIPLE METAMORPHISM

IN discussing the effects of metamorphism, of one kind or another, in rocks of different classes, our general plan has been to take a particular rock-type and endeavour to follow its behaviour when subjected to new conditions of temperature or of stress or of both together. For the sake of simplicity the rock taken as starting-point was conceived as, in a general sense, initially non-metamorphosed. More strictly regarded, it had in every case a certain *past history*, which had left some mark upon it. If a sandstone, it had undergone cementation ; if a limestone, it was often partly recrystallized and perhaps dolomitized. Among argillaceous sediments it would be difficult to find an example which had not already suffered a consider-able amount of mineralogical reconstruction ; and, indeed, the cleaved slates which we have often taken as starting-point in discussing thermal metamorphism already bore the unmistakable imprint of a previous dynamic metamorphism. This subject of successive operations of metamorphism is now to be considered more closely.

Some tracts of the earth's crust have had a varied geological history which included several active episodes, and it would be rash to assume that only the latest of these has left its impress upon the mineralogical constitution and structural characters of the rocks as we now see them. The Alpine chain affords a conspicuous example. Staub [1] had distinguished in the district of Graubünden no less than fifteen distinct and superposed metamorphisms. To inquire into complexity of this order does not lie within our province, and our examples have been drawn, when possible, from some area of less intricate relations. In what follows the subject will be treated only on simple lines, and in such a way as to bring out general principles. The case in which a

[1] *Viert. Nat. Ges. Zürich*, vol. lxv (1920), pp. 323-76.

second metamorphism is of the same general kind as the first introduces no new considerations, and it is therefore unnecessary to discuss it. A special interest enters, however, when rocks, already more or less highly metamorphosed on definite lines, have been subjected to *a new metamorphism of a different kind*. This must often happen. When a plutonic intrusion of some magnitude breaks into the midst of a tract of earlier regional metamorphism, the crystalline schists bordering the intrusion will suffer a new metamorphism of the simply thermal kind. On the other hand, the aureole about a former plutonic intrusion may at some later time become involved (together with the plutonic rocks themselves) in a new metamorphism of the regional kind. So simple thermal metamorphism may be superposed upon regional, or the reverse. We have then to inquire to what extent, and in what ways, the newer metamorphism will modify or obliterate or supersede the effects of the earlier. Changes, both of mineralogical constitution and of micro-structure, are evidently to be expected, and indeed the general nature of these changes may be largely anticipated from what we have already learnt.

REGIONAL FOLLOWED BY SIMPLE THERMAL METAMORPHISM

We proceed to consider, in the first place, the case of *simple thermal superposed upon regional metamorphism*. Abundant illustration of this is to be seen in the Highlands, where the Dalradian and Moine crystalline schists have been invaded by the Caledonian intrusions known as the ' Newer Granites '. The effects to be noted are, in brief, the replacement (or at least a strong tendency to replacement) of stress- by anti-stress-minerals and of the schistose by the hornfels class of structures. This tendency, naturally, is effective in proportion to the grade of thermal metamorphism reached, some characteristic transformations taking effect only in the inner ring of an aureole or in xenoliths enclosed in the igneous rock. The specific mineralogical changes brought about, in rocks of any given composition, must be determined at the first by their actual mineralogical constitution, which in turn depends upon the grade of regional metamorphism to which they belonged.

As regards *argillaceous schists of a low grade*, there is little to be added to what has already been said in preceding pages (Chap. IV) ; for, as already remarked, we there drew freely for illustrative examples upon rocks of the outermost (chlorite) zone of the Highland region. We have seen how the sericite and chlorite are replaced by biotite, cordierite, and andalusite ; then, with advancing metamorphism, either orthoclase or hypersthene normally makes its appearance ;

finally, it may be, sillimanite or perhaps corundum and spinellids. Concurrently the original micro-structure is replaced by some variety of the hornfels type. There is, however, one characteristic mineral of low-grade schists (though not common in the Highlands) which demands more particular notice, viz. chloritoid. As a distinctive stress-mineral, it is naturally destroyed in thermal metamorphism, though it may persist, as in the aureoles of the Skiddaw and Bodmin Moor granites,[1] until after the first appearance of cordierite and chiastolite. It then suffers change ; sometimes into andalusite and magnetite, sometimes, in more chloritic sediments, by contributing to the production of cordierite.

The ordinary *mica-schists*, with light and dark micas, and the garnetiferous mica-schists present more points of interest. Muscovite and biotite are minerals both of purely thermal and of regional metamorphism, the biotite arising in both cases from reaction of muscovite with chlorite and other minerals ; but we have seen that this reaction is deferred by the influence of shearing stress (p. 212). Still more does stress check the reactions by which potash-felspar is formed at the expense of mica. It is natural, therefore, to find that the conversion of a mica-schist to a hornfels is attended by diminution or complete disappearance of muscovite and chlorite and a marked increase in the proportion of biotite ; often also by the formation of orthoclase. The cordierite which is an abundant constituent of such hornfelses is largely a by-product of these reactions. In the most highly metamorphosed rocks of this kind biotite itself often dwindles, and orthoclase becomes a prominent mineral. If silica is deficient, the biotite may also give rise to abundant granular pleonaste. In these metamorphosed mica-schists the setting up of a hornfels structure does not necessarily obliterate the parallelism of elements proper to a schist. If the original rock contained oriented biotite-flakes of some size, these, or finally minerals which replace them, may still preserve something of the characteristic arrangement. Here as elsewhere the strictly limited range of diffusion in thermal metamorphism is the controlling factor. While the smaller structural features are effaced, the larger, such as true foliation, survive, even when translated into terms of new minerals.

A red garnet is a normal constituent of most middle- and high-grade pelitic crystalline schists, but also occurs rather exceptionally as a product of simple thermal metamorphism (p. 54). We have seen reasons for connecting these exceptional occurrences sometimes with a notable content of manganese (or of lime and manganese),

[1] Phillips, *Geol. Mag.*, vol. lxv (1928), p. 553.

sometimes with a high static pressure due to a deep cover. It is to be expected, therefore, that garnet will usually be destroyed in thermal metamorphism ; but we must be prepared to meet with exceptions which, in default of fuller knowledge, may appear as anomalies. The red garnets of the Scottish Highlands, doubtless in general poor in manganese and lime, are in fact found to be destroyed, or partly destroyed, within the aureoles of the Caledonian intrusions ; but the destruction is not always completed at once, and sometimes seems capricious even in a given rock. Cordierite and magnetite are

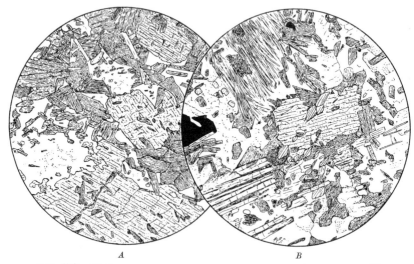

FIG. 170.—THERMALLY METAMORPHOSED CRYSTALLINE SCHISTS ; × 23.

A. A garnet-albite-mica-schist in the aureole of the Garabal diorite complex at Allt Arnan, near Loch Lomond : now an Andalusite-Biotite-Hornfels with cordierite, plagioclase, muscovite and some quartz. The garnet is represented by aggregates of biotite.
B. Sillimanite Gneiss from a patch enclosed in the Ross of Mull granite : showing relics of old acicular sillimanite above and new crystals below ; in the centre new sillimanite is regularly intergrown in andalusite. The other constituents are cordierite, biotite, quartz, oligoclase, and orthoclase.

often the replacing minerals ; in other instances biotite is the chief product, while hypersthene, pleonaste, etc., are less common [1] (Fig. 170, *A*). Sometimes adjacent garnets have suffered replacement in different ways (Fig. 172, *A*). This has been determined by reactions with different contiguous minerals, and may also have been affected by chloritization of the garnet prior to metamorphism. At first there are always recognizable pseudomorphs, but enlarged scope of diffusion at the highest temperatures may cause some dispersal of the new

[1] Some of the best examples are found in the Glencoe district : see *The Geology of Ben Nevis* (*Mem. Geol. Sur. Scot.*, 1916), pp. 195–201.

products, and so obscure the evidence of the former presence of garnet. The literature of the subject, however, furnishes instances in which garnet has been not destroyed but recrystallized in thermal metamorphism. Müller [1] has described the metamorphism of a garnetiferous mica-schist by the granite of Schneekoppe in the Riesengebirge. It is transformed to a rock composed largely of andalusite and biotite with new-formed garnet in well-shaped crystals. The garnet, as analysed, is a fairly normal almandine, and we are probably to see here the influence of high pressure and the 'Volume Law'.

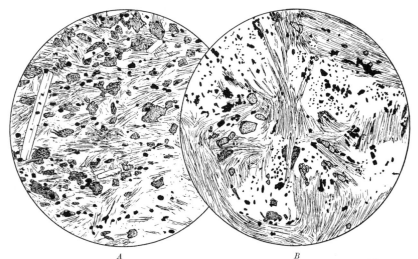

A   *B*

FIG. 171.—THERMALLY METAMORPHOSED CRYSTALLINE SCHISTS ; × 23.

A. Sillimanite-Cordierite-Biotite-Hornfels from Leven Schists near contact with the Ballachulish granodiorite, Argyllshire. Here sillimanite has been abundantly produced in the form of fine needles. Andalusite is present also, in distinct crystals (a large one on the left).
B. Corundum-Pleonaste-Sillimanite-Cordierite-Hornfels, Glen Muick, Aberdeenshire. This is the rock described and analysed in *The Geology of Braemar* (1912), pp. 13–16. It contains SiO_2 42·08 per cent. and Al_2O_3 32·40.

The highest grade of thermal metamorphism in pelitic schists is marked usually by the appearance of new sillimanite, sometimes in slender needles, but often in larger and more distinct crystals. It is remarkable, however, that andalusite is frequently present in addition, the two forms of aluminium silicate occurring in close association and even in regular intergrowth. The other constituent minerals are cordierite, biotite, magnetite, often some acid plagioclase, and in the more siliceous rocks quartz. Some garnet may still remain. In rocks poor in silica other characteristic minerals appear, viz. corundum and pleonaste (Fig. 171, *B*), the status of which as members of an

[1] *Zeits. Deuts. Geol. Ges.*, vol. xliii (1891), pp. 730–3.

anti-stress association has been sufficiently discussed (p. 236). These extreme results of thermal metamorphism are naturally to be seen in rocks close to the contact of a plutonic intrusion and in enclosed xenoliths. When xenoliths have been enclosed in a volcanic magma, another feature is often to be observed. In addition to new minerals and relics of the old ones there may be more or less of a brown glass. Here a high temperature, sufficient to initiate local fusion, has been followed by a cooling too rapid to allow complete recrystallization (p. 27). Good examples are afforded by the ejected blocks at the Laacher See, in the Lower Eifel, as studied by Brauns.[1] In such a case there is often, as a further process, an interchange of material between the xenolith and the enveloping magma, but this is a complication which lies outside our province.

Concerning the effects of thermal metamorphism on the *higher-grade crystalline schists* we possess fewer data. It can scarcely be doubted that the very characteristic stress-minerals staurolite and cyanite must readily give place to more stable products ; but the actual conversion of cyanite to andalusite or sillimanite and the replacement of staurolite by cordierite, magnetite, etc., have not often been recorded from actual observation. Sillimanite stands on a different footing, for we have seen that it characterizes the highest grade both of simply thermal and of regional metamorphism. We find that in a contact-aureole it sometimes remains unchanged, sometimes recrystallizes as new sillimanite or possibly as andalusite. Sillimanite and andalusite of evidently new formation may occur abundantly, the two minerals often in parallel intergrowth, but it is not to be supposed that they come wholly or even mainly from pre-existing sillimanite. An interesting occurrence is that at the Ross of Mull, where sillimanite-gneisses are found contiguous with and enclosed in a large granitic intrusion. Bosworth [2] and Bailey [3] have both arrived at the conclusion that the sillimanite here is wholly due to metamorphism induced by the intrusion. To the present writer it appears that both original and new-formed sillimanite are present, and are clearly distinguishable (Figs. 170, *B* ; 172, *B*). In sillimanite-gneisses poor in silica a high grade of thermal metamorphism brings about a more radical reconstruction ; for sillimanite and garnet alike disappear, being replaced by cordierite, corundum,

[1] *Die Kristallinen Schiefer des Laacher Seegebietes* (1911) and numerous separate papers.

[2] *Quart. Journ. Geol. Soc.*, vol. lxvi (1910) pp. 376–96.

[3] *The Geology of Staffa, Iona, and Western Mull* (*Mem. Geol. Sur. Scot.* 1925), pp. 33–6.

and pleonaste (see equations on p. 236).   Good examples are found in contact with the diorite complex of Clova in Forfarshire.

Thermal metamorphism in crystalline schists of other than argillaceous composition may be dismissed more briefly.   The case of rocks composed largely of iron-silicates has been incidentally noticed above (p. 239).   The psammitic types show, in general, the same mineralogical transformations as the pelitic.   In a quartzite of the commonest variety recrystallization may give rise to very little perceptible change, except that a parallel orientation of mica-flakes is

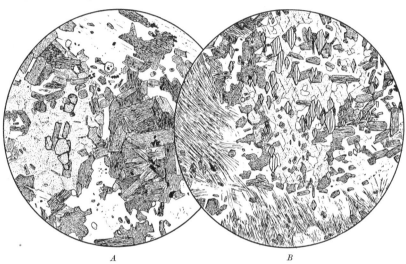

A   B

FIG. 172.—THERMALLY METAMORPHOSED CRYSTALLINE SCHISTS (GNEISSES), ROSS of Mull ;   × 23.

A. Garnet-Gneiss.   A garnet on the left is mostly changed to cordierite (dull from incipient alteration), but shows unchanged relics ; another on the right is completely replaced by biotite.   Other minerals present are plagioclase, quartz, magnetite, apatite, and (elsewhere in the same slice) sillimanite.

B. Sillimanite-Gneiss.   Original sillimanite is in bundles of closely packed needles, and new sillimanite in little crystals intergrown in andalusite with parallel orientation.

sometimes lost in the process.   So too the granulites, which are so widely distributed in the psammitic parts of the Moine Series, often show no noteworthy mineralogical change even in xenoliths.

It is otherwise with rocks which are, or have been, *partly calcareous*. According to their initial composition, regionally metamorphosed rocks of this class present, as we have seen, a wide range of diversity. In addition to the characteristic lime- and magnesia-bearing silicates, there may be more or less abundant residual calcite and often calcite and free silica together.   When such rocks become involved in a metamorphic aureole, there are two points to be observed.   Firstly, there is a strong tendency—more or less effective according to the grade of thermal metamorphism reached—for the replacement of stress- by anti-stress-minerals. Ultimately the epidotes, micas, and amphiboles give place to plagioclase, microcline, and pyroxene. Secondly, the direct reaction between calcite and quartz, formerly inhibited by great pressure, now proceeds freely, giving rise to such minerals as wollastonite, idocrase, and garnet. Both principles are well illustrated by the Deeside Limestone and its associated lime-silicate-schists where they enter the aureoles of the ' Newer Granites ' [1] (Fig. 173). The completely metamorphosed representatives reproduce,

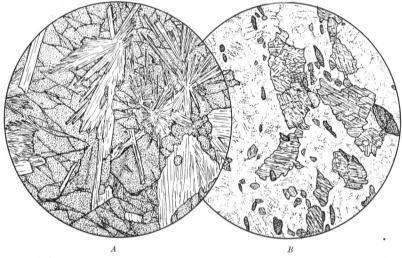

A   B

FIG. 173.—TWICE-METAMORPHOSED CALCAREOUS SEDIMENTS, from a granite
   aureole, Pollagach Burn, near Cambus o' May, Aberdeenshire ; × 23.

   A. This has been a quartzose crystalline limestone with zoisite and minor accessories,
and is now converted to a wollastonite-grossularite-rock.
   B. Formerly a hornblende-plagioclase-rock containing some biotite and quartz,
now a diopside-andesine-rock with orthoclase and sphene.
   The rocks shown in Fig. 35, above, also belong here.

as regards their *mineralogical* constitution, the various types of calcic hornfelses formerly distinguished, with plagioclase, diopside, grossularite, idocrase, wollastonite, etc., precisely as if the original sediments had been directly metamorphosed by the granite without any intervening episode of regional metamorphism. In their *textural* characters, however, and especially in respect of grain-size, these rocks are in strong contrast with the close-textured ' calc-flintas ', etc., which we have seen before. They are of comparatively coarse grain, and have provided mineralogists with fine specimens of wollastonite

---

[1] *Geology of Braemar*, etc. (*Mem. Geol. Sur. Scot.*, 1912), pp. 103–9 ; Hutchison, *Trans. Roy. Soc. Edin.*, vol. lvii (1933), pp. 567–76.

grossularite, idocrase, diopside, sphene, microcline, etc. Here the effects of the previous regional metamorphism are very clearly indicated, foliation and segregation having been especially effective, in these semi-calcareous rocks (p. 207).

The thermal metamorphism of such types as the calc-chlorite-schists and calc-mica-schists again illustrates the same principles, the rather complex initial constitution giving rise often to a varied assemblage of new minerals. Good examples may be studied in the Pass of Brander, where the Ardrishaig Phyllites enter the aureole of the Beinn Cruachan granite. Epidote, hornblende, diopside, sphene, etc., have been produced in abundance, while calcite or quartz (not both) may occur in addition. The original banded structure of the rocks is preserved, so that micaceous seams alternate with others rich in the lime-bearing silicates. With more intense metamorphism the micas themselves would be destroyed, yielding potash-felspar, diopside, magnetite, and spinel.

There remain to be mentioned the *rocks of igneous origin*. In several districts of the Scottish Highlands members of the ' epidiorite ' group of sills (p. 279) may be seen metamorphosed by the later plutonic intrusions. Tilley [1] has described examples from the aureole of the Càrn Chois diorite in Perthshire. Here the sills were in a comparatively low grade of regional metamorphism, being composed essentially of hornblende, a felspar near albite, zoisite, epidote, and chlorite. The early changes include a development of magnetite in the hornblende, partial recrystallization of the hornblende itself, and local formation of biotite, especially round grains of magnetite. Chlorite, epidote and zoisite disappear, and the sodic felspar is replaced by a more calcic vairety. With advancing metamorphism hornblende and biotite give place to augite, which figures at first as a crowd of little granules fringing the hornblende (Fig. 174, *B*). Hypersthene is sometimes formed in addition to augite, and the final product is a pyroxene-plagioclase-hornfels, in which, however, some hornblende often remains. In typical amphibolites, belonging to a more advanced grade, chlorite, epidote, and zoisite have already been taken up into hornblende and plagioclase. Recrystallization of these in thermal metamorphism yields then a hornblende-plagioclase-hornfels with some biotite and minor accessories. There is a strong tendency of the grains of hornblende to gather into complex aggregates, simulating a porphyroblastic structure (Fig. 174, *A*). The replacement, or partial replacement, of hornblende by pyroxene marks a further stage, not always attained.

[1] *Quart. Journ. Geol. Soc.*, vol. lxxx (1924), pp. 65–6.

MECHANICAL GENERATION OF HEAT AS A FACTOR IN METAMORPHISM

A subject which we have not hitherto touched, and one often arbitrarily disregarded, is the mechanical generation of heat as a factor in metamorphism. It will be most appropriately dealt with in this place, inasmuch as the effects to be discussed are usually found superposed upon an earlier metamorphism of the regional kind. The modifications thus induced in the rocks differ, however, from those described in the preceding section, in that they may be only

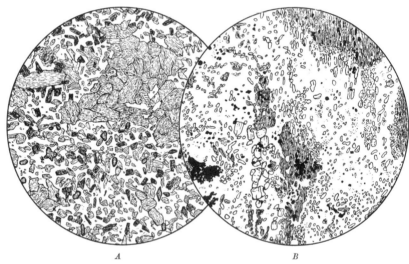

A   B

FIG. 174.—THERMALLY METAMORPHOSED AMPHIBOLITES ; × 23.

A. Near Balmoral, Aberdeenshire : a hornblende-plagioclase-hornfels with some biotite and sphene.

B. Glen Lednock, near Comrie, Perthshire. This illustrates a more advanced stage, in which hornblende is giving place to abundant little grains of augite. (Compare p. 110 and Fig. 41, *B*, above.) The disposition of the minerals still preserves some trace of the former foliation.

partial, and often have a seemingly capricious distribution. There is then a mingling of the two classes of phenomena which· we have distinguished as characteristic of purely thermal and regional metamorphism respectively, and some brief examination of the subject may therefore serve to throw light upon certain apparent anomalies.

When external forces cause deformation and crushing of solid rocks, they thereby do more or less work, in the physicist's sense of the word. An amount of energy, the equivalent of the work done, is liberated within the rock-mass, and must in general assume the form of heat. Mallet long ago pointed out that the quantity of heat generated in the complete crushing down of a hard rock is very considerable. Heat set free within a rock-mass necessarily raises its temperature; but how far such rise of temperature is effective must depend upon the circumstances of the case for the heat, as it is generated, is gradually lost by conduction. Since rocks possess a very low thermal conductivity, the diffusion of heat will be slow; but, on the other hand, the crushing of rocks by orogenic forces is presumably itself a slow process, and conduction may still be adequate to prevent any noteworthy rise of temperature.

If this is the rule, it is not without exceptions. Even sedimentary rocks may be locally fused in the driving of a bore-hole,[1] and crystalline rocks have sometimes suffered fusion under natural conditions, doubtless when crushing has been unusually rapid. The best-known examples are the 'pseudotachylytes' on the borders of the great Bushveld complex in the Transvaal and the Vredefort district of the Orange Free State.[2] There various igneous and sedimentary rocks have been locally reduced to fusion, yielding a black glass, which behaves in the manner of an intrusion. Like phenomena are recorded at numerous localities in the Scottish Highlands; and in the Outer Hebrides [3] fused and partly fused rocks have a wide distribution in association with other crush-effects along a belt of overthrusting.

Since actual melting of rocks is found, though rarely, it is reasonable to expect that less extreme results of the mechanical generation of heat will present themselves more frequently, if sought with due understanding. What is to be looked for is the occurrence of high-temperature minerals at places within an area of merely dynamic or low-grade regional metamorphism. The crucial test is the local distribution of the special phenomena. They will be found localized, not about igneous intrusions, but in relation to the geological structure of the area, and in particular to folding and faulting, overthrusts and crush-belts. On a smaller scale, too, their distribution will be related to the more or less resistant nature of the rocks affected.

The classical area for metamorphism of the kind now in question is the Belgian Ardenne, as described by Gosselet.[4] The country is occupied mainly by Devonian strata, from beneath which emerge

[1] Bowen and Aurousseau, *Bull. Geol. Soc. Amer.*, vol. xxiv (1923).

[2] Shand, *Quart. Journ. Geol. Soc.*, vol. lxxii (1917), pp. 198–219; Hall and Molengraaff, *The Vredefort Mountain Land* (*Verh. k. Akad. Wet. Amsterdam*, 1925), pp. 93–114.

[3] Jehu and Craig, *Trans. Roy. Soc. Edin.*, vol. liii (1923), pp. 430–6, and (1925), pp. 629–33.

[4] *L'Ardenne* (*Mém. Carte Géol. Fra.*, 1888). See also earlier papers in *Ann. Soc. Géol. Nord.*

distinct 'massifs' of Cambrian. Before the deposition of the Devonian, the older rocks were already in the state of phyllites [1] as the result of a regional metamorphism of a low grade. Their highly developed cleavage or schistosity is often seen to be crossed by various structures of the nature of 'false cleavage' (p. 159), connected with the Hercynian crust-movements. Closely bound up with these same movements is a local metamorphism, shown by the production of significant new minerals. The effects are seen at many places in the lower members of the Devonian sequence, especially the Schistes [2] de

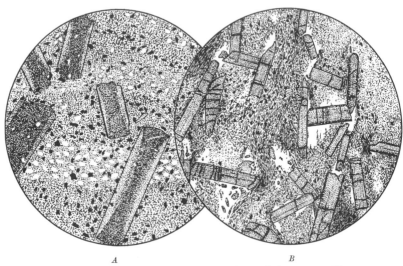

FIG. 175.—OTTRELITE-SCHISTS, in the Belgian Ardenne ; × 25.

A. Ottré, near Viel-Salm. The large ottrelite crystals show the 'hour-glass' structure, very frequent in this mineral.
B. Chateau Seviscourt, near Serpont. The high force of crystallization of ottrelite is shown by the manner in which the crystals lie in all directions, and visibly thrust aside the surrounding matrix.

St. Hubert, and occasionally in the Cambrian phyllites, where these are faulted or overthrust against the Devonian beds.

In places the relations are such that some distinctive new mineral—magnetite, ilmenite, or biotite—characterizes a particular horizon in the Devonian shales over a considerable area. This is the 'méta-morphisme stratique' of Gosselet. More striking is the 'métamorph-

[1] Renard, *Bull. Mus. Roy. Belg.*, vol. i (1882), pp. 215–49 ; ii (1883), pp. 127–52 ; iii (1884), pp. 231–72.

[2] It is to be remembered that the French 'schiste', like the German 'Schiefer' and the 'schistus' of the older English geologists, includes laminated shales and cleaved slates. The now common use by English writers of 'schist', where 'crystalline schist' is intended, is unfortunate.

isme locale ', in which the relation of cause and effect is very clearly displayed. The well-known *ottrelite-schists* have this origin (Fig. 175). They are found on the southern borders of the Cambrian massifs of Stavelot and Serpont, where repeated overthrusts have piled up the Cambrian phyllites and intercalated among them wedges of Devonian strata. Abundant flakes of ottrelite have been developed in both formations, but only in the near vicinity of the overthrusts.

The most interesting localized effects are seen in what Gosselet styles ' métamorphisme par flexion ', which stands in close relation

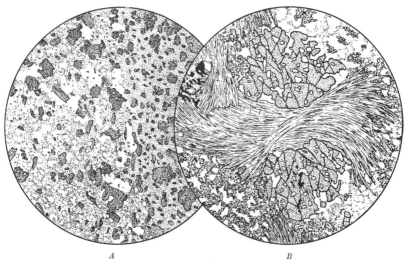

A B

FIG. 176.—HIGHLY METAMORPHOSED SEDIMENTS, Libramont, Belgian Ardenne ;
× 23.

    A. Biotite-Hornfels (' Cornéite ').
    B. Garnetiferous Amphibolite (' Roche amphibolifère '), from a more chloritic and slightly calcareous bed.

to anticlinal or synclinal folds. At Bastogne alternations of grit and shale are exposed in the core of a sharp anticline and similar relations are shown also at Libramont. In each case there has been a development of biotite in the grits and more plentifully in the shales. Moreover, much of the latter has been converted to a hard compact black rock named ' cornéite '. This is composed of quartz, biotite, etc., and shows the micro-structure as well as the constitution of the biotite-hornfels familiar in thermal metamorphism (Fig. 176, *A*). There are varieties containing garnet,[1] and an associated type, which

---

[1] Renard, *Bull. Mus. Roy. Belg.*, vol. i (1882), pp. 1–47. The garnet corresponds with a mixture of equal parts of spessartine, almandine, and grossularite ; it shows often the regular arrangement of inclusions already described (p. 44).

has been partly calcareous, has sheaf-like bunches of a green horn-blende (Fig. 176, *B*). Doubtless, at these localities, the effective crushing, with generation of heat, was in the sharply folded, hard, gritty bands, while the conspicuous effects of metamorphism are shown by the associated argillaceous beds.

The most important overthrust in the Ardenne is the Remagne Fault of Gosselet, which brings into juxtaposition two different facies of the Schistes de St. Hubert, belonging to two distinct basins. At many places along its outcrop it is accompanied by metamorphic effects, which sometimes extend for as much as three or four miles from the fault-line as mapped. There has been a production in different beds of chlorite, ottrelite, haematite, magnetite, biotite, and other minerals. About Libramont, where the metamorphism attains its maximum, the cornéite type is developed, not only in nodules but in continuous bands, with biotite, garnet, and hornblende as characteristic minerals. Even more significant is the occurrence of andalusite,[1] a typical product of thermal metamorphism.

This very summary notice of Hercynian metamorphism in the Ardenne is given here in order to draw attention to an aspect of our subject which has received less consideration than it deserves. It was clearly recognized by Renard, and after him by Gosselet, that there has been at places in this district a notable elevation of temperature in direct connexion with crustal displacements of a pronounced kind, and that it can be attributed only to heat generated by the crushing of the rocks.[2] In so far as such effects can be deemed exceptional, the inference is that crust-movements in general proceed so slowly, that the conduction of heat keeps pace with its generation ; but we are not warranted in assuming without inquiry that a factor in metamorphism, which is so salient in one district, is negligible elsewhere.

A British area which is of interest in this connexion is the Isle of Man, as described by Lamplugh.[3] The Manx slates, with the under-lying grits and flags, present the general arrangement of a complex synclinorium with axis running in a N.E.–S.W. (Caledonian) direction. There is, however, abundant evidence of more than one period

---

[1] Dupont, *Bull. Acad. Belg.* (3), vol. ix (1885), p. 110.

[2] Various alternative interpretations have been put forward by later writers. In particular, Corin has attempted to make out normal zones of metamorphism, while the localized purely thermal effects are attributed to hypothetical intrusions. See especially *Ann. Soc. Sci. Brux.* (B), vol. xlix (1930), pp. 337–48 and vol. li (1931), pp. 57–71 ; *Ann. Soc. Géol. Belg.* (B), vol. liv (1930), pp. 99–115 ; *Bull. Soc. Belg. Géol.*, vol. xli (1932), pp. 340–52.

[3] *The Geology of the Isle of Man* (*Mem. Geol. Sur. Sur. U.K.*, 1903).

of crust-movement. The earlier metamorphism, affecting the whole area, was of low grade (chlorite zone). There is sometimes a true cleavage, but often only some variety of false cleavage or fine corrugation. The latest crust-movement has given rise locally to interesting results, both mechanical and mineralogical. These are found in those places where differential movement and friction reached their maximum, viz. at the junction of the slates with the underlying grits, or in the passage-beds which show alternations of argillaceous and arenaceous bands. On the north-west side of the main axis the

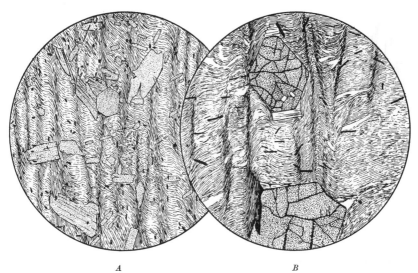

A   B

FIG. 177.—PLICATED AND METAMORPHOSED MANX SLATES, Isle of Man ; × 25.

A. Near Foxdale : composed principally of sericite, with some biotite and quartz, conspicuous crystals of chlorite, and small flakes of ilmenite.
B. On Snaefell : contains the same minerals, with garnet in addition. The flakes of chlorite are developed in places where the resistant garnet afforded protection from the lateral pressure.

effects are merely mechanical, and are seen in a development of crush-breccias or crush-conglomerates, as noted in a former chapter (p. 166). On the south-east side, where the rocks were under a deeper cover, and therefore a greater pressure, the same horizon is marked by a belt of metamorphism, rising in places to a fairly high grade. In the slates, besides recrystallization of the quartz, white mica, and detrital tourmaline, there is a production of biotite, chlorite, ilmenite, and sometimes garnet (Fig. 177). Basic dykes within the same belt of country are not only crushed and sheared, but show a development of new minerals of metamorphism, such as epidote, actinolite, and sphene.

With the exception last noted, the localized phenomena described, both in the Ardenne and in the Isle of Man, challenge attention, because they appear among sedimentary rocks, which have never been subjected to regional metamorphism of any advanced grade. Comparable effects are doubtless of more frequent occurrence in crystalline rocks, which offer more resistance to crushing, and therefore set free more heat when crushed. The British area most interesting in this aspect is the Archaean tract of Western Sutherland, where the Lewisian gneisses and the numerous basic dykes which intersect

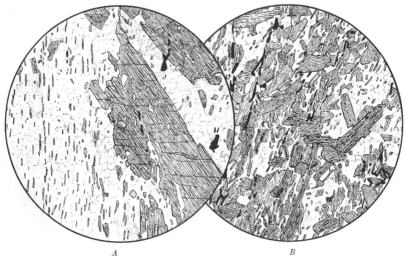

A   B

FIG. 178.—IGNEOUS ROCKS RECRYSTALLIZED IN A SHEAR-BELT, Loch Assynt, Sutherland ; × 25.

A. The coarse pyroxene-gneiss of the district is converted to a porphyroblastic hornblende-biotite-granulite with totally new structure.

B. A dolerite dyke, cutting the gneiss, has been transformed to a plagioclase-amphibolite.

them have been affected in many places by a system of sharply localized disturbances at some pre-Torridonian epoch.[1] Folding, shearing, and disruption assume a variety of forms, which have been fully described by the Geological Survey. Especially instructive are broad belts of shearing, which run vertically in an E.–W. direction. The displacement of dykes shows that the country-rocks on the north side have been moved westward, relatively to their continuation on the south, for a distance of sometimes more than a mile. The two boundaries are sharply defined, and between them gneiss and dykes alike are totally reconstituted in a fashion which indicates an advanced

[1] *The Geological Structure of the North-West Highlands (Mem. Geol. Sur. Gr. Brit.,* 1907), pp. 148–54, 165–70.

grade of metamorphism (Fig. 178). The strict localization of these high-temperature effects points to a local source of heat, which can be no other than the crushing of the rocks themselves.

Another set of displacements in the same area runs in a north-westerly direction, which corresponds with the trend of the dolerite dykes, and movement has in places been localized along the dykes themselves. An example at Scourie in Sutherland was described by Teall,[1] who traced the gradual transformation of the dolerite to a hornblende-schist (plagioclase-amphibolite) (Fig. 179). He pointed

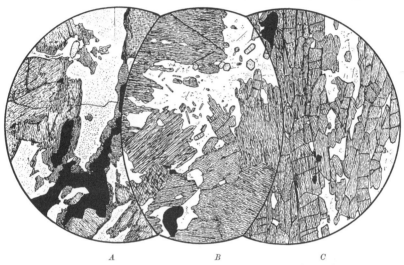

A   *B*   *C*

FIG. 179.—METAMORPHISM OF A DOLERITE DYKE, Scourie, Sutherland; × 25.

A. The augite is partly changed to a green hornblende; felspar and magnetite are unchanged.
B. Here the rock has been totally reconstituted, consisting now of hornblende, magnetite, felspar (andesine), and some quartz, all of new formation. Only the apatite is apparently unchanged.
C. In addition to complete reconstruction, there is now a strongly marked parallel structure.

out that there may be a total reconstitution of the rock without the setting up of any parallel structure. A specimen, taken by itself, will then show metamorphism apparently of purely thermal type, with no *direct* indication of that dynamic factor which is none the less the ultimate cause. The Lewisian gneisses contiguous with the dykes are locally metamorphosed by the heat generated, and sometimes exhibit structures like those shown in Fig. 148, *A*, above.

In respect of their attendant metamorphic effects, these pre-

[1] *Quart. Journ. Geol. Soc.*, vol. xli (1885), pp. 133–44; *British Petrography* (1888), pp. 154–55, 197–200, plates XIX, XX and XXI, Fig. 1; *Geol. Struct. N.W. Highlands (Mem. Geol. Sur.*, 1907), pp. 95–96.

Torridonian crust-movements are in strong contrast with the later (Caledonian) movements of the type of the Moine overthrust. In the latter case crustal displacement, of even greater magnitude but effected more slowly, has caused no serious elevation of temperature, and the accompanying metamorphism is of a very low grade.

SIMPLE THERMAL FOLLOWED BY REGIONAL METAMORPHISM

We go on to consider, though only briefly, the case of *regional superposed upon simple thermal metamorphism*. Evidently we must expect here to find, as the characteristic transformations induced, a replacement of anti-stress by stress-minerals, besides a tendency to the production of dense or high-pressure minerals, and concurrently a replacement of decussate and other varieties of 'hornfels' structures by those which include more or less of the parallel disposition of elements proper to crystalline schists. This is in short the converse of the case already discussed ; but it is easy to see that we cannot now, to the same extent as before, enforce the general principles by direct demonstration. It is of the nature of regional metamorphism that in general it destroys *over large tracts* all trace of the original characters, mineralogical and textural, of the rocks which it affects. If relics survive, sufficient to illustrate the former nature of the rocks and the stages by which they have yielded to the new metamorphism, this is owing to specially favourable circumstances of an accidental kind.

Considering now in particular a hornfels of argillaceous composition, we have to note, as the most conspicuous change in any low grade of regional metamorphism, the production of abundant white mica. Moreover, this comes, not only from potash-felspar, but also from aluminous silicates containing no alkali. The conversion of such a mineral as andalusite to mica demands some source of potash, and this will in general be furnished partly by the chloritization of biotite, partly by the concurrent sericitization of orthoclase :

$$KAlSi_3O_8 + Al_2SiO_5 + H_2O = H_2KAl_3(SiO_4)_3 + SiO_2.$$

So too cordierite, in a low grade of regional metamorphism, yields a mixture of sericite and chlorite, and this is the nature of the so-called 'pinite' pseudomorphs.

In a rock notably rich in iron-ores, however, andalusite may undergo a different change, uniting with ferrous oxide (from magnetite or ilmenite) to yield chloritoid :

$$Al_2SiO_5 + FeO + H_2O = H_2FeAl_2SiO_7.$$

The two minerals andalusite and chloritoid (or the nearly allied ottrelite) are indeed not infrequently found in intimate association. In some occurrences, such as that of Mount Maré in the Transvaal,[1] their mutual relations have been a subject of debate. It seems safe to assume, however, that the equation set down represents a reversible reaction, which under the influence of shearing stress or pressure [2] is driven towards the right, while rising temperature will drive it towards the left. Tilley [3] has described from the Broken Hill district, New South Wales, a very clear instance of the derivation of chloritoid at the expense of andalusite, but this is in an advanced grade of metamorphism.

In rocks of truly argillaceous composition the sericitization of potash-felspar in regional metamorphism seems to be universal, and it may be inquired therefore how any excess of alkali liberated is disposed of in the case of those mica-hornfelses which contained little or no andalusite or cordierite. Its removal in solution, e.g. as carbonate, must be an exceptional process, confined to the shallower levels of the earth's crust. In a garnetiferous hornfels there is probably, as supposed by Flett,[4] a concurrent conversion of part of the garnet to biotite, which takes up the excess of potash. The more arenaceous types of hornfels present a different case; for in such rocks, as we have seen (p. 246), potash- as well as soda-felspar may recrystallize without dissociation.[5]

The British area most instructive in this connexion is that surrounding the granite of Càrn Chuinneag in Eastern Ross-shire.[6] Its intrusion, anterior to the general metamorphism of the Highlands, gave rise to a wide metamorphic aureole of the purely thermal kind. Within this belt the rocks have often been rigid enough to offer effective resistance to the regional metamorphism which came later. There are considerable parts which still retain the hornfels character, and elsewhere the transition may be observed from such rocks to typical mica-schists. The thermal metamorphism was of that deep-seated kind which permitted the formation of almandine [7] (p. 55), and the commonest type of hornfels is composed of garnet, biotite, alkali-felspar and quartz (Fig. 180, *A*). The first effect of

[1] Götz, *Neu. Jb. Min.*, B. Bd. iv (1885), pp. 143–58; Hall, *Tr. Geol. Soc. S. Afr.*, vol. xi (1909), pp. 33–9.

[2] The production of chloritoid by this reaction involves a diminution of volume to the amount of about 43 per cent.

[3] *Geol. Mag.*, vol. lxii (1925), pp. 314–15.

[4] *The Geology of Ben Wyvis* (*Mem. Geol. Sur. Scot.*, 1912), p. 109.

[5] Compare Flett, *loc. cit.*, p. 108.

[6] *Loc. cit.*, pp. 73–88, 102–12; Tilley, *Min. Mag.*, vol. xxiv (1935), pp. 92–7.

[7] The garnet contains only 0·72 per cent. of MnO.

regional metamorphism is seen in the sericitization of the potash-felspar. A dense aggregate of very minute flakes is formed, obscuring the quartz and any residual felspar (Fig. 180, *B*). As the new metamorphism proceeds, quartz, garnet, and biotite are progressively recrystallized, the last-named assuming a regular parallel orientation. In this arrangement the white mica shares, as the minute scales give place to larger and more distinct flakes, the final result being a normal garnetiferous mica-schist (Fig. 180, *C*). The biotite has doubtless undergone some change of composition, the reddish brown colour

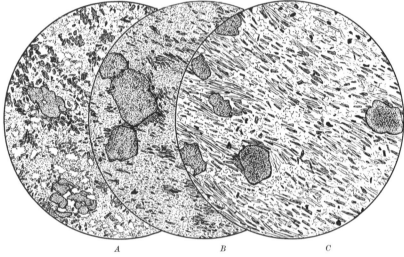

A   *B*   *C*

FIG. 180.—TRANSITION FROM GARNETIFEROUS HORNFELS TO GARNETIFEROUS MICA-SCHIST, in the aureole of the Càrn Chuinneag granite, Ross-shire ; × 23.

   A. Typical hornfels, composed of garnet, biotite, alkali-felspar, and quartz.
   B. The felspar has suffered complete sericitization. The quartz so liberated and the old quartz recrystallized are involved in the dense aggregate of minute scales of white mica. The biotite too is recrystallized, and now shows a well-marked parallel orientation.
   C. Typical mica-schist, with light and dark micas and garnet.

characteristic of the hornfels being replaced by the yellowish brown proper to mica-schist. Whether the garnet also has changed in composition is not determined.. It appears to be somewhat reduced in amount, part having contributed to the formation of new biotite, as already remarked.

   A more aluminous variety of hornfels in the same aureole contains andalusite (chiastolite), and it is interesting to find that, as a result of the succeeding regional metamorphism, this mineral has been transformed to an aggregate of little cyanite crystals, together with white mica. This change has been effected before the breaking down of the hornfels structure. There are also pseudomorphs after cordierite, now composed of bundles of minute needles of cyanite set in a base of muscovite and biotite. With the conversion of the hornfels to a mica-schist the cyanite disappears, being doubtless represented by the abundant muscovite of these rocks. Tilley infers that cyanite ' has only a temporary status in the regional metamorphism '. We may plausibly suppose that here shearing stress continued, or was renewed,[1] down to a temperature at which cyanite was no longer a stable form.

A comparable instance of the direct conversion of andalusite to the stress-mineral cyanite has been described by Eisele [2] in the Black Forest. Here the transformation is sometimes seen in progress, relics of andalusite being enclosed in fibrous aggregates of cyanite.

[1] Clough thought it probable that the shearing was in part later than the intrusion of the (Caledonian) lamprophyre dykes of the district ; *Geol. of Ben Wyvis* (1912), pp. 74–5.

[2] *Zeits. Deuts. Geol. Ges.*, vol. lix (1907), pp. 193–9, 212. Eisele's interpretation has been questioned by Klemm, *ibid.*, vol. lxviii (1916), pp. 86–92.

CHAPTER XX

RETROGRADE METAMORPHISM

Mineralogical Reversals in Thermally Metamorphosed Rocks—Mineralogical Reversals in Regionally Metamorphosed Rocks—Microstructural Rearrangements in Retrograde Metamorphism.

MINERALOGICAL REVERSALS IN THERMALLY METAMORPHOSED ROCKS

THE theme of this final chapter is one to which frequent allusion has been made, though only incidentally, in the preceding pages ; namely, the changes which befall metamorphosed rocks subsequently to the culmination of metamorphism, and which may be viewed broadly as a partial undoing of that process. Like metamorphism proper (as here understood) these changes represent a readjustment, or an essay towards readjustment, to changed physical conditions ; but they are of the nature of *degradation*, as contrasted with the processes of constructive metamorphism. This class of changes includes what Becke has styled ' diaphthoresis ', implying ruin or corruption ; but this rather cumbrous term has not been very widely adopted. It will be more convenient to speak of *retrograde metamorphism*, thus emphasizing its essential nature as a reversal, or partial reversal, of metamorphism proper. It will not be necessary to discuss the subject in great detail. In particular, we shall exclude all those processes of degradation which, taking place in the superficial levels of the earth's crust, are dependent upon an unlimited access of water and atmospheric gases and a free interchange of dissolved material within the rocks.

We begin with the simple case of *thermally metamorphosed rocks*, in which temperature has been the controlling condition, not complicated by the stress-factor. We have seen how, with rising temperature, successive chemical reactions were brought into play, giving rise to new minerals. It was specially emphasized that a new mineral so formed continued to be, potentially at least, a party to any reactions provoked by further rise of temperature, and that in the higher grades the adjustment of equilibrium was in general prompt and complete. It is evident that, when at last the temperature begins to decline, this equilibrium must be disturbed, and increasingly so as the temperature continues to fall. The response to this reversal of conditions is what constitutes retrograde metamorphism. From our study of progressive thermal metamorphism in rocks of different kinds we have learnt generally what mineralogical rearrangements are to be looked for when the process is reversed. Moreover, the evidence of retrograde transformations is usually in this case of a kind easily read. There was in the declining phase much less of that mobility and quasi-vitality which has been noted as a feature of advancing metamorphism. The products generated with falling temperature were for the most part inert, so that the new minerals figure as pseudomorphs after those which they have replaced.

When, however, we turn to an examination of the rocks themselves, the most salient fact that emerges is that, broadly speaking, any effective retrograde transformation is the exception rather than the rule. Indeed, as already pointed out, no systematic study of thermal metamorphism would be possible, were it not that the high-temperature products have remained in great part intact, despite the very different physical conditions to which they are now subjected. The general principle here illustrated is that, in the changes which may affect rocks under natural conditions, the adjustment of chemical equilibrium is much less prompt and complete with falling than with rising temperature. None the less, there are a number of characteristic retrograde transformations which are often to be observed, either partial or complete, and the more important of these we proceed to notice. It will, of course, be understood that they belong to different stages of the cooling down of the rocks. Where there is an exact reversal of some reaction which took place in the original metamorphism at a definite temperature, or through a certain temperature-range, we may assume that such reversal cannot begin until the same temperature conditions are reached in cooling. It may, of course, be deferred later or altogether suspended.

A common retrograde transformation in thermally metamorphosed argillaceous rocks is sericitization of the aluminous silicates. It may often be seen, as an incipient change or more advanced, in the potash-felspars of hornfelses, just as in the same minerals in ordinary igneous rocks ; but it is a more conspicuous feature in the relatively large crystals of andalusite (or chiastolite) and cordierite, which may be more or less completely replaced by an aggregate of fine scales of white mica. Sillimanite doubtless suffers a like change, though this cannot often be verified owing to the usual occurrence of this mineral in slender needles. These aluminous silicates do not, it is true, contain all the material for the making of a mica ; but there is no need for the supposition, in itself highly improbable or in our view inadmissible, that the requisite potash is introduced from some extraneous source. It comes doubtless from the simultaneous alteration of biotite in the rock and, in hornfelses of high grade, from the sericitization of ortho-clase. In all these reactions there is a liberation of silica, and a certain amount of interstitial quartz enters into the micaceous aggregate. In like manner, the magnesian aluminosilicate cordierite yields the aggregate known as ' pinite '. This is, in general, a variable mixture of white mica and chlorite ; but it is sometimes almost wholly mica-ceous, owing to the migration of the more soluble chlorite. According to Clarke,[1] the lateration of cordierite takes place in stages, the first step being the formation of chlorophyllite, $H_4Mg_2Al_4(SiO_4)_5$, by simple hydration.

Biotite is often seen to be partially or wholly chloritized, without losing its shape and cleavage. This illustrates the close relation between forward and retrograde transformations. When biotite is produced in metamorphism, although it is made partly at the expense of sericitic mica, etc., its growth always starts from chlorite. So, when the biotite molecule is broken down, chlorite remains, while the potassic part, as we have remarked, may enter into reaction with andalusite and cordierite to reproduce white mica. The chlorite is of a less ferriferous composition than the biotite from which it derives, and part of the iron comes out in the form of magnetite.

Of arenaceous rocks little need be said. In such as have reached a high grade of thermal metamorphism (with temperatures beyond 575°) there must always have been on cooling an inversion from the higher to the lower form of quartz, but this is a change which cannot be expected to leave any clear indication (p. 67). The former presence of tridymite (which could be produced only in a very high grade) may be indicated by the shapes of the pseudomorphous quartz (Fig. 21, *A*). The less pure arenaceous types may, of course, exhibit the same retrograde transformations that we have noted in rocks of argillaceous composition.

In the instances which have been cited it was possible for minerals more or less closely identical with the constituents of the original sediments (micas, chlorites, quartz, iron-oxides) to be reproduced in retrograde metamorphism without any addition of material from outside, the small amount of water taken up in the process being already present in the rock. In metamorphosed semi-calcareous rocks a reversal to this extent is not possible. The production of silicates

[1] *Bull.* 588, *U.S. Geol. Sur.* (1914), p. 79.

at the expense of carbonates has set free carbon dioxide often in large amount, and doubtless most of this has been eventually lost from the rock, not to be easily recovered. Accordingly, new calcite never figures prominently, and the characteristic retrograde changes are those from higher to lower forms among the silicates themselves.

Very common is a partial, or even total, replacement of grossularite by idocrase ; and, since the opposite process is not found, this may be regarded as a change from higher to lower. It must belong, however, to a very early stage of the cooling ; for we know that idocrase

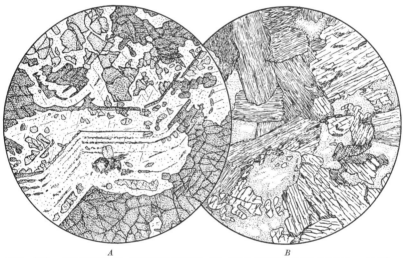

A   B

FIG. 181.—RETROGRADE METAMORPHISM IN SEMI-CALCAREOUS ROCKS ;   × 23.

A. Alteration of a large skeleton-garnet, Balloch Craig, near Braemar, Aberdeenshire. Relics of the original garnet (andradite) remain, chiefly in the upper part of the field, but most of it has been replaced by epidote (shown with heavier stippling) and calcite. Much of the calcite has been enclosed in the original skeleton structure, but that which surrounds the relics of garnet must be new.

B. Altered diopside-microcline-rock, Barnavave, near Carlingford, Co. Louth. The pyroxene is in great part replaced by fibrous actinolite, and the felspar more or less completely sericitized.

very often accompanies grossularite, or occurs alone, as a direct product of thermal metamorphism.[1] The transformation implies no great change in composition, beyond the taking up of a little water, and no important change of volume. A transformation equally common, but belonging to a somewhat later stage, is that by which grossularite gives rise to zoisite or clinozoisite and andradite to epidote (Fig. 181, *A*). As a further degradation some calcite may be produced, and this may also be formed at the expense of wollastonite. Diopside

[1] The derivative idocrase may sometimes be distinguished by its showing anomalous interference-colours ; Osborne, *Geol. Mag.*, vol. lxix (1932), pp. 216, 218.

sometimes shows the change to tremolite or actinolite (Fig. 131, *B*). We have already noted how quickly the periclase in metamorphosed dolomites becomes changed to brucite (p. 77). Probably, too, the serpentinization of forsterite often belongs in part to the time of retrograde metamorphism rather than to ' weathering '. It is needless to remark that those rare minerals of metamorphism which are essentially unstable forms suffer change very readily (p. 96). It may be pointed out, however, that, in accordance with a law of Ostwald, the change is not always directly to the most stable form or association, but may pass through intermediate phases. So, at Monzoni, both melilite and monticellite are found replaced by fassaite, itself a merely metastable form ; while pseudomorphs of grossularite after melilite at the same place represent at least a nearer approach to chemical equilibrium.

Thermally metamorphosed igneous rocks present no features of special interest in this connexion. The retrograde changes which they suffer are merely such as are familiar in normal igneous rocks, which have necessarily gone through the same process of cooling down. It should be remarked, however, that some of the most characteristic changes—sericitization of orthoclase, zoisitization of plagioclase (less commonly with some carbonation), chloritization of biotite —are checked by the limited supply of water (and of carbon dioxide) that can be supposed present. Chlorite requires about three times as much water as mica ; the epidote minerals demand but little ; the zeolites, with their 10–20 per cent. of water, are altogether ruled out in simple retrograde metamorphism.

MINERALOGICAL REVERSALS IN REGIONALLY METAMORPHOSED ROCKS

When now we come to examine the changes of a retrograde kind which may affect *regionally metamorphosed rocks*, the factors to be taken into account are more complex, including not only a gradual decline of temperature, but the possibility of wide variations in the other controlling condition, viz. stress. One obvious difference, as compared with the simpler case which has been considered, is that we must look for important structural as well as mineralogical modifications. To realize the possible complexity, however, demands a more particular discussion.

We have seen reason for believing that, with advancing metamorphism within an extensive region, the shearing stress tends to rise constantly to its limiting intensity ; or, if from time to time it falls short of this, such lapses leave no mark upon the final result. But, if the rising tide can in this way obliterate the effect of fluctuations, it is otherwise with the ebb. In view of the often complex tectonic operations with which regional metamorphism is closely connected, it cannot be supposed that shearing stress merely suffers, like temperature, a steady decline. Rather must we expect incidental revivals of more or less intense shearing stress, recurring, it may be, at various stages of the continued fall of temperature. Moreover, under the conditions supposed, the effects, mineralogical and structural, so impressed on the slowly cooling rocks are likely to be preserved, except in so far as earlier changes of this kind may be obscured or obliterated by later. Any notable renewal of shearing stress during the latest stages of cooling will necessarily leave its mark. The cataclastic and accompanying mineralogical changes in this case will be indistinguishable from such as might be caused by a disturbance of later date, not related to any precedent regional metamorphism ; and it is for this reason that, in the discussion of repeated metamorphism (Chapter XIX), no special notice was given to ' dynamic superposed on regional '.

In crystalline schists, as in rocks resulting from simple thermal metamorphism, high-temperature minerals are liable to be replaced by lower forms ; but there is the further consideration, that, more consistently than before, these products of retrograde reactions will be found in the list of avowed stress-minerals. The aluminous silicates cyanite and staurolite, characteristic of high-grade metamorphism in sediments of argillaceous composition, often give rise, like andalusite and cordierite in the former case, to a confused mass consisting essentially of white mica with some interstitial quartz, the ' shimmer-aggregate ' of Barrow [1] (Fig. 182, A). Unless disturbed by subsequent shearing, the aggregate makes distinct pseudomorphs. Its appearance in this form relegates it to a somewhat late stage of cooling, when only a very limited diffusion was possible. When a like change has taken place at a high temperature, we can sometimes observe the gradual replacement of a cyanite crystal by relatively large flakes of mica (Fig. 182, B). The requisite potash is furnished by the concurrent alteration of orthoclase. The conversion of staurolite to chloritoid [2] is another change which must be referred to a fairly high temperature, probably not much below that at which, in rising metamorphism, chloritoid gave place to staurolite (p. 225). That this retrograde transformation is not often to be observed, is perhaps due in part to its obliteration by further changes of degradation at lower temperatures, when the chloritoid itself ceases to be stable.

The spontaneous breaking down of the common red garnet has

[1] *Quart. Journ. Geol. Soc.*, vol. xlix (1893), p. 340.

[2] Becke, *Forts. Min.*, vol. v (1916), p. 223 : see also Pelikaan, p. 225, above.

usually yielded chlorite. The change is often connected with fracture and crushing of the crystals, and can be seen in all stages of its progress. When garnet is destroyed at an earlier time, and therefore at a higher temperature, it gives rise to biotite ; but this transformation, depending doubtless upon the composition of the contiguous rock-substance, is less frequently found (Figs. 167, *B*, 183, *B*). It may be followed later by chloritization, but the one change is not a necessary step to the other. It should always be borne in mind that, although we may speak of one mineral ' replacing ' another, the reactions in question

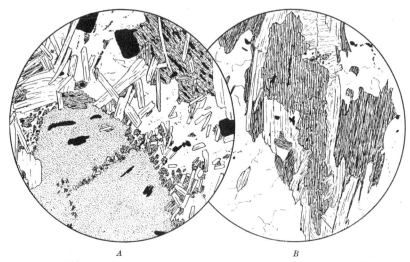

A   *B*

FIG. 182.—RETROGRADE METAMORPHISM IN PELITIC GNEISSES ; × 23.

A. Staurolite-Gneiss, Glen Esk, Forfarshire. The ' shimmer-aggregate ' representing staurolite crystals consists essentially of very fine white mica. Such replacement involves a considerable increase of volume, and this was partly relieved by the expulsion of the more soluble chlorite, seen as an irregular border. Larger flakes of chlorite, with secondary magnetite, represent biotite. The other minerals are muscovite, quartz, a little primary magnetite, and cubes of pyrites, now largely oxidized.
B. Cyanite-Gneiss, Glen Urquhart, Inverness-shire ; showing cyanite in course of replacement by muscovite. The other minerals are biotite, quartz, and magnetite.

are almost always of a complex kind, involving substances other than those named. The garnet produced in metamorphism, taking directly the place of chlorite (Fig. 82), is of course made largely at the expense of that mineral, but also draws upon the iron-ore and probably other constituents of the rock. So, when the process is reversed, the new chlorite is always much less ferriferous than the garnet from which it comes.[1] Chlorite is the common product of degradation of biotite

[1] For comparative analyses see Penfield and Sperry, *Amer. J. Sci.* (3), vol. xxxii (1886), pp. 307–11. The conclusion holds good for chloritized Fe-Mg-garnets in general, in igneous as well as metamorphic rocks : see Lemberg, *Zeits. Deut. Geol. Ges.*, vol. xxvii (1875), p. 534.

as well as of garnet, and in that case its origin is often made certain by the presence in it of numerous needles of rutile.

Among other minerals of metamorphism which are liable to spontaneous alteration may be mentioned corundum, which changes rather readily by hydration to diaspore (Fig. 111, *B*, above). Margarite too, which has been noted above as a frequent constituent of emery deposits, has often a manner of occurrence which relates it to the alteration of corundum, and a like origin has been claimed for the paragonite of certain localities. Of these two minerals, the lime-mica has a composition which can be represented as anorthite *plus* diaspore, and the soda-mica is equivalent to albite *plus* diaspore. Presumably they are the natural end-products of any plagioclase present in a highly aluminous rock.

Rocks of semi-calcareous nature in a high grade of metamorphism usually show some signs of retrograde change. If a lime-garnet is present it is likely to be replaced by zoisite or clinozoisite. The amphibolization of diopside is more common here than in aureoles of thermal metamorphism, being one of those changes which are promoted by shearing stress. As regards the amphiboles themselves, the simple forms remain stable ; but the green aluminous hornblendes, of complex constitution, have a tendency to break down, yielding various products, of which biotite is often one.

These and other common retrograde transformations in regionally metamorphosed sediments of different kinds are to be correlated with the gradual decline of the regional temperature after the climax of metamorphism, and can be referred to different stages of that continued cooling. This, however, is not the only ruling condition to be reckoned with. Highly metamorphosed rocks, such as those exposed in the interior of the Scottish Highlands, have undoubtedly acquired their distinctive characters at very considerable depths beneath the surface. The profound erosion which has brought these rocks to light implies a very great diminution of static pressure ; and the relief of pressure, as well as the lowering of temperature, has been a factor promoting changes of the retrograde kind. It is easy to verify this by applying the ' Volume Law ' to many of the transformations already noticed. In regionally metamorphosed igneous rocks, and such as have been intruded at or near the climax of metamorphism, the influence of this factor is sometimes even more important, especially in rocks of basic and ultrabasic nature. The breaking down of the omphacite of eclogites, described above (p. 309), is an interesting example, and the very general amphibolization of pyroxenes in such rocks falls under the same head.

Concerning rocks of igneous origin in the present connexion not much need be said. We have already made mention of some of the characteristic retrograde changes in a former section dealing with the igneous gneisses. These changes are, broadly, the same that may be observed in ordinary plutonic rocks not related to metamorphism, but with an important difference. The regional decline of temperature with which we are now concerned is a far slower process than the cooling down of an ordinary stock or laccolite intruded among relatively cold country-rocks. This extreme slowness of cooling permits spon-

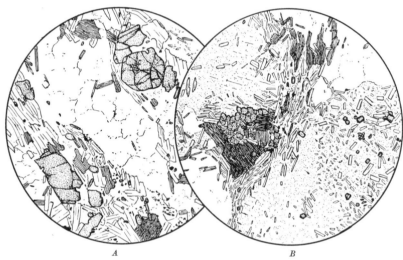

A   *B*

FIG. 183.—RETROGRADE METAMORPHISM IN GRANITE-GNEISS, Eddystone Rock, English Channel ; × 25.

A. The minerals shown are garnet, two micas, two felspars, epidote, and quartz. The abundant flakes of muscovite and the large crystals of epidote in the lower part of the field have been generated at an early stage of the gradual decline of temperature.

B. Here, at a later stage, there has been a plentiful production of muscovite in small flakes at the expense of the orthoclase ; also a breaking down of the garnet, yielding biotite.

taneous reactions to proceed more freely, and the results are more clearly exhibited. So, for example, the ' sericitization ' of a potash-felspar gives rise, not to a swarm of almost invisibly minute scales, but to very evident crystals of white mica ; and the ' saussurite ' change in a lime-bearing felspar yields, instead of an obscure cloudiness, distinct crystals of zoisite or epidote. The only check is imposed by the limited amount of water present, since most of the new minerals contain constitutional hydroxyl. It is to be observed also that such retrograde changes as those just cited are possible through a wide range of declining temperature, and may be repeated at different stages in the history of a given rock. The earlier-formed products, generated at a high temperature and with a prolonged growth, may build relatively large crystals, while a later generation of the same mineral is in smaller flakes or grains (Fig. 183 ; compare also the two generations of epidote in Fig. 165, *B*, above). Incidentally it should be remarked that in the more sodic plagioclases the production of zoisite or epidote does not involve a breaking down of the crystals, as in the more calcic varieties, but merely a change to a still more sodic composition.

Distinct from definite chemical reactions, giving rise to new minerals, is a class of changes connected with solid solution in the felspars and other groups of minerals. Here, again, the principle involved is a simple one, viz. the contracted limits of solid solution consequent upon lowering of temperature. Vogt long ago pointed out the importance of this in relation to perthitic intergrowths. According to his calculations, potash-felspar at magmatic temperatures can take up 28 per cent. of albite in solid solution, but at ordinary low temperatures not more than 15 per cent. A crystal of orthoclase or microcline, saturated with dissolved albite at the temperature of crystallization, is therefore no longer stable when cooled. Whether it remains in a metastable (supersaturated) state or is able to disgorge its excess of dissolved albite in the form of inclusions, depends upon the rate of cooling. The former case is represented by the soda-sanidine of volcanic rocks, always rapidly cooled. In an ordinary granite stock or laccolite microperthitic structure often indicates at least a partial discharge of the excess of dissolved albite,[1] and still more in associated pegmatites, but it is probable that complete equilibrium is not always attained. Only within a tract of regional metamorphism is this result ensured by the extremely gradual decline of temperature, and here accordingly perthitic intergrowths are of very common occurrence. Probably any recurrence of stress-conditions during the time of cooling will promote the change, the effect being analogous to that of stirring a supersaturated liquid solution. The relative infrequency of antiperthite seems to indicate that the solubility of potash-felspar in albite is less affected by temperature. Other examples of the effects of ' exsolution ' are afforded by intergrowths of magnetite and ilmenite and of haematite and ilmenite.[2] The setting up in an igneous rock of a myrmecitic structure is more probably a late-magmatic than a metamorphic effect ; but this, too, involving diffusion, must be facilitated by a very slow rate of cooling.

[1] This secondary perthite is to be distinguished from the primary intergrowth, regarded by Vogt as a eutectic of approximate composition 40 Or+60Ab.

[2] Ramdohr, *Neu. Jahrb. Min.*, B.B. liv, A (1926), pp. 335–58.

From what has been said concerning retrograde metamorphism in rocks of various kinds, it is apparent that we cannot in general expect much regularity in the incidence and distribution of changes of this kind. Whether a particular retrograde reaction does or does not take place must depend upon a number of factors—the rate of cooling, the stress conditions, the water content, the contiguity in the rock of minerals capable of reacting together. In particular cases an approach to regularity may perhaps be realized. In the Sulitelma district T. Vogt [1] found it possible to lay down on the map the approximate limits of chloritization of biotite in the metamorphosed sediments. In most of its course the line does not differ much from that which marks the coming in of almandine in advancing metamorphism, thus indicating a considerable ' lag ' in the retrograde change.

MICROSTRUCTURAL REARRANGEMENTS IN RETROGRADE METAMORPHISM

Having dealt sufficiently for our purpose with the more characteristic changes of a mineralogical kind to which rocks are liable in the waning stage of regional metamorphism, we shall conclude our treatment of the whole subject with some general remarks concerning the associated changes in microstructure.

Of directional structures there is little to be said. Schistosity in rocks of sedimentary origin may be accredited wholly to direct metamorphism. Foliation, however, of the kind which depends upon segregation of the more mobile elements of the rock, is essentially a cumulative process, and the freedom which these minerals enjoy at high temperatures will not be lost in the earlier stages of the decline. Again, the corrugation which is often found modifying the schistosity of phyllitic schists has doubtless in many cases a relatively late origin. Thus a simple type of rock may consist of parallel flakes of muscovite set in quartz. The trains of mica-flakes have merely been thrown into a system of minute folds, perhaps with some bending of the larger flakes, while the quartz has accommodated itself to the deformation of the rock by recrystallizing. A certain freedom of migration, too, is shown by its tendency to collect in the bends of the folds (Figs. 94, B ; 119). Often strain-shadows give evidence of a later renewal of stress, when the quartz had lost its facility of recrystallization.

In the case of igneous rocks intruded in relation with the metamorphism, as we have seen, other considerations enter. Foliated and allied structures may arise in connexion with the intrusion itself, or may be impressed at a somewhat later stage, or again congenital

---

[1] *Norges Geol. Und.*, No. 121 (1927), plate XXXIX.

features may become further accentuated. Schistosity proper, in such rocks, belongs to an early stage of declining temperature, and is therefore logically a retrograde effect. Instead of a schist passing up into a gneiss, a gneiss, under the appropriate conditions, passes down into a schist.

We proceed now to those characters which depend on the shape and dimensions of the individual elements of a rock—in Grubenmann's terminology ' structure ' as distinguished from ' texture '. During the prolonged decline of temperature which follows the climax of regional metamorphism two opposing agencies may be called into play, the one destructive and the other recuperative. Any recurrence of the orogenic forces, if sufficiently intense, will cause fracture and displacement of crystals, the different minerals giving way or resisting according to their specific properties. On the other hand, such innate power of crystallization as the minerals possess, if any, will be exerted to repair their injuries and obliterate the effects of the crushing.

The normal micro-structure of metamorphosed rocks is that which we have defined as crystalloblastic, and, as a rule, is easily recognized. This characteristic structure is modified by any fracture or displacement of crystals, and in a complete crushing of the rock is quite superseded by structures of the cataclastic kind. It is the part of recrystallization to rehabilitate, so far as is possible, the shattered fabric. If the temperature is still sufficiently high, all the essential minerals of the rock may be able to respond to the call, and the crystalloblastic type of structure may be perfectly restored, though probably with some diminution of grain-size. At lower temperatures this is no longer possible, and the cataclastic effects remain in permanency.

In conclusion, it will serve to enforce the paramount influence of temperature in metamorphism, retrograde no less than direct, if, at the cost of some repetition, the matter is set forth more precisely. It is fundamental to a true conception of advancing metamorphism that a new mineral, generated at the appropriate temperature, does not thereafter remain passive and lifeless. If not consumed in reactions producing other minerals, it retains at all higher temperatures the power of rejuvenation. This power it still continues to enjoy during the waning or retrograde phase, until the temperature has fallen to approximately that at which the mineral made its first appearance.[1] For each mineral of metamorphism, then, if it has not already given place by retrograde change to other and lower minerals, there comes a time when it is left inert and dead. The ebb of metamorphism

---

[1] The fact that temperature is not the sole ruling condition may imply a certain latitude.

M.—23*

leaves the several minerals stranded in turn, in an order the reverse of that in which they first appeared in the advance of metamorphism.

It is easy now to understand why cataclastic effects are most frequent—i.e. most frequently preserved—in the high-grade minerals of metamorphism. Those characteristic of ordinary argillaceous rocks will afford a sufficient illustration. Cyanite is a mineral eminently susceptible of strain-effects, a certain amount of yielding along gliding-planes being very general (Fig. 105, *B*, above). Under more severe stress the gliding becomes accentuated, and actual fracture takes place,

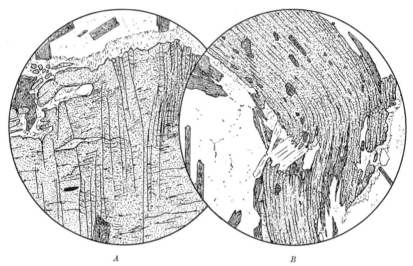

A   *B*

FIG. 184.—CATACLASTIC EFFECTS IN CYANITE, from cyanite-gneiss, Loch Assapol, Ross of Mull ; × 23.

A. This crystal has yielded chiefly by gliding-lamellae, but actual shattering has begun in places (above, left). A thin crust of ' shimmer-aggregate ', the beginning of chemical break-down, is of later origin.

B. Here a large crystal has been sharply bent, with the result of opening cleavage-cracks and initiating a more complete shattering. The fissures have been occupied by recrystallized biotite, with some muscovite, proving that the fracture took place at a somewhat high temperature.

beginning often with bending of the crystal and the opening of cleavage-cracks (Fig. 184). Staurolite, with its perfect cleavage, suffers fracture even more readily (Fig. 105, *A*, above). These two minerals have seldom, if ever, been able to recover from cataclastic accidents. The common garnet is a brittle mineral, and its crystals, except the smallest, are seldom free from fracture. Under a simple type of stress-distribution it often, as a first result, develops cracks with a regular parallel arrangement, at right angles to the direction of maximum tension [1] (Fig. 84, *A*, above). When there has been actual stretching, the

[1] This is common also in the garnets of igneous rocks : see Fig. 160, *B*.

fragments of a crystal are slightly parted, the intervals being filled by newly crystallized quartz. More prolonged movement results in a train of scattered fragments. Garnet, however, is capable of recrystallizing at a temperature only moderately elevated, and then speedily recovers its crystal shape (Fig. 185, *A*). Crushed at a lower temperature it remains in a shattered state, unless indeed it also breaks down chemically, yielding new products (Fig. 185, *B*, *C*).

The point to be especially noted is that these products of high-grade metamorphism are often seen broken and displaced, while micas,

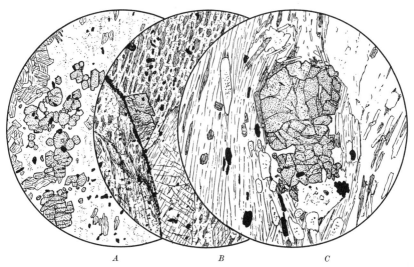

FIG. 185.—CATACLASTIC EFFECTS IN GARNET ; × 23.

A. Garnetiferous Hornblende-schist, Kerrysdale, near Gairloch, Ross. A large garnet has been shattered and dragged out, but at a high temperature which permitted recrystallization.
B. Garnetiferous Hornblende-Biotite-schist, from the same neighbourhood. A garnet and a large porphyroblastic hornblende have been partly cut away by a small crush, along the line of which are biotite and magnetite derived in part from the destruction of the former minerals.
C. Garnetiferous Albite-Mica-Gneiss, Loch Dochart, Perthshire. The garnet has been fissured and partly shattered, and newly crystallized albite and quartz are moulded on the broken surface.

tourmaline, magnetite, sodic felspars, and quartz in the same rock seem to have been immune from attack. Clearly these lower-grade minerals must have suffered, each after its kind, under the same powerful forces which crushed the garnet, etc. ; but they possessed a power of rejuvenation which enabled them to obliterate the traces of the crushing. That they have in fact recrystallized is often apparent from the manner in which they have filled fissures in the high-grade minerals and moulded themselves upon the broken surfaces.

The same principle is illustrated by the relations of the different low-grade minerals to one another. Albite and quartz are seen moulded upon and partly embracing bent flakes of mica or cementing the fragments of a broken crystal of tourmaline. These accidents happened when the temperature was such that the micas and tourmaline had lost the power of recrystallization, which albite and quartz still retained. At a still lower temperature a crystal of albite may be broken and its interstices occupied by newly crystallized quartz (Fig. 115, *B*, above).

These considerations are emphasized here because, while essential to a clear view of the nature of metamorphism, they have also a bearing on some questions relative to geological history. One such question is that of the supposed existence of more than one important system of regional metamorphism in the Scottish Highlands. Clough,[1] impressed by the abundance in many of the Argyllshire schists of albite porphyroblasts free from any sign of fracture or deformation, was led to the conclusion that they indicate a distinct and later metamorphism.[2] Cunningham Craig,[3] adopting the same conception, has attributed the formation of albite to a special hydrothermal type of metamorphism. In this Bailey [4] concurs, with the stipulation that ' the two metamorphisms probably continued in operation side by side for a considerable period, although eventually the albitic survived the garnetiferous '. It is certain, however, that, if the term ' hydrothermal ' implies merely the presence of water in conjunction with a more or less elevated temperature, this condition is necessary for the crystallization of both minerals. Further, we have seen that, when in the same rock albite crystals are seen intact, while garnets are shattered, this is because albite was the earlier, not the later, mineral to form. The two minerals are, in fact, found together in many rocks in the garnet-zone, and whether one or other or both be produced must depend upon the composition of the rocks. Here, then, we have to note one more illustration of how a clear understanding of the processes of metamorphism may be vital to a true reading of geological history.

---

[1] *Geology of Cowal* (*Mem. Geol. Sur. Scot.*, 1897), pp. 39–43.

[2] Clough further suggested an introduction of soda from some external source : see above, p. 212.

[3] *Quart. Journ. Geol. Soc.*, vol. lx (1904), pp. 26–7.

[4] *Geol. Mag.*, vol. lx (1923), pp. 326–7.

INDEX

359